The Environmental Challenge

The Environmental Challenge

Edited by

Willis H. Johnson,
Wabash College

William C. Steere,
New York Botanical Garden

Holt, Rinehart and Winston, Inc.
New York Chicago San Francisco Atlanta
Dallas Montreal Toronto London Sydney

Library of Congress Catalog Number: 73-3947
ISBN: 0-03-084185-2
Printed in the United States of America
9 8 7 6 5 4 3 2 1 090 7 6 5 4

Preface

During the past few years much attention has been focused on the problems that man faces with respect to the environment. A number of books dealing with these problems have been published—many of them collections of previously published articles. It occurred to the editors of this volume that a series of chapters written expressly by experts in the field to follow a preconceived outline could be a worthwhile contribution.

Four chapters are devoted to basic background material—The Structure and Function of Ecosystems, The Diversity of Life, Energy Relations between Plants and Animals and Their Environment, and The Regulation of Populations.

In any consideration of the problems of man and the environment the question of human population must be considered. The next chapter considers the question: Can We Control Human Population?

The major aspects of environmental pollution of all kinds are treated in four chapters: Air Pollution; Water; The Land Resource Base; and Land Use and the Land Ethic.

The greatest problems of both population and pollution are found in our large cities. The next chapter is an essay entitled Toward an Ecology of the Urban Environment.

It is not enough to state what the problems are. It is also necessary that we indicate what is being done and what should be done to alleviate these problems. This is undertaken in the next two chapters: The International Biological Program and Government and Academia—Roles and Responsibilities in Man's Environmental Crisis.

With the exception of the chapter on the urban environment, which was written by a political scientist, the other chapters were written by biologists. It seemed to the editors that the voice of a humanist should be heard also. Chapter 14 is a humanist writing on American Institutions and the Ecological Ideal. The editors have written an introduction and a summary.

Each author has written with the college student in mind. Thus, the book can be used by students in a special course on man and the environment, or as collateral reading by students in a general or introductory biology course, or by students in other courses where environmental problems are considered, or by the general reader with an interest in environmental problems.

All of the chapters were written especially for this volume except Chapters 3 and 14. William C. Steere was recently involved in the writing of a chapter on *The Diversity of Life* for *Biology and the Future of Man,* edited by Philip Handler, Oxford University Press, 1970. This chapter covers the field so well that we obtained permission to reprint it here. The material in Chapter 14 was originally presented in an AAAS symposium in Boston in December of 1969 and was later printed in *Science.* Since it so clearly states the viewpoint of a humanist we asked the author to include it here with only a few modifications.

The editors were gratified by the responses of the authors to their invitations to write the chapters, and we want to thank each one of them for joining in this venture. A brief biographic account of the author is placed at the beginning of each chapter.

Crawfordsville, Indiana
New York City
February, 1974

W. H. J.
W. C. S.

The Environmental Challenge

The
Environmental Challenge

1

Introduction

WILLIAM C. STEERE

New York Botanical Garden and Columbia University

William Campbell Steere was born in Muskegon, Michigan, in 1907. After graduation from the University of Michigan in 1929, in plant physiology, he carried on graduate studies in cytology at the University of Pennsylvania for two years. In 1931 he became an instructor at the University of Michigan, receiving the Ph.D. degree there in 1932; he became Professor of Botany in 1946 and Chairman of the Department in 1947. Between 1950 and 1958 he was Professor of Biology and, from 1955 on, also Dean of the Graduate Division at Stanford University. Since 1958, he has held the position of Director, and more recently, President, of the New York Botanical Garden, as well as Professor of Botany at Columbia University. Other professional activities have been, while on leave, Exchange Professor at the University of Puerto Rico, 1939-1940; Senior Botanist, Board of Economic Warfare, with quinine-procurement missions in Colombia, 1942-1943, and Ecuador, 1943-1944; and Program Director for Systematic Biology at the National Science Foundation in Washington, D.C., 1954-1955. He is a past president of the Botanical Society of America, the American Bryological Society, the American Society of Naturalists, the American Society of Plant Taxonomists, the California Botanical Society, and the Torrey Botanical Club. In 1959, during official ceremonies of the IX International Botanical Congress, he received the degree of Docteur es-Sci. (honoris causa) from the University of Montreal; in 1962, the University of Michigan awarded him the degree of D.Sc.

His research interests center about the geographical distribution of plants, especially mosses, and he has investigated problems involving their systematics, ecology, cytology, and paleobotany. These studies, based on extensive field work in many parts of the Western Hemisphere, have resulted in more than 350 published research papers, of which the following are illustrative: "Chromosome Number and Behavior in Arctic Mosses," *Botan. Gaz.*, (1954); "The Mosses of Porto Rico and the Virgin Islands" (with H.A. Crum), New York Acad. Sci., *Sci. Survey P. R. and the V. I.*, 1957; "A Preliminary Review of the Bryophytes of Antarctica," Nat. Acad. Sci., *Nat. Research Council Publ.*, 1961.

At no time in the history of mankind has there ever occurred the popular concern and apprehension over the problems of population pressures and consequent damage to our environment that we find today. The growth of the world's population has demanded the exploitation of natural resources, many of which are irreplaceable. The increased production of food and other materials required by man goes hand in hand with the need to dispose of more and more waste materials. Industrial wastes pollute the air and the waters; man's wastes of all imaginable kinds are strewn over the land.

The problems are vast and chaotic, and entirely of our own making. Much of our present concern is based on emotional grounds, and has been generated by exaggerated alarms spread by self-elected prophets of doom. However, the problems are critical and their solution depends on increased willingness of government, industry, and the general public to contribute energy and funds, as well as a scientific approach and technical skill.

The primary purpose of this book is to take stock of all the components of our environment so that we may have an orderly assessment of the nature and magnitude of environmental problems. Emotion may give us the motivation to attack and even to define these problems, but their solution requires knowledge and understanding. It is our hope to set the stage here for clearer and more logical thinking simply by furnishing the requisite background information.

During the past decade, more than in any other, we have become aware of the unity, essential fragility, and uniqueness of planet Earth. The U.S. space exploration program has brought home to us as nothing else could do that the earth itself is like a space capsule, and the world's population its passengers. Nothing can be added, nothing can be subtracted. However, some of the complex ecological systems can be interrupted or diverted. Before the age of glass, metal, and plastic, man used natural products that soon returned their component materials and energy to the soil through decay by microorganisms that for their own metabolic processes liberated the chemical energy that had been bound in wood, skins, and other natural materials. Today, many of our waste materials, especially those derived from synthetics, are not biodegradable, that is, microorganisms just do not have the special enzymes necessary to decay such substances.

Our planet consists of a central lithosphere which is surrounded by a thin layer of soil and water, enveloped in turn by the atmosphere. This thin outer layer provides the only habitats for living organisms and is therefore called the biosphere. The biosphere, then, is a unitary concept, in which all living organisms occur. Life on earth began long ago and the date of the origin of the first simple green plants, the blue-green algae, has been pushed back very recently to 3.8 billion years ago. Through this long history, millions of kinds of organisms have evolved; far more are extinct than still exist today. The diversity and complexity of life on earth reflects both the long period of time through which it has evolved and the interactions of the living organisms with both each other and with their environment. The ecology of organisms, then, is as complex as their evolution, and the two scientific approaches are wholly and inextricably interrelated.

The science of ecology, which for decades was a little-recognized field, has now become a household word, which, we are afraid, means different things to different people. For the record, ecology, which has already been defined in the preceding paragraphs, concerns itself with all interrelations and interactions between organisms, both of the same kind and other kinds, and their external environment. Ecology is a transcendental science today, as it draws upon the knowledge and the techniques of all fields of biology, as well as from the other sciences. The fields of botany, zoology, microbiology, and the other life sciences, chemistry, physics, geology, geography, geophysics, hydrology, systems engineering, soil science, meteorology, climatology, and paleo-climatology have all contributed information to the various chapters of this book.

Within the concept of the earth's biosphere, we find that many ecosystems occur; these are the basic functional units of ecology and concern the interaction of all the organisms of a large ecological system or community among themselves and with their environment. Large and characteristic ecosystems occur in different climates, as determined especially by mean annual precipitation and mean annual temperature. Major habitat types that are populated by individual types of plants and animals are called biomes—the importance and validity of this concept is easily recognized when one reviews in his mind's eye the various kinds of biomes to be found in North America, such as the deciduous forest, the coniferous forest, the tropical forest, as well as grassland, desert, and arctic and alpine tundra.

The detailed ecology of any single biome is incredibly complex. In its most simplistic definition, an ecosystem or biome may be analyzed in terms of its trophic levels, or levels of energy transfer within the nutritional system. The elements of a biome, in a greatly oversimplified way, are as follows: (1) The *producers* are the green plants which are able to combine water and atmospheric carbon dioxide into simple carbohydrates in the presence of sunlight and chlorophyll. To be more specific, the process of photosynthesis just described transforms the kinetic energy of sunlight into the chemical energy of sugar, through a long and fascinating series of biochemical and physiochemical steps, each governed by one specific enzyme. Without the ability of green plants to trap and store the sun's energy, an ability that evolved surprisingly early in geological history, the course of evolution would have been in directions that are difficult to imagine today—and yet which may have been explored by the very earliest living things. (2) *Consumers* are animals and non-green plants that depend for their metabolic energy on green plants, or other organisms that have derived energy from green plants, thus establishing their place in a food chain. Food chains at their simplest are the cow eating grass and small marine crustacea engulfing green phytoplankton. However, some food chains are unbelievably long and complex between the original producer and the ultimate consumer. In any event, it should be kept in mind that all the energy that drives the metabolism of all living organisms was derived in the beginning from the sun. (3) The *decomposers* (also called reducers) are microorganisms, mostly bacteria and myriad varieties of fungi that live in the soil and on or above its surface. Actually, the supreme importance of decomposition in the dynamics of an ecosystem has not yet been fully realized, and it may even turn out to be a controlling rather than a subsidiary factor. As energy is released and utilized by the enzymes of the microorganisms that carry on decomposition, it is then made available for the nutrition of higher plants in an ancient nutritional cycle. (4) *Abiotic factors* are those non-living physical aspects or components of the ecosystem that nevertheless have a definite influence on it, such as weather, climate, soil geology, water temperature, and hydrogen-ion concentration. After all, it should not be forgotten that all energy used by all living organisms for metabolism comes from the sun. To sum up, it appears that the most critical steps in the flow of energy in an ecosystem are the initial capture and transfer of the sun's energy into the chemical energy of the carbohydrates, proteins, and oils of plants, the utilization of the energy-containing materials of plants by animals in one or several steps, and the eventual release and utilization of energy-rich compounds of larger plants and animals by microorganisms. Thus, photosynthesis derives energy from the abiotic environment, and respiration contributes energy to drive metabolic processes, as well as to the external environment.

In view of the foregoing, it can be seen that energy flow through food chains and through ecosystems has been far more influential on the course of evolution of living organisms than has been formally recognized. As our environment gradually and irrevocably changes in response to man's ever-increasing activities and population pressures, we must learn more and more about the transfer of energy from organism to organism, as well as the man-created processes that have interrupted normal cycles of energy transfer, either by inhibiting the metabolic activities of the decomposers or by presenting them with energy-rich synthetic materials which they cannot metabolize.

Much of what might otherwise be said here has already been included in the summary chapter by Willis Johnson; in fact, there might be some difficulty in deciding which of these chapters should be the introduction and which the summary!

As already stated, we have attempted to present in this book a factual background for the intelligent understanding of our environment, which will aid greatly in the evaluation and assessment of environmental problems. We are all faced with potentially catastrophic problems and dilemmas; what we do about them must be decided by information and not by hysteria. Unfortunately, even now we do not possess a fully adequate body of basic information in the form of quantitative measurements to determine the long-range effects of man-made changes in ecological systems. However, an increasing tempo of research devoted to the understanding of how large ecosystems function has been launched under the auspices of the International Biological Program. By developing suitable models, and by the computer processing of enormous masses of data, eventually it may be possible to predict the behavior of these systems, and even to forecast how they will react to the stresses that modern man imposes upon his environment.

2

The Structure and Function of Ecosystems*

GEORGE MASTERS WOODWELL

Brookhaven National Laboratory

A native of Cambridge, Massachusetts, George Masters Woodwell attended the Boston public schools, graduating in 1946 from the Boston Public Latin School. He holds an A.B. degree from Dartmouth College (1950) and A.M. and Ph.D. degrees from Duke University (1956, 1958). Between 1950 and 1953, he served as an Ensign and Lieutenant j.g. in the U.S. Navy on oceanographic survey ships and for one year as Electronics Officer and Underway Watch Officer on a cruiser. He has held appointments and lectured at various universities. Between 1957 and 1961, he taught in the Department of Botany at the University of Maine in Orono, leaving the University of Maine as Associate Professor of Botany to join the staff at Brookhaven National Laboratory in 1961. In addition to his appointment as Senior Ecologist in the Biology Department at Brookhaven, he holds an adjunct appointment as Lecturer in Ecology at Yale University.

His research interests at Brookhaven have been focused on the structure and function of ecological systems, having worked specifically on the ecological effects of ionizing radiation and on the patterns of movement of the nutrient elements and toxic materials through the various biological, geological, and chemical cycles of the earth. He has been especially interested in the cycling of DDT over the surface of the earth and has recently been involved in efforts to use the courts to bring a closer relationship between what we know of ecology and what we do through government in the management of resources.

It is interesting to contemplate a tangled bank, clothed with many plants of many kinds, with birds singing on the bushes, with various insects flitting about, and with worms crawling through the

*Research carried out at Brookhaven National Laboratory under the auspices of the U.S. Atomic Energy Commission.

damp earth, and to reflect that these elaborately constructed forms, so different from each other, and dependent upon each other in so complex a manner, have all been produced by laws acting around us.

CHARLES DARWIN,
The Origin of Species

The Unity of Nature

The unity of nature that Darwin saw in a tangled bank is the subject of this essay. The unity extends not only to the plants and animals whose evolution is governed by the laws Darwin and Wallace and others set forth, but also to the nonliving aspects of environment: to the soils, water, air, and local climate whose characteristics are also changed by living systems. Phrases such as "plant community," "animal community," and various types of "associations" have long been used for assemblages of species to distinguish these arrays from the species themselves that are the principal units considered by Darwin and most contemporary evolutionists. These assemblages, however, were recognized early to be segments of a still larger unit, which includes both living and nonliving components, all of which change as evolution progresses. The word used for this totality of environment is "ecosystem," a contraction first set forth by the British botanist A.G. Tansley (1935), who saw the need to express the unity of nature more precisely than his contemporaries were doing with their definitions of "community" and "association."

The meaning of ecosystem has been broadened in recent years from what Tansley had in mind. It is common now to speak of "the human ecosystem" or an "agricultural" or "urban ecosystem" as though these units were immediately obvious, clearly delimited, and universally defined. They are not, of course, and there is reason to question whether any universal definition of the limits of such systems is possible. It is, however, possible to recognize units of landscape that are more or less intact, partially isolated physically and perhaps biologically from other units. Drainage basins are a good example. Some others are lakes and streams; forests, grasslands, tundra; oceans and segments of them that are also more or less isolated such as estuaries. At the other end of the scale we must consider the entire earth as an ecosystem, reflecting in its qualities the sum of all of the influences of life including the influences of man himself. And we can think of this larger, ultimate unit as comprising many smaller interacting ecosystems, each of which represents some segment of the earth's surface. Each segment is in constant, vital interaction with each other segment. These interactions are similar to those of the diverse species of Darwin's tangled bank. And we must recognize that the ecosystems themselves are in large degree a product of the same type of evolution that builds species. Ecosystems are the basic unit of ecology and our objective is an examination of the principles governing their structure and function.

There are two principles that we must set forth in the beginning and will refer to throughout the discussion. Both treat the development of ecosystems toward "maturity." But they treat development on different scales of time, one evolutionary, the other successional. First, let us consider evolutionary maturity. We know from careful observation around the world that evolution tends continuously to develop new ways of supporting more life. When resources are available, species are evolved that use them. As possibilities for further evolution diminish and rates of change slow, we say that an ecosystem is approaching "maturity." Tropical forests and the animal communities of the oceanic depths are mature on this scale. A clear, simple definition of evolutionary maturity is less important than an understanding of the pattern of changes leading toward maturity. A considerable body of evidence and opinion suggests that as evolution progresses, existing species become more sharply differentiated and new species are formed. The new species are themselves new resources allowing further evolution, and the capacity of the site for supporting life is enhanced (Odum, 1969, 1971; Margalef, 1963, 1968). The structure and degree of organization of the total living system tends to increase. The structure may be in the form of species with larger bodies, such as trees, or it may be in the form of more complex feeding relationships, with higher and more intricate forms of carnivores.

Succession refers to the short-term changes that follow any disturbance such as fire or wind-throw or harvest by man, and leads in years or decades to an ecosystem that changes very slowly. We call this stage "stable," "climax," or "successionally mature." The difference between evolutionary changes and the short-term changes of succession is frequently overlooked in attempting to analyze important environmental questions. The difference is in time; evolutionary changes may require hundreds of thousands of years while successional changes are short term, spanning decades or perhaps a century or so. Successional changes are of practical interest to man in altering the structure of natural ecosystems in the short term; evolution is of interest in general only over much longer periods, too long to be of immediate significance in correcting year-by-year man-made changes in environment. Our interest is in both evolutionarily and successionally mature ecosystems that are the matrix within which man builds his own domestic ecosystems.

Measurement of Structure and Function

Energy and Carbon

For an example of mature natural ecosystems we turn most commonly to forests, which are the product of both evolution and succession in much of the habitable portion of the temperate zones. Here certain aspects of structure are conspicuous. Trees obviously provide the form of the forest; they provide shade, habitat for other organisms, and major segments of nutrient and hydrologic cycles. Other less obvious aspects of structure are the quantitative relationships between populations of plants and animals, or between individual plants and animals, yet such relationshps are an essential part of the structure of natural ecosystems. The discovery of details of the structure and function of such ecosystems and the expression of these details in a simple, comprehensible, and useful way is a major objective of ecology. It is also a major challenge for science because such systems are extraordinarily complicated. The subject can only be sketched in a brief outline here. What can be said of progress to date?

Analyses of the structure and function of ecosystems are often based for simplicity on the use and distribution of energy or carbon (Odum, 1971; Morowitz, 1968). The advantage is that the technique can be applied to all ecosystems and gives a basis for making comparisons among the earth's different systems. The flows of energy and carbon are linked directly to the cycles of mineral elements such as nitrogen, phosphorus, potassium, calcium, magnesium, sulphur, and iron that are necessary for life. Once we know the pools and fluxes of carbon, we can complete major segments of the cycles of other elements. The disadvantage of this approach is that we have not accounted for the intricate relationships between the organisms that make the energy or carbon cycles work.

The fluxes of energy, dry organic matter, or mineral nutrients are widely recognized now as extraordinarily complex, so complex in fact as to require models of various types to help us understand them. Various models have been used, ranging from simple visual schemes to complicated mathematical models designed for computer use. Most such models emphasize the flux of energy or carbon because these two factors integrate the total function of ecosystems. The models help to clarify the function of each ecosystem, its interactions with other systems, and its potential for support of man. One of the simplest models (Figure 2.1) was designed to show the effects of disturbance on a forest (Woodwell, 1963). The basis was energy although it just as well could have been carbon. The model was designed to show simply and diagrammatically how disturbance shifts the structure by changing the sizes of major pools of energy and the ratio of inputs (photosynthesis) to outputs (respiration). Such a model gives little information about the internal structure of ecosystems.

A more complex model takes into account the flux of energy not only into the plants but also into the animal populations, including both the grazing or browsing food chains, and into decay. Much of the original insight for such analyses came from the work of a brilliant young scientist at Yale named R.L. Lindeman, who in 1942 published a paper in the journal *Ecology* entitled "The Trophic-Dynamic Aspect of Ecology," setting forth qualitatively his view of the relationships that must exist between plant and animal populations and among animals that use different types of food. His work started a new scheme of

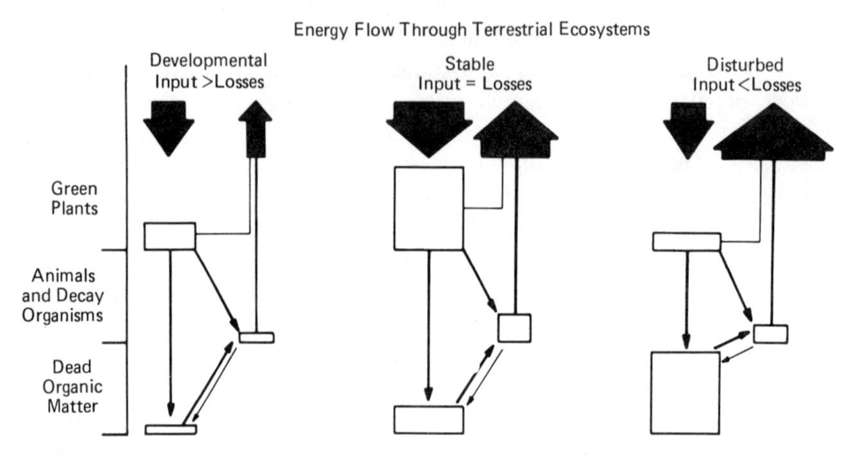

Figure 2.1 A Simple Model of a Forest Based on the Distribution of Energy. Disturbance Shifts the Relationships among the Various Reservoirs, Increasing the Amount of Respiration Temporarily.

analysis that led to the contemporary approaches to ecology outlined here. The essential point of all such schemes is that the flux of energy or carbon follows two basic routes: a grazing food chain and a decay chain, which have many interconnections. Both chains are obviously complex and are probably better called "webs" to emphasize this complexity. The pathway through decay is further complicated by a major split between the type of respiration that proceeds in the presence of oxygen (aerobic) and the type that has no requirement for oxygen (anaerobic). Thus there are three major routes that carbon, or energy, may follow through an ecosystem (Figure 2.2). Unfortunately for students, these three routes are not simply laid out in nature, but exist totally interlaced by virtually every organism.

Much research has been aimed at resolving the question of the efficiency of the transfer of energy from plants to animals and from one population of animals to a predator. There is no simple generalization that applies universally because the transfers are usually not simple. Herbivores such as the gypsy moth, the hippopotamus, or cattle do not digest all of the food they take in, but excrete an appreciable fraction intact or only slightly changed. Efficiency of such transfers can vary very greatly. A reasonable assumption for our purposes appears to be that 10 to 20 percent of the energy entering any trophic level is available for transfer to the next higher trophic level in a grazing food chain without substantial disruption of the populations. This means that 10 to 20 percent of the energy fixed by plants is available for transfer to herbivores directly. And 10 to 20 percent of that energy is available to support carnivores that prey on the herbivores. While this is an attractive simplification, it does not account explicitly for transfers of energy to the decay routes, nor does it account for the flux from decay back into the grazing chain. Both fluxes are significant, although

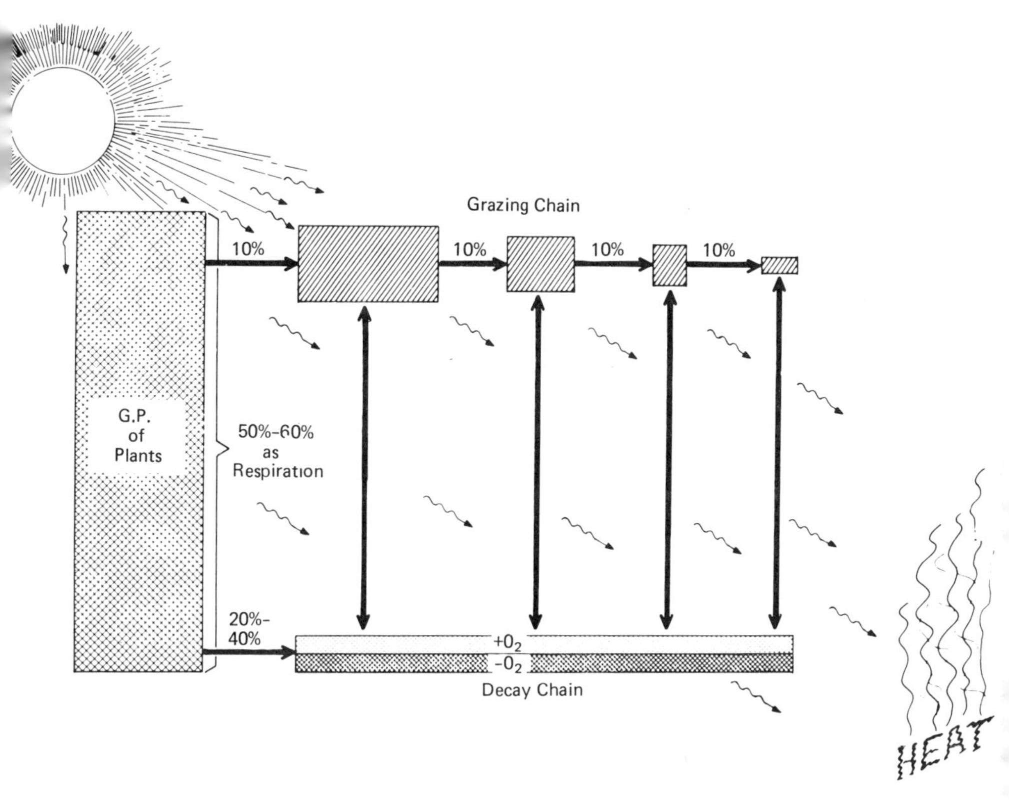

Figure 2.2 The Trophic Structure of Natural Ecosystems. The Flux of Energy Is through Plants either into a Grazing Food Chain or into Decay. With Severe Disturbance the Grazing Chain is Lost and That Energy Is Diverted to Decay.

variable. Their total magnitude is generally less than 30 to 40 percent of the total energy fixed in photosynthesis, although as ecosystems are simplified by chronic disturbance, including exploitation by man, the amount following the decay route increases.

The total energy fixed (Gross Primary Production) by the plant community of a temperate zone forest has been shown to be divided approximately as outlined in Table 2.1 (Woodwell, 1970). The forest in this instance is considerably simplified inasmuch as segments of the grazing food chain have been eliminated by man; the flux of energy into decay increases to almost 25 percent of the gross production. As the forest approaches successional maturity, the fraction of gross production stored decreases with a parallel increase in the amount flowing into decay. At maturity total decay will obviously exceed 40 percent of gross production. An intact grazing food chain in its maximum development would probably not change this picture appreciably.

The relationships outlined in Figure 2.2, while grossly simplified, show

Table 2.1 The of Gross Production (Total Energy Fixed) in a Temperate Zone Forest

	Organic Matter *Oven Dry Weight* (grams)	Percent Gross *Production*
Respiration of plants	1450	55
Growth (in immature forest)	550	21
Herbivores	30	1
Decay	620	23
Total gross production	2650	100

SOURCE: G.M. Woodwell, "The Energy Cycle of the Biosphere." *Scientific American*, 1970, **223**.

certain limits on the structure of natural ecosystems. First, there is an inexorable commitment of gross production to support of respiration of the plant population. This commitment for a temperate zone forest is about 55 percent of the gross production, leaving less than half the total as net production for support of all consumer populations and for growth of the plant community. Second, grazing food chains, no matter how well developed in an evolutionary sense, probably do not use more than 20 percent of the gross production; storage and the decay chains receive the remainder. In most instances grazing chains receive very much less than 20 percent as indicated for the forest of Table 2.1. Third, losses between trophic levels place a limit on the amount of trophic structure that can be developed. If transfers along a grazing food chain are restricted to 10 percent efficiency, then a third level of carnivores could receive not more than one-thousandth of the total gross production. Clearly, the third level carnivores are going to be limited in size and abundance. Fourth, in an ecosystem such as the forest of Table 2.1, a major fraction of the gross production passes through decay food chains.

One final point is essential to our discussion. Disturbance, whether short term or chronic, changes these aspects of structure in one direction; it tends to reduce grazing food chains, shifting the flux of energy into the decay routes. In terrestrial ecosystems the balance between aerobic and anaerobic decay may not be affected greatly. In water, however, disturbance and especially pollutions of various types tend to shift the decay chains toward the anaerobic route, accentuating the series of changes now recognized as "eutrophication."

Thus far the emphasis has been on terrestrial ecosystems because they are the most familiar. In lakes, streams, estuaries, and the oceans the flux of energy and carbon is quite parallel to what has been outlined for a forest. In aquatic ecosystems, however, the structure of the standing crop of plants and animals is usually less conspicuous than for a forest; the plants are smaller, frequently unicellular, shorter lived, and more numerous. The ocean is probably the oldest and most stable habitat—and the most mature in an evolutionary sense. It supports large numbers of different species of plants and animals. Food webs tend to be more complex; the grazing chain is more important in most segments of water bodies that are regularly aerobic, including oceanic and coastal waters. Here food webs start commonly with smaller plants, planktonic or free floating algae, that exist with almost incredible diversity. Such plants are not conveniently harvested or used directly by man. Generally, human harvests of the sea are taken several trophic levels above the plants, perhaps as many as five or more under certain circumstances, but commonly as many as three. The amount of energy or fixed carbon available by this route is obviously limited, not only by the limits of production of the algae, which are low (these yields have been compared recently in Woodwell, 1970), but also by the limits set by losses in passing through several trophic levels. Recent analyses of the potential of world fish production suggest that maximum yield can probably not be greater than about 100 million metric tons, or less than twice the yield in 1968. These estimates did not take into account the losses of fisheries through pollution or mismanagement, now so common in the oceanic fisheries.

The Chemistry of the Biosphere

The flux of energy and carbon through natural ecosystems is so complex and fascinating in itself that we tend to forget that this flux is sustained by a continuous and predictable flux of physical conditions and chemical substances. While there is no question that throughout the earth's history there have been many abrupt, even cataclysmic, changes in these conditions and many much slower but equally revolutionary modifications of climate, it is also true that the earth's present biota has had many millions of years of conditions not greatly different from the range of conditions now existing on earth to reach some degree of evolutionary maturity. Ecologists usually think of evolutionary maturity as having reached a maximum in the tropics, especially in the tropical rainforests, but recent studies of the ocean floors remind us that the seas are even older than the land and that here we may find the highest degree of evolutionary maturity on earth (Sanders, 1969). The important point, however, is the principle that the interactions of organisms and environment are reciprocal; not only does environment determine the conditions for life but the organisms themselves influence the conditions of their environment.

The evolution of the atmosphere is the most profound example. The atmosphere is the result of the evolution of a system of using solar energy to obtain hydrogen from the water molecule. The hydrogen is used to reduce carbon dioxide. Molecular oxygen is released. As this type of photosynthesis became common over many millions of years, a new atmosphere was formed on the earth, one containing about 80 percent nitrogen, .03 percent carbon dioxide, and about 20 percent oxygen. The oxygen made a new, more efficient type of respiration possible, with a whole new array of plants and animals, and our modern biota began its rapid evolution, speeding the evolution of the atmosphere.

There are countless other clear examples. A. C. Redfield, in a splendid essay in the *American Scientist* in 1958, showed that the chemistry of life in the oceans is adjusted with precision to the chemistry of the oceanic waters. The relationship is hardly surprising when considered against the evolutionary background, but to generations reared to understand that it is possible and necessary to adjust the chemistry of crops and soils to produce a good yield in agriculture, there is some novelty to the idea that the survival of oceanic fishes depends on the maintenance of the chemical conditions in the oceans that have prevailed on earth throughout the past several million years. Thus, there is virtually no possibility of effecting a change in chemistry that would improve the fisheries at all!

Such statements are widely challenged these days as unproved and unprovable within the ordinary context of science. This does not make them false as many of the more exuberant technologists would suggest; it simply places them in a special category of probable truths that must be tested in myriad ways with time. We gain further insight at the moment by returning briefly to consider the models we have used above. Energy alone is obviously inadequate as a basis for examining nutrient relationships of natural ecosystems. The fluxes of energy and carbon indicated in Figures 2.1 and 2.2 must be governed in some degree by the availability of nutrients such as nitrogen, phosphorus, and potassium, each potentially affecting evolution, succession, and the structure and function of any ecosystems. While we might like to think that Liebig's law of the minimum should apply in all such cases, it has a much more restricted application in natural ecosystems than it has in agricultural systems, where it was developed. Liebig's law says simply that the element available in least abundance governs the function of the system much as the weakest link in a chain affects the use of the chain. The law does not apply, in a simple way at least, to natural ecosystems because their fluxes into and from the system and their separate routes through the species are all integral to the system. The species, races, strains, genotypes peculiar to that place are the current survivors of an age-old evolutionary scramble for the resources of that site. The resources have been divided among those species through the intricate interactions of evolution acting on the hereditary material of the species themselves, producing an array of plants and animals that not only can survive in that physical and chemical setting but have become a part of it, enhancing the resources of the site by building on them, recycling nutrients, and ameliorating extremes of weather and climate. Under these conditions no *single* nutrient is "limiting." *All* nutrients are limiting. An addition or loss from the stock of any nutrient will cause changes in the system depending on the nutrient and the time and place of the change.

If we were to attempt to add the cycles of essential nutrients to our models of Figures 2.1 and 2.2 we might simply substitute the nutrients for carbon using the same pattern of flow because the nutrients travel in one form or another the same routes as carbon, although the ratios of one nutrient to another may vary considerably. Such an approach is simple enough and is frequently used. On the other hand, we might see a need for a separate model to incorporate the cycling of nutrients into the carbon flux. The nutrient flows would be connected through a myriad of cross connections among themselves and to the flux of carbon, and the model would become more complex in proportion as more nutrients were added. Complexity would increase further as other factors became important, such as the rate of successional change. Clearly such models are not simplifications; they will not be discussed further here. They exist in trial form as parts of varied efforts at interpreting details of the structure and function of ecosystems. The complexity of such models is, however, not an invention of man; it is, rather, a characteristic of nature and the models are an attempt to simplify nature into some scheme that will allow further understanding.

The two types of maturity we have discussed are part of the nutrient problem as well. Succession, proceeding as it does at a rate measurable year by year, can indeed be limited in rate and probably changed in direction by the availability of nutrients. Succession is probably also affected by the ratios in which nutrients become available, although there is very little research on such topics bearing on natural systems. As succession progresses, changes in the structure of ecosystems occur less rapidly, nutrients are caught in more complex cycles and tend to be regenerated within the system in increasing degree. At maturity or climax, inputs to an ecosystem are presumably balanced by losses. One might wonder whether there is a significant evolutionary tendency for climax ecosystems to expand, absorbing, holding, and recycling more and more nutrients in time as evolution progresses. Probably such changes occur in evolutionary time, measured for the systems we are discussing in hundreds, thousands, or hundreds of thousands of years, not in periods that are of short-term significance to man.

With this background the question of what happens in response to changes in the physical or chemical conditions of the earth becomes clear. Evolution offers no possibility of providing short-term adjustments to new conditions. The systems we must work with are those that are extant now, having reached their present height of complexity through evolution and the much shorter-term changes of succession. Disturbance of the conditions under which these systems have evolved can only disrupt the systems. In the simplest case of small but chronic disturbance, the ratios of one species or group of species to another may be changed in barely measurable degree. A more severe change might eliminate a species, affecting in the process all of the other species of the ecosystem. Such an effect might not be catastrophic; other species replace them just as oaks replaced the American chestnut of the eastern forests of North America in the first half of this century. Such a change, however, constitutes a profound shift in the structure of nature, for specialized herbivore and carnivore food chains that were based on the chestnut were lost, replaced by the food webs based on oaks, which already existed in many of the forests. Clearly the loss of this one species constituted a loss of structure from the forest, the structure being measured both as the loss of one of the largest and most majestic trees and as the loss of those segments of the trophic structure that were specialized for the chestnut.

We have an even more dramatic example in the eutrophication of shallow waters. Here the addition of phosphorus and nitrogen, but especially phosphorus, shifts both the total nutrients available and the ratios of nutrients to favor the growth of coarse, filamentous green algae instead of the normal flora that is frequently characterized by 50 to 100 species of diatoms. The filamentous algae do not support the complex food webs previously supported by the extremely diverse original flora; fisheries disappear or are replaced by ''trashfish'' or detritus feeders. Frequently, a major segment of organic production flows into the detritus food chains, overloading them with organic matter causing deficiencies of oxygen and causing release of all of the noxious products of anaerobic decay. The classic example of this series of changes occurred in the estuarine bays on the south shore of Long Island where wastes from a hundred or more duck farms were allowed to accumulate in the bays during the 1940s and 1950s causing an extreme case of eutrophication. During the hot periods in summer large sections of Moriches and Bellport Bays became anoxic, releasing H_2S in such quantities as to make boaters ill and to cause white paint on boats and houses to turn yellow. The decline of duck farming coupled with improved flushing of the bays brought relief. Such problems, however, are becoming increasingly common, as estuaries and other shallow waters are degraded by accumulations of nutrients, sewage, toxins, or by simply dumping waste heat from power generators.

Crisis of Environments

The Earth as a Series of Interacting Ecosystems

What seems clear from even such a superficial consideration of the structure of natural ecosystems as this is that the world has functioned through all of the history of man as one single living system, a biosphere, whose principal interacting units are major ecosystems: forests, grasslands, lakes, estuaries, and oceans. The important characteristics of each of these is not size alone, but the fact that they are living systems, adapted by an age-old evolution to a particular set of physical, chemical, and biotic conditions. They build themselves, repair themselves, stabilize their habitat, and regulate themselves—all on solar energy without help from man. The natural ecosystems are self-sufficient matrices potentially capable of supporting themselves indefinitely, supplying clean water, clean air, sewage treatment, and diverse natural resources, including fisheries and other food for man.

We are concerned at the moment that there may be limits on the largesse of these systems and on our freedom to exploit the biosphere for the support of man, a single species out of millions. One set of limits must lie at the point where human activities change those characteristics of the biosphere that are essential to the support of life as we know it, where the structure of natural ecosystems is changed worldwide, shifting the flux of energy from the grazing food chains that are the principal chains tapped by man and are characteristic of mature and healthy systems into the decay chains.

We are without question at or beyond the threshold of such changes. There are many signs. Perhaps the least controversial sign is a worldwide increase in the CO_2 content of the atmosphere due to the consumption of fossil fuels. While the evidence available at the moment suggests that the increase holds no immediate threat to the earth's biota, indeed may increase slightly the rate of fixation of carbon and the amount of energy available in support of animals, including man, the fact that human activities are now large enough to affect the worldwide pool of a substance as essential to life as carbon should be ample proof that human activities have passed one threshold of safety. Thus it should not have taken a decade to convince the world that a worldwide pollution with radioactivity posed a serious threat, and two decades to convince us that we had a worldwide problem with DDT residues that was serious, growing, and threatening to man in many ways. Nor should there be surprise in discovering that other chlorinated hydrocarbons travel the same routes and have similar effects—and are also worldwide pollutants. Nor should we be surprised if the list of toxins grows exponentially in the next years as our technology expands.

These are not simply examples of corporate misbehavior and callousness. They are evidence that the world is already overcrowded, that there is not now room in the world for all of the people and the technological trappings of the second half of the twentieth century without major changes in patterns of human activities. It is evidence that human ecosystems must suddenly adopt as limiting characteristics certain of the same features that mark the structure of natural ecosystems. And what are these?

Mature natural ecosystems are integral physical and chemical units that have well-defined interactions with one another. Chemically they are "tight," retaining and recycling their essential nutrients, not normally dumping them to degrade themselves and other systems. The problems we see as the "crisis of environment" arise because man-dominated systems and most of the natural systems that modern technology has domesticated are the antithesis of the functionally tight systems of nature; they leak nutrients and toxins that travel freely and degrade the other systems nearby that should be available to support man and his ecosystems. The point has been reached now where the effects are very far-reaching, perhaps worldwide; fisheries are threatened not simply by overharvests, but also by destruction of the estuaries; forests are affected widely, and many believe that the earth's potential for support of man is diminishing.

Man may survive on an earth that is impoverished, eutrophic, dirty, and dull; but must he?

The solutions are simple enough, if viewed in the abstract. They are to reduce the effects of man to a point where they are small in proportion to the capacity of the earth to repair itself, to acknowledge that living systems will always run the world and that safety, comfort, and purpose for man lies in simple restraint. The solutions are complex only if we acknowledge that avarice is intractable, that man will expand to occupy and use every square inch of the globe, that the growth of population and the growth of unbridled technology are inevitable and that this growth will bring its own solutions. These are risky assumptions, propounded by starry-eyed optimists, who believe that man can run the world . . . or don't care if he can't.

The effects of man can be reduced by recognizing, first, that man's ecosystems—urban, suburban, agricultural—have been designed to be parasitic on mature natural systems close by. New York uses the Hudson River, Jamaica Bay, the New York Bight, and a large segment of the Atlantic Ocean to clean its sewage. This is, perhaps, an appropriate use for those waters, but we recognize now that the same waters cannot also be used for bathing, fishing, for waste heat from reactors, and for treatment of the wastes of the growing millions of Long Island and New Jersey. For the man-dominated systems to continue to expand without inexorable degradation, they must absorb more of the "costs" of their existence, reduce their interactions with the natural systems they depend on. They must become more closely integrated within themselves, recycling their "wastes," limiting their degree of dependence, mimicking in structure and function the subtle and still poorly understood details of the interactions that Darwin saw so clearly among the plants and animals of a tangled bank.

References

MARGALEF, R., "On Certain Unifying Principles in Ecology." *American Naturalist*, 1963, **97**; pp. 357–374.

MARGALEF, R., *Perspectives in Ecological Theory*. Chicago: University of Chicago Press, 1968.

MOROWITZ, H.J., *Energy Flow in Biology*. New York: Academic Press, 1968.

ODUM, E.P., "The Strategy of Ecosystem Development." *Science*, 1969, **164**; pp. 264–270.

ODUM, E.P., *Fundamentals of Ecology*. Philadelphia: W.B. Saunders, 1971.

ODUM, H.T., *Environment, Power, and Society*. New York: Wiley-Interscience, 1971.

REDFIELD, A.C., "The Biological Control of Chemical Factors in the Environment." *American Scientist*, 1958, **46**; pp. 205–221.

SANDERS, H.L., "Benthic Marine Diversity and the Stability-Time Hypothesis." *Brookhaven Symp. Biol.*, 1969, **22**; pp. 71–81.

TANSLEY, A.G., "The Use and Abuse of Vegetational Concepts and Terms." *Ecology*, 1935, **16**; pp. 284–307.

WOODWELL, G.M., "The Ecological Effects of Radiation." *Scientific American*, 1963, **208**, pp. 40–49.
WOODWELL, G.M., "The Energy Cycle of the Biosphere." *Scientific American*, 1970, **223**, pp. 64–74.

3

The Diversity of Life

ERNST MAYR, RICHARD D. ALEXANDER, W. FRANK BLAIR, PAUL ILLG, BOBB SCHAEFFER, and WILLIAM C. STEERE

Chapter 3, The Diversity of Life, was written by a Committee (Panel 13) selected by the National Academy of Sciences, which consisted of Ernst Mayr, Harvard University, Chairman; Richard D. Alexander, University of Michigan; W. Frank Blair, University of Texas; Paul Illg, University of Washington; Bobb Schaeffer, The American Museum of Natural History; and William C. Steere, New York Botanical Garden. Each Panel member composed an extensive draft statement covering his areas of special interest and competence; William C. Steere brought these drafts together into a single first draft which was then completely revised and rewritten by Ernst Mayr, who produced the final report.

The Oxford University Press has very kindly given permission for the reprinting of this chapter, which appeared as Chapter 13, pages 498–530, in *Biology and the Future of Man* (1970), edited by Philip Handler. In our opinion, this book is by far the most comprehensive and up-to-date review of the broad scope of biology yet to be published.

Wherever we study nature, be it the microscopic life in a drop of pond water, the flowers on an alpine meadow, the birds and trees in our woods, or the marine invertebrates in the intertidal zone, we are overwhelmed by the number of different kinds of organisms we find. When we go north, south, east, or west, we encounter new kinds of animals and plants, and if we go to another continent, South America, Africa, or Australia, the entire animal and plant life will be different.

Only he who studies it with great care can appreciate the magnitude of this organic diversity. Physicists deal with about fifty kinds of elementary particles. Three of these combine to form just over a hundred different elements, and the number of naturally occurring simple molecules is a matter of a few thousand. In contrast, biologists estimate that five to ten million different species of organisms exist today, each differing significantly from its closest relative. Within a single species most of the millions of members are genetically different from all the others. They are unique. And yet currently living species are only a small fraction (far less than one percent) of all the different kinds that have lived and died since life began on earth, some three or four billion years ago. The totality of biological diversity is almost incomprehensible to the human imagination.

The biological science that deals with this diversity is *systematics*. It deals with the elementary classification of animals and plants (taxonomy) and studies the evolutionary, ecological, and behavioral aspects of their diversity. In these broader concerns it joins with all other biological disciplines.

The study of diversity is important for many reasons. Organic diversity is one of the most conspicuous aspects of the world we live in. It is part of man's nature to be interested in more than purely utilitarian things, to concern himself with the stars in the heavens and with all his fellow creatures. As early as 1694, the great British botanist John Ray charmingly discussed "why the good Lord had created such a vast multitude of insects as the world is filled with; most of which seem to be useless, and some also noxious and pernicious to man." He gave a number of reasons. "One might be to exercise the contemplative faculty of man; which is in nothing so much pleased as in variety of objects; . . . new objects afford us great delight, especially if found out by our own industry." The modern biologist sees still other reasons.

Every species of organism is a unique genetic system, differing from all other species in structural, physiological, behavioral, and other characteristics. Every biologist must determine the identity of the organism with which he is working to make sure that his observations and experiments are meaningful. A classification of the diverse species is an indispensable information retrieval system. The name of an organism is like the index number of a file; it gives immediate access to all the information existing about it. The scientific name is the key to the entire literature that deals with a particular species or higher taxon. Evolutionary relationships among organisms, revealed by classification, represent an important source of information for every kind of biological study.

The high specificity of different kinds of organisms is also of great practical importance. When prickly pear was introduced into Australia by a cactus fancier, it spread unchecked and overran fifty million acres of valuable sheep pastures. Only after 145 organisms found feeding on prickly pear in North and South America were carefully studied was a small and inconspicuous moth finally located in Uruguay which flourished in the Australian climate, and whose larvae soon destroyed the prickly pear plants, restoring the pastures to their former value. Hundreds of similar examples can be cited from the biological control literature.

Water pollution is now rightly considered one of our foremost problems. There is no more accurate or quicker way to determine the degree and the nature of pollution of a given body of water than to study the algae and other microscopic plankton organisms found in the water. Each species has its own requirements for oxygen, nitrogen, and other organic and inorganic additives to the water, and their study permits a rapid and inexpensive monitoring service. Monitoring for pollution through numerical surveys for key species plays an increasing role in the study of soils and polluted waters. Other areas where a precise knowledge of organic diversity is of great importance will be discussed later.

Aspects of Diversity

Life appears to have originated on earth only once. The diversity of life now found is due to a continuous process of speciation, the process wherein one species yields several descendant species. Since each of the major groups of organisms—the seed plants, the insects, the birds, the mammals, the primates, the hominids, and so on—go back to one original founder species, speciation is the most important single event in evolution. How does it come about?

Darwin's great work was entitled *On the Origin of Species.* Yet, curiously, Darwin actually failed to solve the problem of the multiplication of species, the problem of how one species splits into several daughter species. He failed because he did not fully understand the nature of the biological species. Species of organisms cannot be defined by degree of difference or by their essences, as can "species" of inanimate objects or the species of the philosophers. A species of animals or plants is characterized by two properties: by representing a gene pool adapted to utilize a particular niche in nature, and more importantly by the possession of protective mechanisms that prevent mixing with other gene pools. The genetic mechanisms by which crossbreeding with other species is prevented are called *isolating mechanisms.* Each living species is a harmoniously adapted genetic system, and the isolating mechanisms operate to prevent the mixing of two incompatible genetic systems which could lead to the production of disharmonious and selectively inferior hybrids.

It is now possible to pose the problem of speciation more precisely. How do new gene pools orginate protective isolating mechanisms that work efficiently? The most frequent process to achieve this is "geographical speciation," the temporary spatial separation of a population from the gene pool of the parental species and the gradual building up of isolating mechanisms in the geographically isolated population. After the isolating mechanisms have been reasonably perfected through steady process of mutational change, the extrinsic (mostly geographical) barrier can break down, and the daughter species can now coexist with the parental species, protected by its isolating mechanisms.

Another process of speciation is by chromosome doubling (polyploidy) a process to which about half of the plant species owe their origin, but which is very rare in sexually reproducing animals. As noted in the earlier discussion of the origin of new proteins, in an organism possessing a duplicate set of chromosomes, one set is free to mutate without threat to viability. If the products of the mutated genes afford selective advantage in a specific environment, an evolutionary step has been taken.

Although the process of speciation is now understood in its simplest outlines, there is still much uncertainty about the factors that control the rate of speciation in different organisms. What influence does the size of the gene pool have on this rate? There are indications that small "founder populations" speciate far more rapidly than large populations. Most of our knowledge is based on birds, a few other groups of vertebrates, as well as some insects, molluscs, and flowering plants; speciation in lower invertebrates and lower plants has hardly been explored. Most available data concern organisms that live in the middle latitudes of the Northern Hemisphere. Here it is evident that the climatic changes during the Pleistocene fragmented the ranges of many species, greatly accelerating the process of geographical speciation and that here, as well as on islands, are numerous instances of incipient species, many of them highly localized. Incipient species provide an invaluable opportunity for elucidating the rate of geographical speciation and the factors that contribute to it. It is imperative that such studies are carried out before man's misuse of the environment destroys this opportunity forever. The nature of the biotic environment and other ecological factors basic to speciation have been studied insufficiently. There are, as yet, no answers to such questions as:

1. How does the coexistence of populations of other, closely or distantly related, species affect the rate of speciation?
2. When several coexisting species have similar environmental requirements, does this facilitate shifts into other environmental niches?
3. Under what conditions does a limited amount of hybridization between sympatric species result in a gradual breaking down of the reproductive isolation between these species, or conversely lead to the perfection of the isolating mechanisms through the selective elimination of inferior hybrids? Do plants differ in this respect from animals?

Answers to these questions can be provided only in part by a perceptive study of nature's own experiments; it will surely also require some deliberate experimentation—that is, the manipulation of populations under essentially natural conditions. A particularly great need exists for expanding speciation research into the tropics. The tropics, for example, the Amazon Basin, are notable for their extraordinarily rich faunas and floras. These areas are also remarkable for their highly stable environments, and the question arises to what extent is the high number of tropical species due to a high rate of speciation and to what extent due to an increased survival of species in the equitable tropical environment. The richness of the tropical biota does not necessarily prove an elevated rate of speciation. Sophisticated work on rates and mechanisms of speciation in the tropics is remarkably scarce. Tropical ecosystems are characterized by containing very many species, each with relatively few individuals, in contrast to nontropical ecosystems containing many individuals of relatively few species. Is there any relationship between these differences in ecosystem structure and the rates and processes of speciation?

There are indications that geographical isolation may be unnecessary for speciation in certain ecological specialists, such as plant-feeding insects that are highly host-specific. After a population has become established on a new host plant, can selection pressures overcome the leveling effects of gene flow between populations on the old and the new plant hosts? How large a part of the genotype participates in the genetic basis for host specificity and isolating mechanisms? These unanswered questions are wide-open areas for future experimental research.

We have every reason to assume that no qualitative difference exists between speciation in marine organisms and the processes established for terrestrial organisms. But we do not know what the geographical barriers in the oceans are, particularly for pelagic organisms. Many species are restricted to certain water masses with well-defined physical-chemical properties, and this preference may reinforce the geographical barriers. As with freshwater organisms, the presence of particular species in a given body of water is often as useful an indicator of the history and the physical properties of this water mass as certain physical constants.

The shoreline, particularly the intertidal zone, presents some fascinating problems. The total width of the species range at a given locality may be just a few meters, and yet may extend for five hundred or a thousand miles. The essentially linear nature of the distribution of the rich intertidal fauna and flora poses problems of climatic gradients, geographical variation of niche preference, and changes in the competitive interaction with the changing biota of other localitites. Many intertidal genera are rich in sympatric sibling species, and their precise identification is a prerequisite of ecological studies.

The determinants of the actual number of species found in a given locality are still incompletely understood. Why is a tropical reef ten times as rich in species as a rocky reef in cool waters? Why is a tropical forest so rich in trees, insects, and birds? What are the respective contributions made by competition, predation, richness of resources, or mitigation of seasonal contrasts? We cannot begin to answer these questions until we have better factual information, based on improved census methods. The study of the causes for changes in species diversity is at present one of the most active areas of population ecology.

Geographical Variation

A species is represented at every locality by a single population, and it is this population that interacts with the coexisting populations of other species. Yet each species has a more or less extensive geographical range, and most consist of thousands or even millions of such local populations. Even though they share the isolating mechanisms and other physiological and developmental components of the genetic system of the species, they will, nevertheless, differ in minor genetic ways from one another, not only owing to chance phenomena (e.g., mutation, recombination, errors of sampling), but also because the population of each locality is exposed to a slightly different set of selection pressures. For example, where the environment changes abruptly, as may occur in soil color, the protectively colored coat of mice may change equally abruptly, while other attributes of these mice remain unchanged. Where the environment changes gradually from one end of the species range to the other, one generally finds only a gradual change in those characters that vary geographically.

Any feature of an organism that adapts it to its physical or biotic environment may vary geographically. This is as true for physiological and behavioral characters as for structural ones. Populations of a grass may flower in March along the Gulf Coast, while other populations of the species in Kansas may not flower until May or June. The breeding season of populations of a species of toad that lives in the Great Plains is regulated by rainfall, while in populations of the same species in the Great Basin, with permanent runoff water from the mountains available, the breeding season is regulated by temperature and/or photoperiod.

No matter how well populations respond to the demands of local environments, there is a line, the geographical species border, which is the limit of tolerance of the genetic system of a species. Insufficient research has been done to answer the tantalizing question of why the frontier populations of a species do not respond to the selection of new genotypes that would permit expanding the species border at a regular annual rate. To answer this question would require a far greater knowledge of the cohesive factors in the genetic systems of species than we have at present.

There are rather severe limits to the tolerance of a gene pool. A phyletic line can expand its utilization of the environment only by budding off new gene pools (speciating). Yet, the entering of new niches can be effected by a minor reconstruction of the genotype. Birds illustrate excellently how a type of organism can become adapted to a multitude of different environments without any basic change of its original morphology. No matter how diverse the niches of woodpeckers, penguins, hummingbirds, albatrosses, ducks, swifts, etc., are, these birds are nevertheless anatomically, physiologically, and in most aspects of their life histories, remarkably similar.

Among plants we find that some families have become almost entirely adapted to certain special environmental situations, such as the cactus family to arid country and the water-lily family to aquatic situations. Other families are much more flexible and flourish in many different kinds of environments without giving up their basic family identity. For example, the sunflower family runs almost the whole gamut of plant forms, ranging from annual and perennial herbs, including some of our most familiar weeds, to large woody trees and shrubs, herbaceous and woody climbing and twining vines, the strange plants of high mountains in Africa and South America that consist of a tall, thick stem with a crown of leaves at the top, and even to cactus-like succulents.

Evolutionary convergence, with similar selective pressures driving unrelated and rather different genotypes into similar ecological niches, is another exciting but inadequately investigated phenomenon. The "cactus" growth form has appeared in several distinct families of plants, and some of these succulents are so similar in appearance that a nonspecialist has difficulty in identifying the family. Old World tree frogs and New World tree frogs are so similar in adaptation to life in trees that it is necessary to examine the skeleton to determine which is which, yet they have originated as independent radiations within separate families. In Australia, where ordinary frogs (ranids) are virtually absent, a tree frog has evolved habits, body size, and even shape and appearance of our common leopard frog (a ranid). These and many other examples of convergence imply that there is a finite number of ways in which an organism of a given basic genotype can "make a living." However, the study of convergence is only now beginning to shift from a purely descriptive natural history phase to a sophisticated and statistical analysis. We do not know yet what part of the genotype responds to such highly specific selection pressures and how the harmony of the gene pool is maintained during the acquisition of the new adaptation.

Origin of Higher Groups

In addition to species, we distinguish broader assemblages in the diversity of nature, such as birds, beetles, ferns, and snails. How did they come about? Some students of evolution formerly favored large-scale mutations as the cause of the origin of new types. But all the available evidence indicates that higher groups of organisms arise through an extension of the same summation processes that give rise to species. The interpretation of the fossil record by paleontologists, combined with an understanding of the genetic and evolutionary mechanisms, permits the evolutionist to paint a convincing picture of how macroevolution occurs.

The first step is always the entrance of a population or species into a new niche or adaptive zone. It seems as if some individuals of such a population are genetically predisposed to make such a change. In other cases the mere formation of a new habit without an initial genetic change may be the first step. Once this step is taken, a new selection pressure will arise, favoring all those individuals which by their genetic constitution are superior in utilizing the new niche. This is apparently how the earliest ancestors of all terrestrial animals became amphibious and how the earliest ancestors of all flying animals began to glide. The probability is, of course, extremely low that any particular genotype has the potential to initiate a new major group, but with millions of species simultaneously in existence, this must have happened many times in the course of the more than three billion years of life on earth. Even where the first adaptations were behavioral, like air swallowing in an ancestor of the lungfishes, they set the stage for additional structural and functional adaptations which make the occupation of the new adaptive zone more effective.

How do entirely new organs or structures come into existence? Darwin knew that there are two answers to this question, even though much of the detail has been worked out only in recent years. Sometimes merely the intensification of the function of a previously existing organ is involved, such as when lungs developed as sinuses of the esophagus in air-gulping fishes living in stagnating swamps. The other answer is that a preexisting structure takes on a second function without interference with the original function until the new function becomes the primary one. The feathers in birds, for instance, are believed to have been acquired by the avian ancestors to facilitate maintenance of a constant body temperature and acquired only secondarily the new function as enlargements of the gliding surface of the wing. Evolution is quite opportunistic, selection making use of whatever variable structures are available. Floating in water is tremendously important in all pelagic animals. Yet each group has solved this problem in a somewhat different way: some by air-filled swim bladders, others by storing oil drops, and still others by enlarging the body surface (to increase friction) or by methods for active swimming. It is this very diversity by which evolution solves an adaptational challenge which indicates that the Darwinian interpretation, variability and selection, is indeed consistent with all the known facts.

The fossil record poses numerous challenges to the evolutionist. Why do certain groups remain virtually unchanged for hundreds of millions of years, while others simultaneously undergo radical changes, seemingly in the same environment? Why do certain groups undergo almost explosive diversification leading to a simultaneous invasion of all sorts of new adaptive zones—referred to as adaptive radiation—while others show no indications of branching and remain within the ancestral niche?

Some of these evolutionary happenings, long known to the paleontologist, seem to be going on right under our noses. Beautiful contemporary cases of adaptive radiation are known from tropical archipelagos. An example is Darwin's finches in the Galápagos Islands. Here a finchlike ancestor has given rise to grosbeaks with enormous bills and to other forms that fill the ecological niches of warblers and woodpeckers. Even more extreme is the case of the Hawaiian honeycreepers (Drepanididae), in which just about every bill type has evolved that is known among songbirds, ranging from extreme grosbeak types to the most slender, long, curved sicklebills (adapted for nectar feeding). There is every reason to believe that cases of adaptive radiation are extremely widespread, but only a few have been thoroughly studied up to the present.

One of the most interesting challenges of evolutionary research is the reconstruction of the intermediate stages leading from one kind of organism to another one. The discovery of key fossils, so-called missing links, has been a great help, such as *Archaeopteryx* between reptiles and birds, the mammal-like reptiles (therapsids) between reptiles and mammals, and *Ichthyostega* between rhipidistian fishes and amphibians. Many other such transitions are still shrouded in complete mystery. The higher plants (angiosperms), now the dominant group of plants in most of the world, have not yet been clearly traced to an ancestral group among the lower plants. Some 25 major phyla are recognized for all the animals, and in virtually not a single case is there fossil evidence to demonstrate what the common ancestry of any two phyla looked like. Representatives of nearly all the animal phyla with preservable hard parts appear in definite form at the time of the Cambrian (or latest Precambrian) from 500 to 600 million years ago, when the first animals turn up in the fossil record. A more careful study of the earliest fossils, a study of the chemistry of the preceding strata to determine the reasons for the sudden occurrence of so many hard-shelled forms, a far more extensive search through late Precambrian deposits, are among the needs of further research. This has to be combined with a new look at the axioms of comparative anatomy. What structures of the primitive invertebrates are genuine indicators of relationship, and what others might have originated repeatedly as the most logical responses to functional demands? What is the meaning of the facts of embryology in relation to phylogeny? A large area of biology that seemed an exhausted mine is suddenly becoming productive again.

PROCESSES OF EXTINCTION

Of all evolutionary phenomena, extinction, the termination of an evolutionary line without descendants, is perhaps the most poorly understood and yet one of the most remarkable. More than 99 percent of all evolutionary lines that once existed are now extinct. Considerable progress has been made in recent years in explaining the extinction of species owing to changes in the physical or biotic environment. Yet, all explanations for the decline and final extinction of entire major groups, as has occurred many times in the geological past, are still totally unsatisfactory.

At the close of the Cambrian period, of sixty trilobite families, the dominant group of animals in this era, nearly forty disappeared from the subsequent fossil record, and after another period of flowering in the Ordovician, the entire phylum disappeared before the end of the Paleozoic. Near the end of the Permian about twenty-four orders and half of the then known families of animals became extinct; and again, during the last part of the Cretaceous, one quarter of all the animal families were eliminated. Such periods of extinction have been equally devastating for dominant groups in the oceans, like the ammonites, and on land, like the dinosaurs. This great group of reptiles included relatively small species and gigantic ones, carnivores and herbivores, running species and sluggish, amphibious dwellers of swamps, and yet they all succumbed. Not one of the many theories advanced to explain this catastrophe is convincing. By contrast, plant extinctions have been more gradual; the important floristic changes have not coincided with the major extinctions of animal groups. Most major groups of plants persist for long periods, and many are seemingly immortal.

The Permian, Triassic, and Cretaceous extinctions involved both marine and terrestrial animal groups, many of them distributed throughout the world in a wide range of habitats. Competition from evolutionary newcomers, changes in climate, or modifications of breeding habits can hardly explain the disappearance of such dominant and widespread groups as the dinosaurs or the marked decline of the ammonites at the close of the Permian, during the Triassic, and their final extinction at the end of the Cretaceous; nor can these factors explain the extinction of the Devonian placoderm fishes and the abundant and diversified Mesozoic marine reptiles. With so many species in existence, one would have expected at least some of them to shift into new adaptive zones and thus escape the fate of their relatives. Why was natural selection so powerless to prevent these wholesale extinctions?

Competition is sometimes a factor, and the absence of it is undoubtedly a primary reason for the survival of *Sphenodon,* the last representative of the rhynchocephalian order of reptiles, on an island off New Zealand. The prosperity of marsupials in Australia and the Papuan region may well be due to the scarcity of placental mammals in these areas. It is much harder to believe that the diurnal terrestrial reptiles of the Mesozoic were actually competing with the presumably nocturnal Mesozoic mammals, although it is correct that the extinction of the Mesozoic reptiles vacated broad niches and habitats that were subsequently filled by the diversifying mammals.

A family, an order, a class, or a phylum does not become extinct at once. Instead, the most reasonable general explanation for extinction is to relate it to changes in the entire ecosystem. The rise of the angiosperm plants in the Cretaceous undoubtedly had an adverse effect on the previously ruling herbivore reptiles, and this in turn brought about the decline of the carnivorous types. The important point is that the entire biota at any one time is interrelated and interdependent in an extremely complicated manner. Whatever factor affects the primary producers in such a system will have a profound and selective effect on the primary and secondary consumers that may lead to partial and complete extinction and to the formation of new ecosystems.

The History of Life

Students of fossil life have succeeded to a remarkable extent in reconstructing the sequence in which various forms of life first turned up in the various geological ages. Establishing this chronology opens up many intriguing biological problems to investigation, particularly trends and rates of evolution, and the many questions raised by the origin of evolutionary novelties. Without a knowledge of the fossil record the evolutionist cannot begin to solve these problems; indeed, he would be unaware of some of the more fascinating phenomena of evolutionary biology. All more ambitious studies must start with the precise description and reliable classification of the fossil forms. To draw the proper inferences from the seriously incomplete fossil record is no easy task. How incomplete this record is, particularly for soft-bodied organisms, is indicated by the estimate that only one out of 5,000 to 10,000 formerly existing species is preserved in the known fossil record. Organisms with preservable hard parts have left a far better record. For example, the history of the vertebrates is now understood in its major outlines, even though there is still much uncertainty about the earliest history of the amphibians and the more primitive types of fishes. The origin of the major groups of plants is far more shrouded in mystery, well illustrated by the fact that we still search for the ancestor of the dominant group of modern plants, the angiosperms. The gaps in our knowledge are still enormous, but there is hope that some will be filled through further research.

Organic Diversity and the Conceptual Framework of Biology

The life sciences encompass a broader field than the totality of the physical sciences. Within this area, each branch of biology makes its particular contribution to the sum of biological understanding. Neglect of any part of biology leads to a weakening and an imbalance of biology as a whole. The continued well-being of systematic biology is of such great importance to biology as a whole because the study of diversity with its emphasis on evolution, the creativeness of the selective process, the history of genetic programs, the uniqueness of individuals and species, and the statistical properties of populations and taxa places emphasis on perhaps the most strictly "biological" of all phenomena. At the molecular and cellular levels, the biologist is very close to chemistry and physics; indeed it is his avowed objective to reduce his findings to phenomena that can be expressed in terms of chemistry and physics. The evolutionist, on the other hand, is constantly aware that organisms are the products of individual genetic programs carefully adjusted by hundreds of millions of years of natural selection. All the components of these organic systems are explicable in terms of physics and chemistry, but each system as a whole, with its evolved genetic program and all of its homeostatic control mechanisms, is more complex by at least one order of magnitude than anything found in inorganic nature. Stated otherwise, granted our understanding of DNA, enzyme mechanisms, membranes and nervous conduction, these still do not permit prediction that there would be butterflies, orchids, or porpoises, much less man. Philosophy has not yet digested the biologist's way of looking at living nature, free of all vitalistic and finalistic ideas; but in the wake of the spectacular advances of chemical and evolutionary biology, one day there must emerge a new philosophy of science, based largely on the findings of biology rather than those of physics.

Systematics, including evolutionary biology, has produced fundamental generalizations of concern to every thinking human being. Perhaps the most important contribution after the theory of organic evolution is the development of "population thinking." The populationist stresses that all classes of objects and phenomena in the living world are composed of uniquely different individuals and that all statements concerning such populations of individuals must be taken in a statistical sense. Almost all statements which one makes in biology except at the molecular level are relative or statistical. One cannot understand the working of natural selection or the phenomenon of race or of fitness unless one fully understands the populational nature of biological phenomena. The gradual replacement of the ideology of essentialist philosophy, dominant from Plato and Aristotle to the Kantians, by population thinking during the past century has been one of the most important, even though hardly noticed, conceptual revolutions in the history of biology, a revolution originating in systematics and evolutionary biology.

These developments have decisive bearing on our understanding of man. Although man is unique, he is part of the evolutionary stream and cannot be understood if treated as an isolated phenomenon. He must be compared with the remainder of the organic world; systematic biology has supplied much of the information and the conceptual framework to make such a comparison feasible. This comparison has revealed in which respects man resembles other organisms as a consequence of his evolutionary heritage and in which respects he is indeed unique. *Evolutionary biology has succeeded in dealing with man objectively and scientifically, rather than as an object of ideology or dogma.*

The evolutionary process applies to man in the same manner as to all other organisms. Every problem faced by man, whether it be disease prevention, increasing the length of life, population control, ethical systems, the possibility of eugenics, etc., will be better understood as our comprehension increases of the evolutionary process by which man evolved and acquired his present characteristics. As man's knowledge increases, and as he acquires more and more control over his own fate and his planet, he takes on an increasing responsibility to predict the consequences of his actions farther and farther into the future and to adjust his actions accordingly. He can do this only by understanding the present in terms of the past. It would be a fatal mistake to consider man as something static. He is the product of evolution, and he is continuing to evolve as an inevitable consequence of genetic variability and differential reproduction. Evolutionary biology gives us major tools with which to conduct the studies that are necessary for a scientific approach to the future of man, and will contribute to creation of the badly needed bridge between biology and the social sciences.

Applications of the Study of Diversity

The study of diversity is of great practical importance to man. Perhaps the most important lesson we have learned is the high degree of specificity of every species in nature. It is not nature that produces drying oils, fibers, or drugs, but it is sheep that produce wool, cotton plants cotton, pigs bacon, cows milk, *Hevea* trees rubber, a specific mold penicillin, and so forth. The general laws of biology are important, and the study of the basic unit processes and macromolecules is fascinating and useful, but biology would remain one-sided and ineffective if it did not simultaneously develop a study of individual organisms on a broad scale.

Animals and Plants

Whether we focus on beneficial organisms or noxious ones, it is most impressive to consider the impact of certain species on man. Let us start with some noxious species. The cotton boll weevil requires the annual application of tens of millions of dollars worth of pesticides to prevent destruction of the cotton crop. *Complete control of one major pest, the rice stem borer of Asia, would immediately provide food for 120 million people, without additional effort or labor or updating of agricultural methods.* Consider the impact on the North American landscape and agriculture of the inadvertent introduction of such pests as the European corn borer, a multimillion dollar pest, or the gypsy moth, also from Europe, which has destroyed tens of thousands of acres of forest, or the notorious Japanese beetle, or the fungi responsible for the chestnut blight, or the Dutch elm disease! Past experience has shown that chemical controls of such pests have numerous drawbacks. They are expensive, usually result in the development of resistant strains, and tend to do much damage to the associated fauna and flora. This use of dangerous poisons must be eliminated, so far as possible. Biological control is far preferable, as demonstrated by the control of the prickly pear, the Klamath weed, various citrus scales, and the screwworm. Yet this requires an imaginative and painstaking analysis of the life cycle of the pest, its diseases, predators, and competitors, its wintering habits, etc. Control of the screwworm was not possible until it was shown that two sibling species were confused with each other, only one of which was noxious. Highly beneficial species can exist undiscovered right under our noses, with their ecological role unknown and unappreciated. For instance, attempts to grow figs in the United States failed repeatedly until a tiny wasp, previously not known to be necessary for pollination of the fig flowers, was imported.

It is obvious that permanent, effective control of some of the world's most destructive insects has not yet been achieved. In considerable measure, this reflects lack of a broad systematic and biological knowledge of the diverse potential parasites and predators that would carry out the job for us.

A thorough knowledge of the diversity of organisms is even more essential when we try to harness additional beneficial species for our purposes. The world's oceans have been called the last "great unexplored and unexploited frontier," and it has been suggested that the world's protein deficiency could be covered by the productivity of the sea. There has been a great increase in the harvesting of fish, particularly by automation, but the development of fundamental background knowledge has not kept up with purely commercial exploitation. All the fishing pressure at present is on comparatively few species. Undoubtedly, other species might be added to the pool of potentially exploitable species. How large are their geographical ranges, and when are their breeding seasons? What restrictions must be placed on fishing to maintain the breeding stocks? Decisions based on many kinds of biological information will surely have to be made. How to divide up the oceanic pool of fish derived from continental breeding stocks (such as salmon) has already become a question beset with ancillary economic and political complications. Fishing pressure cannot be regulated wisely until much more research has been carried out. For the Northern Pacific salmon populations it has become necessary to identify individuals and populations correctly. It has become necessary to apply immunological procedures, sophisticated morphometric analysis, and other modern systematic approaches. Even the parasites have been useful in difficult cases, and extensive lists of organisms associated with the salmon have been compiled for different geographical regions.

Crustaceans furnish rich yields of high-grade protein, but remarkably few species are so far exploited. A substantial proportion of the annual catch in the United States consists of one or two species of shrimp, even though hundreds of unexploited species of crustaceans of comparable size exist, as well as thousands of smaller species, which could be subjected to processing of some kind. Why should it not be possible to develop harvesting and processing methods for the krill (*Euphausia superba*), the whale food, which is obviously highly nutritious and occurs in rich abundance? Protein extracts from fish are now being produced routinely; crustaceans could be used just as well, and have an additional tremendous potential as producers of oils and of vitamins and their precursors. A small, ancient industry in southeast Asia produces "shrimp paste" from a variety of crustaceans, not only highly palatable and nutritious but also promising as a processing approach for a great variety of crustaceans now completely bypassed, yet in large part readily available the world over.

The possibility of direct harvesting of plankton has often been discussed. Here the problems are decidedly taxonomic, since repugnant and indeed toxic species exist in many plankton assemblages. The identification of directly usable plankton might eventually lead to the possibility of specialized culturing of such organisms. Pilot activities indicate the manifold problems that still have to be overcome, particularly an exhaustive study of the total requirements of the relevant species.

Aquatic farming for many kinds of organisms is rapidly expanding. Raising carp and carplike fishes is an ancient custom in Europe and Asia. Fish rearing in tropical lands has long passed its pioneer phase, but its full potential has nowhere been reached. Many additional species that might be suitable for freshwater farming have yet to be exploited.

The farming of marine organisms is even less developed. Its greatest success has been the farming of shellfish, such as clams and oysters, here an even richer protein source can be made available through further development of technology. At present only a few species are used the world over, but with the systematic knowledge now in hand, shellfish farming could unquestionably be vastly expanded through imaginative and ingenious exploration, unless prevented by the increasing pollution of our estuaries and coastlines. The production of pearls and the harvesting of many products of marine plants, as the alginates, agar, and other compounds, involve procedures basically similar to those of food harvests. Commercial pearl production has also expanded to previously unexploited species of molluscs because they produce products that are in special demand. Many further species of marine organisms with a wide diversity of applications could be successfully subjected to various farming procedures. In addition to their food value, they are undoubtedly an important potential source of drugs, antibiotics, and other complex products.

In the exploitation of wild plants for drugs and other natural products, the identity of the plants concerned, as well as their systematic relationships, becomes especially critical. For example, those species of *Cinchona* related to *C. calisaya* give a high yield of quinine and quinidine, and are the source of highly selected varieties long cultivated in Java and Sumatra. *Cinchona pubescens* and its relatives, however, as well as species of the related genera *Remijia* and *Ladenbergia*, produce primarily the alkaloids cinchonine and cinchonidine, which do not have nearly as high an antimalarial value. *Rauwolfia*, from which the first successful tranquilizer was obtained, has been well known in India for millennia. However, because the genus was relatively well known systematic botanists predicted immediately that certain related tropical American species would likewise be active—as proved to be true. Yet only three species are so far being exploited commercially, in spite of a hundred that need to be studied. Another relatively recently introduced drug is *curare*, whose source was a mystery well kept by South American Indians. Now, however, we know that the primary genera concerned are *Strychnos* and *Chondrodendron;* here, again, both the Indian who makes curare for hunting and the individual who collects these plants commercially for industrial purposes must be aware of the various species and which are the best ones for the purpose.

Not only individual species, but whole genera and even families have highly specific characteristics. For example, the Rubiaceae are rich in genera that produce alkaloids: in addition to *Cinchona*, we find coffee and ipecac, which was known to the South American Indians as a specific for amebic dysentery long before it was added to the European and American pharmacopoeia. The Asclepiadaceae and Apocynaceae, very large families especially well developed in the tropics, contain few members that are not toxic. Because of their abundant production of alkaloids—our common roadside weed of the temperate United States, *Apocynum cannabinum*, for good reason is called the "dogbane"—many tropical and subtropical ornamental plants, belonging to these families, are dangerously poisonous, including such familiar plants as the oleander. This knowledge of widespread properties is useful in many families of plants and gives a very useful guideline for drug companies and others who are interested in natural plant products.

An excellent illustration of the utility of systematics is given by two closely related plant families, some of whose members are difficult to distinguish by the usual morphological methods: the Solanaceae (or nightshade family) and Scrophulariaceae (or snapdragon family). Most members of the first family, even the so-called Irish potato, are rich in alkaloids, many of them of medicinal importance, such as atropine, whereas the second family has no alkaloids at all, only saponins, some of which, like digitalis, are also of great medicinal value. By a survey of higher plants, then, one can identify certain families that have, through evolution, specialized in the production of certain useful materials. To find new sources and new products, one need only extend the study within a promising family. Already published systematic research has provided an extremely useful shortcut in such searches, whereas poorly known plant groups have had to be investigated laboriously for potentially useful products.

Other evidence of the economic value of systematics comes from a survey of the cultivated plants—and it is astonishing to see how few plant families provide so much of man's food and fiber. All our major cereals belong to the grass family, in the tropics as well as in temperate areas. The hundreds of kinds of peas and beans of the legume family found in every continent not only form a staple food of man but provide the greatest source of protein of any major plant family. The Rosaceae are the greatest producers of fruit cultivated by man or harvested in the wild, from strawberries through raspberries, blackberries, and their relatives, to stone fruits as peaches, plums, and cherries, and apples, pears, and their relatives. Most of our root crops are produced by two families, the carrot family and the turnip family, both of them excellent examples with many related forms that immediately come to mind. The carrot family provides such edible leaf stalks as celery, fennel, and parsley, whereas the turnip family provides us with such edible stems, leaves and flowers as cabbage, brussels sprouts, cauliflower, and many others. Other families of importance to man, especially for food, are the lily family (onions, garlic, and relatives), the squash family (including melons, cucumbers, pumpkins), the nightshade family, with many edible forms, such as potatoes, tomatoes, and peppers (as well as poisonous ones), the heather family with cranberries, blueberries, lingonberries, and many other kinds of berries—and so on. Some families are so important that whole cultures are based around them, for example, the palm cultures of tropical India and elsewhere, and the bamboo cultures of southeast Asia, where the plants provide not only basic food but also the primary source of shelter.

In summary, the families of higher plants that are rich in useful species can have their usefulness extended by further systematic study, and the many plant families that are now little used may well turn out to have unexpected and surprising products of great use to man if we have but the wisdom to seek them.

FUNGI AND MICROORGANISMS

Yeasts, molds, and other fungi, bacteria, and other microorganisms form a vast and almost untapped reservoir of natural resources which cannot be fully utilized until their classification has been further advanced. Yet their economic potential is highly diverse:

1. As food sources for the future: Although only few species of mushrooms are used in commerce today for food, many thousands of species are edible and untold further thousands are still undescribed. Yeast, known to be an excellent source of food protein and vitamins, can be produced very economically from waste products of many industries to supplement the nutrition of man, livestock, and poultry. The use of fungi as food needs further study, as man cannot afford to overlook any source of food for the future, in view of the population explosion. Bacteria that can be grown on petroleum have a high potential for tomorrow.

2. As sources of drugs: The discovery of antibiotics in fungi revolutionized the drug business, as well as public health activities and practice. In spite of the importance of antibiotics, it is possible that only a small beginning has yet been made. The recent discovery of tumor-inhibiting substances in puffballs has created an urgent need to collect, identify, and test this group. Other fungi, taken by primitive peoples for their hallucinogenic effects, are a rich source of alkaloids related to mescaline.

3. As sources of toxins: The poisonous products of nature are legion, from mushrooms that are deadly poisonous to the botulinus bacterium that occasionally makes canned and stored foods so extraordinarily toxic to humans. Although inimical to human life and welfare, some poisons may be beneficial in very small quantities.

4. As pathogenic organisms that affect man, other animals, and green plants: For this reason they are of the utmost importance to mankind and warrant intensive study. Extensive attention has been given to those organisms recognized as pathogenic for man, but much remains to be done. The biochemistry of parasitism poses some fundamental problems, since it sometimes determined how one organism interacts with another. The fascinating series of races of *Fusarium* wilt, for example, were perhaps the first group of "cryptic" taxa known. Although exactly alike morphologically, they differ in the particular host they harm. For example, the race of *Fusarium* specific for watermelons will rapidly kill watermelon plants, especially at the seedling state, while other crop plants, each with its own specialized *Fusarium* race, are resistant in varying degrees to the watermelon wilt. This is an area that deserves research as much for basic scientific as for economic reasons. The fungal diseases of the tropics are poorly understood. As vast areas are cleared for a one-crop agriculture, there is real danger of new diseases appearing, which could overnight make the whole enterprise uneconomical, as happened at one time to the banana industry.

5. As agents of deterioration: The tropical deterioration of electrical

equipment, leather products, paper goods, optical equipment, and textiles, etc., caused great problems during and after World War II. As time goes on, we will become more and more aware of damage by highly specialized saprophytic fungi to materials that would have seemed completely resistant. Certain fungi and bacteria are able to ferment aliphatic hydrocarbons such as petroleum, thereby helping to keep highways clean.

6. Industrial uses: Through fermentation of solutions containing carbohydrates, such metabolic products of fungi and bacteria are produced commercially as alcohols, citric and other acids, glycerol, enzymes, steroids, fats, plastics, growth-promoting substances, vitamins, and other compounds of great importance. Curiously enough, few species have so far been tested, and even fewer are used industrially. There is probably no chemical transformation that cannot be carried out by some microorganism, under proper cultural circumstances, even though it might not be economically worthwhile at the moment. Some of the metabolic products are so complex and so biologically active that it seems almost impossible that they could be produced in plants—for example, the production by some mushrooms of polyacetylenes, which are actually explosive, and of other compounds so energy-rich they give off light. It is safe to predict that the industrial use of fungi in chemical transformations will increase as natural resources become less abundant.

7. Major transformations in nature: Without the destruction of organic wastes by bacteria and fungi, all life, and especially the higher forms, would soon disappear. These organisms are vital for the removal of organic debris, the concomitant return of carbon dioxide to the atmosphere, and humus formation, essential for the growth of plants. By the same process, of course, fungi cause wood to rot and food to spoil, to mention two among hundreds of detrimental phenomena. The fixation of nitrogen by bacteria, and many other transformations in our environment are given little recognition considering their great importance.

8. Mycorrhiza: The relationship between fungi and the roots of trees is poorly understood, even though this symbiotic relationship is vital to the development of many forests and for the growth of many cultivated plants, including food crops and the commercial production of orchids.

9. Biological control: Fungi are effective agents in the control of other organisms in nature; many fungi are parasitic on insects in nature, yet their potential for biological control has not yet even been touched, except for the use of bacteria for beetle larvae that damage lawns. It is possible that fungi can be used for the control of other fungi, especially pathogenic ones, as well as of nematodes and other organisms harmful to man's activities. The great potentiality of this kind of biological control depends largely on our systematic knowledge of the organism concerned. All in all, the study of the diversity of fungi and other microorganisms is eminently justified by the extraordinary range of activity of these organisms.

The Task Ahead

What are the most urgent tasks of the student of diversity? This question, which every scientist should ask himself occasionally, is not easy to answer. The sheer magnitude of organic diversity poses problems of research strategy not found in other areas of biology. The increase in our knowledge has been so rapid that the fraction of all kinds of organisms which a single individual can handle is steadily becoming smaller. During the 1700's, Linnaeus was able to treat both animal and plant kingdoms in their entirety. But he was the last person to do so. In 1818, the famous French naturalist Lamarck published an account of all invertebrate animals, again the last person to be able to do so. By now, two centuries since the great works of Linnaeus, systematists have discovered, described, and classified about a million and a half different species of organisms, and yet each year more than ten thousand previously unknown species are discovered and described. At that, the formal description of a species is only the very first step in the task of the systematist (to be discussed). How much of his research time shall he devote to it? How endless is the task ahead of him? Is it really worth doing? In order to answer these questions, we must give a short report on the current state of the inventory of the world's biota.

Inventory of the World's Animals

The inventory of the animal and plant life on earth has progressed very unevenly. This is evident from a cursory survey of our knowledge.

Vertebrates. Such large and conspicuous diurnal animals as the birds are relatively well known. The two main senses of birds, vision and hearing, are the same as in man, and all communication among birds is based on these senses (plumage coloration, displays, and songs). Not a single special technique is required for the recognition of all birds in a given district, merely the keen sense of observation characteristic of the naturalist. Consequently, only two or three new species from very remote regions are added annually to the 8,600 species of birds already known. Other less conspicuous or nocturnal vertebrates, even mammals, lizards, and snakes, are far less known. Fishes, in spite of their great potential for exploitation for food and the important role they play in aquatic ecosystems, remain the most poorly known group of vertebrates. By a conservative estimate, at least one third of the living species of fishes remain unknown to science. A leading authority stated in 1931 that there were probably 20,000 species of fish known at that time, "and one hundred or more new forms seem to be discovered every year." That the number of new forms described every year has not decreased during the nearly forty since this statement was made is indicative of the remaining task of inventorying the world's fishes. Many of the undescribed fishes are marine. The most spectacular illustration of our incomplete knowledge of diversity of life in the oceanic depths was the discovery in 1938 of the coelacanth fish, *Latimeria*, representative of a group thought to have been extinct since the days of the dinosaurs, more than 70 million years ago. The need for a continued exploration of the diversity of marine fishes is indicated by the fact that as recently as 1966 a new suborder (Megalomyceroidei), containing two new genera, was described from material collected off Bermuda. Poorly known continental areas probably equal the oceans in number of unknown fishes, as, for instance, the huge Amazon Basin. One specialist of Amazonian fishes has estimated that although well over 1,000 species of Amazonian fishes are already known, the figure will ultimately reach 2,000.

Invertebrates. Compared to the inventory of vertebrates, that of the invertebrate animals is much less complete; indeed, it is certain that far more species are still undescribed than those that have already been made known to science. To illustrate this by just one example: a group of soil ecologists at the University of Chile found it necessary to describe 700 new species during an examination of the first 10 percent of the soil samples collected during a transect of Chile.

Insects. Insects far outnumber all species of plants added together, known and unknown. There are three species of insects known for every other animal species of any kind. And yet, in spite of the 750,000 species already described, insects are among the most poorly known of all animal groups, and 6,000 or 7,000 previously unknown species are being discovered every year. Undiscovered species are by no means restricted to such out-of-the-way places as tropical jungles in underdeveloped countries. Only a few years ago an entomologist discovered, within a few miles of the U.S. National Museum, a whole series of new species of fireflies which had previously been thought to be a single species. By analyzing cricket songs with the help of new sound-recording equipment, two other entomologists found that more than 40 percent of the 108 cricket species of the eastern United States had been unknown prior to their research. And yet crickets are highly conspicuous animals, and this is one of the most intensively studied regions of the world. Another entomologist estimates that of the approximately 500 species of ichneumon wasps collected in traps on his lawn in Michigan during the past few years, half are unknown species. Based on the proportions of undescribed species in collections that he has examined in various parts of the world, he estimates that three fourths of the ichneumon wasps, a group with over 15,000 described species, are still unknown. All of them are insect parasites, and many of great potential in the control of injurious insects. The rate of discovery of new species of insects is determined almost entirely by the number of investigators available for the search at any given time. For this reason an estimate of the total number of insect species is extremely difficult. Guesses of experts have ranged as high as 10 million, and this may not be far off, although we can accept 2 or 3 million as more conservative estimates.

Mites. The rapid enlargement of the inventory of species of mites is illustrated by advances in the knowledge of the chiggers (Trombiculidae), a family of mites of great medical importance as vectors of scrub typhus and other rickettsial diseases: 3 species were known in 1900, 33 in 1912, 517 in 1952, and about 2,250 in 1966. Athough thousands of species have been described in medically unimportant families of mites, it is estimated by specialists that less than 10 percent of the total number of species existing on the globe have been identified. What an enormous task is ahead for the specialist concerned!

What is true for insects and mites is equally true for many other kinds of invertebrates. For instance, the nematodes, a group of mostly small worms, some free-living, others parasitic, are still largely undescribed, in spite of their great importance in the ecosystem and for man's welfare.

Marine invertebrates. The inventory of marine invertebrates is still in its early stages, and much of the world's oceans remains unexplored. Most of their shores have only been sampled, although intensive studies have been carried out in Europe and along the coasts of the United States. The coastlines of the other continents of the world, largely unexplored, harbor countless species of marine organisms unknown to science. The exploration of new habitats and the development of novel collecting procedures will open whole new worlds of marine organisms that will provide a mass of new information.

Marine microorganisms, particularly bacteria and fungi, are poorly known, but procedures for obtaining them in sufficent quantity and in appropriate condition for taxonomic treatment are now being developed. We may be sure that microorganisms from plankton, benthos, and other communities and habitats will contribute much information and material for description.

Rather recently, an entire new fauna, called *psammon*, was discovered to live between the sand grains and mud particles within the substrate of beach and ocean floor. This fauna, at a size level well above that of microorganisms, contains many previously unknown types of organisms, including even new orders and classes. It will clearly add many new species to our inventory, belonging to many different classes of invertebrates. It is estimated that only about one percent of this fauna has so far been described, because the onshore forms, being highly sedentary, are very localized geographically. The portion of the psammon that lives below the tidal zone has been so little sampled that a one percent sampling figure is probably also true, even though this fauna may have a wider distribution.

Every deep-sea expedition has returned with an abundant harvest of new species. Even though the deep-sea fauna of all major oceans of the world has been studied, the difficulties of collecting suggest that less than half of the species have so far been described. Our biological knowledge of this fauna is virtually nil, and much work remains to fill in such essential information as the stages of the life cycle, place in the ecosystem, relative abundance, dispersal pattern, mode of speciation, reproductive pattern, structure of communities, and other biological information of vital importance for the biological oceanographer.

High-speed sampling is still yielding a high proportion of forms new to the systematist, and a vast area of the ocean still remains to be studied by this technique, so that the potential of further new forms of midwater fauna to be discovered and studied is great.

Although some 150,000 species of marine organisms (invertebrate and vertebrate) have now been described, fewer than 15,000 species parasitic upon them have been recognized. Since the parasitic population of a single fish can run to tens of species of assorted organisms, and many invertebrates have numerous known parasites, it is clear that a vast descriptive task still remains to be done in this area of systematic biology.

Inventory of the World's Plants

Even though the number of species of higher plants is considerably smaller than that of higher animals, our knowledge of them is still very incomplete, because many species are highly localized. Even so, the individuals of a single species may be highly scattered. In the tropical rainforest, for example, only one or two specimens of any one tree species generally occur in the same square mile, so that botanical collecting has had to be largely a kind of random sampling.

The Amazon Basin possesses an especially rich flora; it probably surpasses any other area of similar size in the world in the number of species of plants present. Richard Spruce, who spent from 1849 to 1860 in the Amazon Basin, wrote in 1858 that by moving one degree of latitude or longitude, he found approximately half the plants to be different. He calculated that there yet remained to be discovered in the great Amazonian forest, "from the cataracts of the Orinoco to the mountains of Mato Grosso," some 50,000 or even 80,000 new species. In the hundred years since Spruce left South America, many of these species have been discovered, but others remain undescribed. The magnificent forests of the Pacific coast of Ecuador and Colombia are little known and almost equally rich.

Approximately half a million higher plants are now known, but probably a quarter-million species remain undetected in remote tropical areas. Ferns, with perhaps 5,000 species, hepatics, with approximately 10,000 named species, mosses, with some 25,000 named species, represent primitive groups which are of special interest because they represent very ancient lines of evolution in the plants. Yet they are highly vulnerable to extinction when their habitat is changed through man's activities.

Another large group of plants that needs a more rapid tempo of investigation is the fungi. Although of great economic importance, systematic mycology lags far behind other areas of systematics. Our knowledge of fungi in vast areas of the world, particularly in the tropics, is virtually nil. Little is known about the ecology, genetics, physiology, chemistry, and behavior of the species that have already been described. Their total number will certainly reach at least half a million, even though only 100,000 species have so far been described.

Considering that fungi can carry out almost any conceivable chemical transformation, it may be expected that many more will be used in laboratories and in industry as their synthetic capabilities are made evident by research.

Inventory of the World's Fossil Organisms

Paleobotany and paleontology, the sciences of fossil plants and animals, produce information of great importance for evolutionary biology, historical geology, and other branches of science. It has been estimated that of all the species that have existed, only 1 out of 5,000 to 10,000 fossil species have so far been discovered and described. Considering the fragmentary nature of the fossil record, it is remarkable how well paleontologists have succeeded in reconstructing the probable history of life. "Missing links" are now available to fill the gaps between many of the major types of vertebrates. Some of the gaps in our knowledge stem from inadequate exploration and sampling. It is urgent that these gaps be closed before man's further alteration of the landscape irretrievably destroys the record. Some groups of organisms, such as the early invertebrates, are poorly represented in the fossil record because they lacked hard parts, or because such parts are loosely associated or fragile. However, our knowledge increases steadily through new discoveries and more thorough investigation of previously described forms. Modern techniques in collecting and preparation have made it possible to obtain and study many rare and poorly known groups. The development of acid etching and celloidin "peel" techniques for fossil plants has enabled the paleobotanist to study microscopic structure. Numerous large groups of invertebrate animals, as the corals, sponges, and radiolarians, and of lower plants, especially bryophytes, have so far been rather neglected by paleontologists. The exciting field of paleomicrobiology is on the verge of a flowering that could rival in excitement the discoveries of any branch of science.

We may safely predict that remarkable and startling discoveries about evolutionary history will be made through the study of fossil plants and animals. An example is the recent discovery of what may be the earliest evidence of life in the form of primitive algae and bacteria deposited in rocks that may well be over three billion years old. This work makes us realize more than ever before the drastic differences in rates of evolution in different organisms. Close relatives of some of the blue-green algae found in these early rocks are still in existence and have undergone astonishingly little morphological evolution in several billion years.

Is an Inventory of All Species Needed?

The magnitude of our ignorance about kinds of animals and plants is obvious. Should we really try to describe them all? If 5 million species of animals remain to be described, and if taxonomists throughout the world annually described 10,000 species, it would require another 500 years to complete the task of naming these animals and thus making them known to science. And yet, describing is only the first step: the scientist must also study the totality of properties of these organisms; arrange them in a meaningful classification, investigate their speciation and biological evolution, and so forth.

A single systematist can become expert on a group containing perhaps 1,000 to 3,000 species. The number of groups of organisms is so great that frequently one man during his lifetime is the only person investigating a particular group. Many groups (taxa) are at present entirely "orphaned": not a single specialist in the entire world is studying them. In large groups, such as most insects, the number of new species described at any time is limited almost entirely by the number of investigators searching for them during that period, in contrast to the situation in birds and mammals where the number of species that remain to be discovered is steadily diminishing.

Large groups of invertebrates remain undescribed because the techniques for their study are too demanding, and this along with practical considerations places those who work on these taxa at a disadvantage. Protozoa, for example, are so difficult to collect and preserve that it is impossible to establish adequate type collections or extensive museum reference collections, making their descriptive taxonomy a formidable task. The taxonomy of sponges (Porifera) requires the study of anatomical features that are only disclosed by special techniques of preservation and of microtechnical procedure. The various jellyfish types (Coelenterata and Ctenophora) present considerable problems in both collection and preservation, so that few satisfactory collections are available for study in museums. Sea anemones require such extensive and detailed microtechnical processing for a determination of the basic taxonomic characters that hardly a specialist for them is left in the whole world. Similar difficulties exist for virtually all soft-bodied invertebrates, including flatworms, nematodes, internal parasites, and such marine groups as Phoronida, Enteropneusta, Pterobranchia, Nemertea, Sipuncula, Echiura, and Priapuloidea. The result is a great shortage of specialists, even though many of these groups are important as components of ecosystems, as parasites, or as a major food source of food fishes. For example, at most 10,000 species of nematodes are now known, even though competent authorities estimate that 100,000 free-living species still remain undescribed.

The investigator facing such problems is tempted to abandon all further work on an inventory of species and concentrate on what may appear to be more rewarding aspects of the study of diversity. In the face of this situation, it seems advisable that the taxonomist should continue to devote part of his research time to the description of species for some of the following reasons.

1. Every species has unique properties. Knowing all about the bluebird will not tell us all about the nightingale. Every species of organism is a unique genetic system. Each species represents a separate line of evolution, and each species is characterized by different biochemical specificities. The exact identity must be known for every organism that is used for experiments in genetics, physiology, behavior, or ecology; otherwise results cannot be compared and repeatability,

the essence of science, is lost. The recognition and description of species is thus important for basic as well as applied biology. As one striking example, this kind of knowledge led to the development of a synthetic sex attractant that has proved efficacious in the trapping of the gypsy moth, a major insect pest, and may prove to be valuable in its control. In 1964 the highly injurious oriental fruit fly was eradicated from the Pacific island Rota with the use of eugenol, an organic chemical highly attractive to males of the insect. Small pieces of fiberboard saturated with a combination of the attractant and an insecticide were uniformly scattered about the 33-square-mile island. In less than six months the entire fly population was destroyed. Similar knowledge of other species is yet to be acquired, and its acquisition constitutes a profitable field of investigation.

2. The characters of natural groups of organisms cannot be determined until a minimal number of species in these groups has been described. Every organism is a unique product of millions of years of selection in a complex and fluctuating physical and biological environment. Every property of every organism, whether structural, functional, chemical, or behavioral, is thus the result of its past history, and its study is basic for understanding the biotic world as a whole.

3. Our great ignorance about the properties of most organisms precludes determination of their usefulness for man. Who would have guessed the medical usefulness of molds until penicillin, streptomycin, and other antibiotics were discovered? Who knows the potential mèdical significance of the many organisms that have not yet been taxonomically described? Certainly, many species would be more suitable as biological material in laboratories and industry than the ones now employed. The relatively recent introduction of *Neurospora*, *E. coli*, hamsters, etc., is indicative of the untapped reservoir.

4. Understanding of ecosystems depends on a knowledge of the individual components of the system. Whenever new ecosystems need to be studied, whether in the temperate zone or tropics, on land or in water, an inventory-taking of the taxonomic components is essential.

How urgent is the task? In a stable world, one might well recommend declaring a moratorium on the further describing of new species, in order to concentrate instead on a thorough study of those already known. Unfortunately man's technological progress has released forces that lead to an ever-accelerating destruction of natural habitats. Dozens, perhaps hundreds of species are annihilated each year, species that required hundreds of thousands or millions of years to evolve. They cannot be replaced. Wherever man transforms the landscape for his own purposes, he destroys most of the native populations, usually causing their replacement by a few species that thrive in man-made environments. Highly localized species are most vulnerable, and this class includes an amazingly high percentage of all species. Cave species are destroyed by the exploitation and pollution of caves, as are stream endemics through pollution, silting, anf the conversion of streams into water reservoirs. Estuaries, marshes, and swamps are being drained or polluted. The native vegetation is disappearing at a rapid rate in many parts of the world, especially in the tropics, as the result of the pressures of expanding populations and intensified agriculture. Many species of potential practical and scientific interest have undoubtedly already been exterminated. The existence of others is acutely threatened at this very moment, but it is only for a limited number of large mammals, birds, and reptiles that we have any information whatsoever concerning their vulnerability. The proverbial dodo, the mammoth, the New Zealand moas, and several hundred species of large mammals and birds have already become extinct since man became a hunter. The near extinction of David's deer, the Hawaiian goose, the whooping crane, the California condor, and the Arabian gazelle illustrate the ongoing process. Once a population is reduced below a certain level of size, its genetic variability and flexibility become so low that extinction is almost inevitable. Where an organism is rigidly specialized for a single habitat, it is doomed to extinction as soon as this habitat is threatened. Our knowledge of almost all groups of animals and plants is so limited that we cannot enumerate the species of which only few indivdiduals still survive. The rate of extinction at the present time is greater in subtropical and tropical areas than in the temperate zone. Here, the population explosion is particularly rampant and the destruction of the native vegetation vegetation takes place at a threateningly rapid rate. Even in the vast Amazon Valley, large international corporations are planning to clear huge areas for the large-scale cultivation of single tree species (for pulp and veneer) with the inevitable destruction of the vast local fauna and flora.

Extinction has been going on through the geological ages, witness the dinosaurs and ammonites. The present range of the *Sequoia* (''big tree'' and redwood) is only a small remnant of the range of this genus in the recent geological past, when it occurred widely in North America and across Asia into Europe. In Japan, China, and other parts of the Orient, some plants were so rare and unusual that they became objects of religious or superstitious attention, and we owe to this fact the survival of many species known today only in temple gardens in the Orient: they have become extinct in the wild. An outstanding example is the *Ginkgo,* a very primitive type of gymnosperm with motile sperm, which at one time had an extremely broad geographical distribution. Botanical gardens, in an analogous manner, have brought some particularly interesting species into cultivation. But what fraction of the estimated 750,000 species of higher plants can any botanic garden maintain, indeed all botanic gardens of the world together?

There is an urgent need for the conservation of as many types of original biota as is possible. Even though many areas still appear to be wild and untouched, they have remained so only because local population pressures have not yet reached the level to demand their exploitation. As yet we do not even have the information to determine where such sanctuaries should be and what proportion of the vanishing faunas and floras they would preserve. Where they cannot be preserved in the wild, at least an effort should be made to preserve samples of these unique representations of life in scientific collections. Once they are destroyed, they are irreversibly gone. We have an obligation to posterity to prevent this whenever possible.

Classification. The description of species, the inventory of the units of diversity in nature, is only the first task of systematics. Another task is the arranging of the chaotic mass of species into groups of related units, the so-called higher taxa. The 5,300 species of songbirds, for instance, are much easier to understand after they have been classified into thrushes, warblers, flycatchers, wrens, etc. Such classification is even more urgent for large groups, like the 750,000 species of insects or 25,000 species of fishes. The members of a higher taxon are descendants of a common ancestor, and each taxon is thus the product of evolution. The taxa fit into a hierarchy of larger and larger groupings of less and less closely related forms, with the kingdoms of animals, plants, and microorganisms forming the most comprehensive groups. The nearest relative of a group of organisms is by no means always evident, and indeed the determination of relationships is one of the most challenging areas in taxonomy. The findings of comparative biochemistry, cytology, physiology, and behavior are used to shed light on relationships, each new finding requiring scrutiny and sometimes correction of the still inadequate existing classifications largely based on a few morphological characters.

Conversely, the whole field of comparative biology depends on sound classification. A continuous reciprocal feedback between classification and other branches of biology must be cultivated for the health of biology as a whole. Every classification is a scientific theory. Its explanatory value is that each taxon is a phylogenetic unit (descendants from a common ancestor), and its predictive value is that it predicts most of the characters of newly discovered species as well as the taxonomic distribution of newly studied characteristics. Physiologists, cytologists, and biochemists not only contribute by their findings to the improvement of classifications but themselves make use of the zoological and botanical classifications in the planning of their comparative researches.

A classification is also an information retrieval system, and the superior taxonomist delimits his taxa to increase their value for this objective. The existence of millions of species, each with sexual, age, and other variability, each with information on biology, physiology, structure, and biochemistry, poses enormous problems of information storage and retrieval which even the largest computers cannot yet handle.

The Species as a Unit of Biology

The original description of a new species, often based by necessity on the morphological characters of a few specimens, represents only a meager beginning. Species are populations that coexist with one another in highly complex ecosystems, and have many other attributes far more important than the few obvious morphological characters. A species is not merely an item in an identification key, but a biological system. The student of diversity must determine and describe for every species all stages of the life cycle, differences between the sexes, internal anatomy, physiological characteristics, behavior and ecology, pathology and parasites, and the innumerable interactions with other inhabitants of its geographical range. There are only few species on which we have reasonably complete information of this kind. For well over 90 percent of the 1.5 million "known" species of animals and plants on this globe, nothing more is known than a name and a few diagnostic characters. This is why taxonomists increasingly study living animals and plants in their natural environment, and this is why new techniques for the study of new characters are being used, particularly all sorts of experimental approaches. Biochemical systematics, a rapidly expanding field of investigation based on the study of molecules, has provided a new set of tools for obtaining information on relationships. The new methods of analysis, such as chromatography, electrophoresis, gas chromatography, and amino acid analysis, have come into wide use in both plant and animal systematics.

Hybridization experiments provide still another important source of information. By attempting to hybridize species, the experimental taxonomist can determine whether or not species populations that occur together in nature, or that might possibly come together at some future time, are capable of influencing one another through the interchange of hereditary factors. Since all systems of an organism do not necessarily evolve at the same rate, failure of individuals to cross sometimes reveals them to belong to "cryptic" species with similar morphology but with reproductive incompatibility. The hybridization experiments thus aid in improving the classification at the species level. Since the degree of compatibility indicates the degree of genetic divergence, hybridization experiments also help in the reconstruction of evolutionary history. Cytological studies of chromosome number as well as of "marker" chromosomes that can be identified in various groups give important information about relationship. Chromosome numbers and patterns have been determined for only a small fraction of existing plant and animal species.

One of the most exciting areas of investigation of species is the study of the causation of reproductive isolation between coexisting (sympatric) species. Contrary to popular belief, this is, at least in animals, rarely a sterility barrier. Behavioral barriers that prevent mating are far more important than the genetic incompatibility which comes into operation after mating. It is obvious that an arrangement which prevents an unproductive mating between species is of greater selective advantage than one that permits the wastage of the gametes of the hybridizing pair through genetic incompatibility. Two main classes of premating isolating mechanisms are commonly involved in any particular instance. Signals, by which individuals attract and identify other individuals as being reproductively ready members of the opposite sex, are most effective, because they are misunderstood only rarely. In birds, frogs, and many insects, vocalizations are important as signals. The availability of high fidelity portable tape recorders and audiospectrographs permits accurate measurement of these signals and experimentation on the effectiveness of differences in vocalization as isolating mechanisms. This type of work has already revealed many cryptic species in all the vocalizing groups, and future work will certainly contribute to the more accurate recognition and classification of species.

In other kinds of animals, as mammals, salamanders, many insects, and other invertebrates, species-specific secretions (scents, odors, sexstuffs, etc.) serve as means of communication among potential mates. Very few of these compounds have so far been identified, but present-day methods of analysis, as by gas chromatography, make this now feasible, and this particular area of research should much expand in the immediate future. In still other kinds of animals, such as insects, fishes, reptiles, birds, visual signals involving color patterns and ritualized displays and postures provide premating isolating mechanisms. Here again, a great deal of interesting and important research is to be done.

The second major class of premating isolating mechanism involves separation of the reproductive activities of sympatric species by space or time. They may reproduce at different times—as in one species of elm that flowers in spring and a sympatric species that flowers in late summer—or they may select slightly different breeding sites, as do two sympatric species of toads, one of which breeds in running water and the other in rain pools of the same small valley. Premating isolating mechanisms of this category are widespread both in plants, where they take the place of behavioral isolating mechanisms, and in animals, where they tend to augment them. Up to the present, however, only a minute fraction of the species of plants and animals are adequately known with respect to either of these classes of premating isolating mechanisms or with respect to their postmating mechanisms.

In short, the study of biological diversity, based on all properties of the living organism, has hardly begun. This work is necessary for the taxonomist himself, to eliminate errors in classification based strictly on morphology. More important, it is imperative that we know as much as possible about *all* the attributes of those species that significantly affect man and his resources.

References

ALSTON, R.E., and B.L. TURNER, *Biochemical Systematics.* Englewood Cliffs, N.J.: Prentice-Hall, 1963.

BLAIR, W.F. (ed.), *Vertebrate Speciation.* Austin, Texas: University of Texas Press, 1961.

CRONQUIST, ARTHUR. *The Evolution and Classification of Flowering Plants.* Boston: Houghton Mifflin, 1968.

DAVIS, P.M., and V.H. HEYWOOD. *Principles of Angiosperm Taxonomy.* Edinburgh and London: Oliver and Boyd, 1963.

FINGERMAN, MILTON. *Animal Diversity*. New York: Holt, Rinehart and Winston, 1969.
HYMAN, L.H., *The Invertebrates* (vols. I–VI). New York: McGraw-Hill, 1940–1967.
MAYR, ERNST. *Animal Species and Evolution*. Cambridge, Mass.: Harvard University Press, 1963.
PENNAK, R.W., *Fresh-water Invertebrates of the United States*. New York: Ronald Press, 1953.
ROMER, A.S., *The Vertebrate Story*. Chicago: The University of Chicago Press, 1959.
SIMPSON, G.G., *Principles of Animal Taxonomy*. New York: Columbia University Press, 1961.
SOKAL, ROBERT R., and PETER H.A. SNEATH, *Principles of Numerical Taxonomy*. San Francisco: W.H. Freeman, 1963.
SOLBRIG, OTTO T., *Principles and Methods of Plant Biosystematics*. New York: Macmillan, 1970.
SWAIN, T. (ed.), *Chemical Plant Taxonomy*. London and New York: Academic Press, 1963.
YOUNG, T.Z., *The Life of Vertebrates*. New York: Oxford University Press, 1962.

4

Energy Relations between Plants and Animals and Their Environment

DAVID M. GATES
University of Michigan

David M. Gates was born in Manhattan, Kansas, May 27, 1921, the son of Frank C. and Margaret T. Gates. He was educated at Kansas State University and The University of Michigan (B.S. 1942, M.S. 1944, Ph.D. 1948). Dr. Gates is currently Director of the Biological Station and Professor of Botany at the University of Michigan. He was Director of the Missouri Botanical Garden and Professor of Biology at Washington University from 1965 to 1971. During the academic year 1964–1965 he was Professor of Natural History at the University of Colorado. From 1957 to 1965 he was Consultant to the Director of the Boulder Laboratories of the National Bureau of Standards and Assistant Chief of the Upper Atmosphere Space Physics Division of those laboratories. He was Scientific Director of the London Branch of the Office of Naval Research at the American Embassy in London from 1955 to 1957. Dr. Gates began his teaching career at the University of Denver where he was Assistant and Associate Professor of Physics during the period 1947 to 1955.

Dr. Gates originated a new branch of ecology called biophysical ecology. His research contributions involved atmospheric spectroscopy, solar radiation, upper atmosphere physics, and energy exchange relationships between plants and animals and their environment. Recent research papers include: "Toward Understanding Ecosystems," *Advances in Ecological Research*, 1968; "Transpiration and Leaf Temperature," *Annual Review of Plant Physiology*, 1968; "Thermodynamic Equilibria of Animals with Environment," *Ecological Monographs*, 1969; "Animal Climates (Where Animals Must Live)," *Environmental Research*, 1970; "A Model Describing Photosynthesis in Terms of Gas Diffusion and Enzyme Kinetics," *Planta*, 1971. In addition to his numerous scientific papers, Dr. Gates is the author of *Energy Exchange in the Biosphere* 2d ed., Harper & Row, 1969; *Atlas of Energy Budgets of Plant Leaves*, Academic Press, 1971; *Man and His Environment: Climate*, Harper & Row, 1971.

The world we live in is a world of energy flow; it is a thermodynamic world. If it is too cold or too warm, all life ceases. The Earth is at the proper distance from the sun to receive the amount of heat necessary to give it a temperature compatible with life. Our planet rotates about its axis at a rate which makes the days and nights of a length such that the daytime side does not get too hot nor the nighttime side too cool. If it rotated too slowly the sunlit side would get so much heat from the sun that its temperature would be very hot and temperatures as high as 130°F or greater would be commonplace. The opposite side would radiate toward the cold of space during the long night and temperatures there would plunge to very low values.

The Earth, in orbit around the sun, is immersed in the tenuous extension of the solar atmosphere and is swept over by the solar wind. Our planet is bombarded by solar protons, electrons, and ions, most of which are deflected by the Earth's magnetic field or captured into the Van Allen Radiation Belts surrounding the Earth. Some penetrate the Earth's atmosphere and create showers of ionizing particles, known as cosmic rays, which cascade downward toward the ground. The atmosphere attenuates some of these particles and shields us from the more hazardous environment of space. Ultraviolet radiation from the sun, if penetrating to the surface of the Earth, would destroy most organic molecules, and life as we know it would not exist. We receive some ultraviolet sunlight at the Earth's surface, but no wavelengths short of 2900 Å reach the ground because of the absorbing property of ozone in the stratosphere. Ozone is a triatomic molecule, O_3, formed from the recombination of $O_2 + O \longrightarrow O_3$ after the dissociation in sunlight of molecular oxygen O_2 into atomic oxygen O. This blanket of invisible ozone with its pungent odor floating in the atmosphere 15 miles above the ground shields us from the intense ionizing ultraviolet radiation from the sun. The irony of this protective phenomenon is the fact that oxygen in the atmosphere is a product of photosynthesis and the ozone layer would not exist if green plants did not cover the surface of the Earth.

Presumably, at the time of the primordial Earth, the atmosphere made up largely of nitrogen was transparent to ultraviolet sunlight. The first unicellular plants grew in the sea where they received the protection of the water and of the overhanging rocks along the edges. As these pioneer plants produced oxygen as a by-product of photosynthesis, the oxygen dissociated in the atmosphere to form ozone, and some attenuation of the incident ultraviolet afforded partial protection to the plants against these actinic rays. The sensitive green plants could then emerge onto the land where they spread abundantly, photosynthesized rapidly, and added greatly increased amounts of oxygen to the atmosphere. More ozone became concentrated in the stratosphere to afford more protection and the plants continued to proliferate across the land. Life on the land and the atmosphere above evolved together during millions of years of slow evolution.

Today man impulsively adds insult after insult to the atmosphere in the form of pollutants, paves and plows the green surface, and in short order upsets the intricate and intimate balance between life and the delicate semi-transparent gaseous atmosphere.

Life on Earth is possible not only because of the temperature or energy level of the surface but because energy flows into the system from the sun and escapes from the surface to the cosmic cold of space. Thus we have a source and a sink of energy with the Earth sandwiched between as one gigantic transducer. High temperature energy or light comes in to the Earth and low temperature radiant heat leaves to space. The Earth remains in temperature equilibrium and has a mean surface temperature of about 12°C or 53°F. As the energy streams through the Earth's living interface, between ground and sky, organic molecules are built, work is done, and life continues unabated. The secret of it all, of life on Earth, is the Sun and cold radiation sink of space itself.

Solar Radiation

Solar heating of the atmosphere, ground, and oceans drives the general circulation of the air and waters of the world, evaporates water from the land and sea, and is essentially the primary force for all climate and life on Earth. Because of the enormous significance of solar radiation with respect to most processes on Earth it is of utmost importance to know the amount of sunlight incident upon the Earth's outer atmosphere. This has been a difficult measurement to make because previously the value has had to be inferred from observations made at the ground at the base of the atmosphere. Now, with the advent of high altitude balloons and satellites, direct measurements of the incident solar radiation above the atmosphere can be made. The solar constant is the flux of solar radiation outside the Earth's atmosphere received on a surface which is perpendicular to the solar rays at the mean distance of the Earth from the sun (1.5×10^8 km). The solar constant is 1.94 cal cm^{-2} min^{-1}. The solar constant is not a true constant since it undergoes some variation as the result of changing solar activity. The fluctuations of the solar constant are about ± 1.5 percent.

The Earth receives about 2790 cal cm^{-2} day^{-1} outside the atmosphere. The Earth, with a diameter of 1.27×10^4 km, intercepts a total of 3.67×10^{21} cal day^{-1} on the average. If this number is considered as unity we can compare other sources of energy with it. The total world use of energy by people is about 0.01, an average hurricane consumes about 10^{-4}, the Krakatoa volcanic eruption consumed about 10^{-5}, a thermonuclear weapon about 10^{-5}, and the average tornado about 10^{-11}.

Of the solar radiation irradiating the Earth's outer atmosphere, about 24 percent is reflected back to space by clouds and about 6 percent is reflected by atmospheric dust for an average total reflectance of about 30 percent. Additional sunlight, which is reflected back to space, is the fraction of sunlight reflected from the ground. The 30 percent given here is just the percentage reflected by clouds and atmospheric dust. The average reflectance of the Earth as a whole (clouds, dust, and ground) is 36 percent of the total incident sunlight, which means the ground reflects an additional 6 percent that gets through the atmosphere to space. But here, for the moment, we are concerned with how much sunlight reaches the ground surface. Hence, high in the atmosphere we have lost 30 percent by reflection from clouds and dust. Of the remaining 70 percent of the incident solar radiation above the atmosphere, about 17 percent is absorbed by clouds and atmospheric constituents, 22 percent reaches the surface as scattered skylight, and 31 percent as direct beam solar radiation. Thus, on the average, about 53 percent of the solar radiation intercepted by the Earth reaches the ground surface.

Man's activities are dumping into the skies about 12 million tons of particulates per year in the United States alone. This dust and dirt is greatest over our metropolitan areas but also persists in the air for hundreds of miles downwind. The dirty aerosols of New York City are now being noticed by airline pilots in the air flow above the Atlantic Ocean almost to the coasts of Europe and Africa. In Washington, D.C., where very accurate records of incident solar radiation have been kept for more than 60 years, comparison between now and 40 years ago or more shows that on the sunniest days there is nearly 16 percent less sunlight today than earlier. More serious than this is the fact that observations of solar radiation at stations remote from industrial contamination are showing reductions of 1.5 to 3.0 percent between now and a few decades earlier. This may not seem like a very significant change, but when continued decade after decade it becomes an extremely serious modification of our atmosphere and its effect on our primary source of energy input, that is, sunlight, may be disastrous. Generally an increase of dust in the atmosphere would reflect more sunlight to space, but if the dust is black, such as from carbon particles, then instead of reflecting light it will absorb more radiation. Whether the result is to reflect more, or to absorb more, it is nevertheless a manmade, inadvertent modification of our environment which is striking at the ultimate source of all physical and biological processes on Earth, that is, at the solar radiation budget of this planet.

The character of life on Earth varies enormously from pole to equator to pole. The number of species and the density of biomass is greatest in the warm tropics, less in temperate latitudes, and least at the cold poles of the planet. These features are primarily the consequence of sunlight and season. This can be understood by inspecting Figure 4.1 for the seasonal variation of the amount of sunlight incident on a horizontal surface outside of the Earth's atmosphere as a function of the time of year. The irradiation of the equator varies relatively little throughout the year and is greater at the time of the equinoxes than it is during midsummer when it is minimal. It is interesting to note that the maximum irradiation of the equator is less than it is of other latitudes. The maximum irradiation is of the pole and high latitude stations during the summer solstice when in the land of the midnight sun the sun's rays slant across the landscape to illuminate the surface. Here the daily total of radiant input is great because of 24 hours of sunshine. The amounts shown in Figure 4.1 are for the sun's rays beyond the Earth's atmosphere; when penetration through the atmosphere to the surface is considered there is then a considerable reduction. Here at high latitudes the winters are totally dark when the sun never appears above the horizon for approximately four to six months. The irradiation by the sun of middle latitudes is low in the winter and high in the summer. Here there are distinct summer and winter seasons, as there are also at higher latitudes, whereas in the tropics there is relatively little distinction between winter and summer except with respect to other climatological patterns such as rainfall.

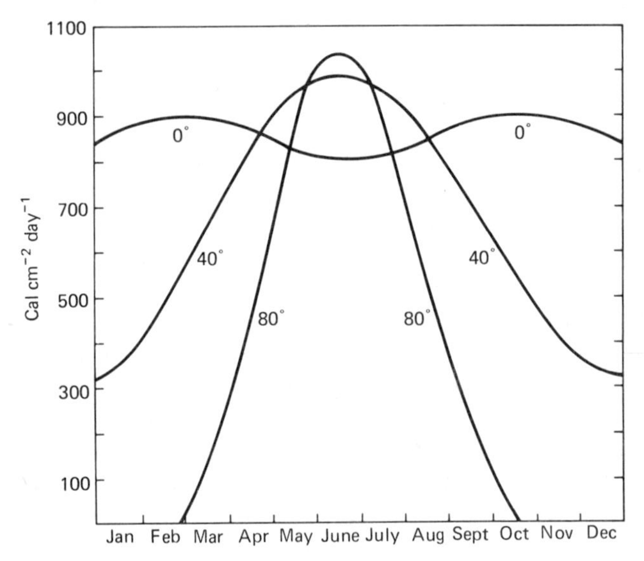

Figure 4.1 Daily Totals of Undepleted Solar Radiation Received on a Horizontal Surface Outside of the Earth's Atmosphere for Various Latitudes as a Function of the Time of Year.

The ground in the tropics is always warm and the seasonal pulse of heat flow to and from it is small. At middle and high latitudes the ground is a source of heat going into the atmosphere as winter is approached and a sink for heat from the atmosphere as summer comes on. The tremendous abundance of plant and animal species in the tropics would appear to be related strongly to the fact that there are no winters there to hold organisms in check. Growth and decay are rapid in the tropics, insects are incredibly abundant, and the forests are teeming with life. The lack of winters allows for more reproductive cycles during the year in the tropics than in temperate and arctic regions.

Since so much of the character of the landscape, and of life itself on the Earth, depends upon the amount of solar radiation at the surface, it is well to consider the distribution of sunlight throughout the world. The average annual solar radiation on a horizontal surface at the ground is shown in Figure 4.2 in units of kilocalories per cm^2 per year. Maximum amounts greater than 200 kcal cm^{-2} yr^{-1} are associated with the major deserts of the world. In these regions as much as 80 percent of the sunlight incident at the top of the atmosphere reaches the surface as compared with 53 percent average for the whole world. Solar irradiation amounts less than 100 kcal cm^{-2} yr^{-1} are found poleward of 40° latitudes over the oceans and poleward of 50° latitudes over the continents. Curiously enough, small amounts of sunlight reach the tropical forests of West Africa because of a large amount of persistent cloud cover throughout the year.

In the United States the maximum amounts of solar irradiation are 750 to 800 cal cm^{-2} day^{-1} in July over the southwest desert. Typical average amounts of solar radiation incident on a horizontal surface in July elsewhere are New York, Washington, St. Louis, New Orleans, and Seattle about 550 cal cm^{-2} day^{-1}, Chicago and Detroit 500 cal cm^{-2} day^{-1}, while Los Angeles and Denver receive about 650 cal cm^{-2} day^{-1}. During the winter the maximum amount of solar radiation at the surface is over the southwest desert where the January monthly average is 350 cal cm^{-2} day^{-1}. In January, Seattle is 100, New York and Chicago 150, Washington and St. Louis 200, New Orleans and Los Angeles 250, and Miami 350 cal cm^{-2} day^{-1}. One gets some general idea of the distribution of sunny and cloudy climates from these figures.

The daily cycle of sunlight is shown in Figure 4.3 for a typical middle latitude station (about 55°) in the northern hemisphere in June. Sunrise is a little before 0400 and sunset a little after 2000. The direct sunlight and skylight together exceed 1.1 cal cm^{-2} min^{-1} at noon. There is sunlight reflected from the ground surface, which in this case is grass covered. The reflected radiation at noon is 0.2 cal cm^{-2} min^{-1} and, of course, follows the same trend as the direct sunlight does throughout the day.

There are other streams of radiation in the normal environment out of doors which are not as apparent to a person as are sunlight and skylight. There is a law in physics which states that any surface emits energy in the form of radiation by virtue of its surface temperature. It goes on to state that the intensity of the radiation emitted is proportional to the fourth power of the absolute temperature of the surface. The absolute temperature of any object is its temperature above absolute zero, which is −273° C. Hence the absolute temperature is the centigrade temperature plus 273 degrees. Most surfaces emit radiation with a high degree of efficiency; however some surfaces do not radiate as "blackbodies" but as a certain fraction ϵ, called the emissivity of the surface.

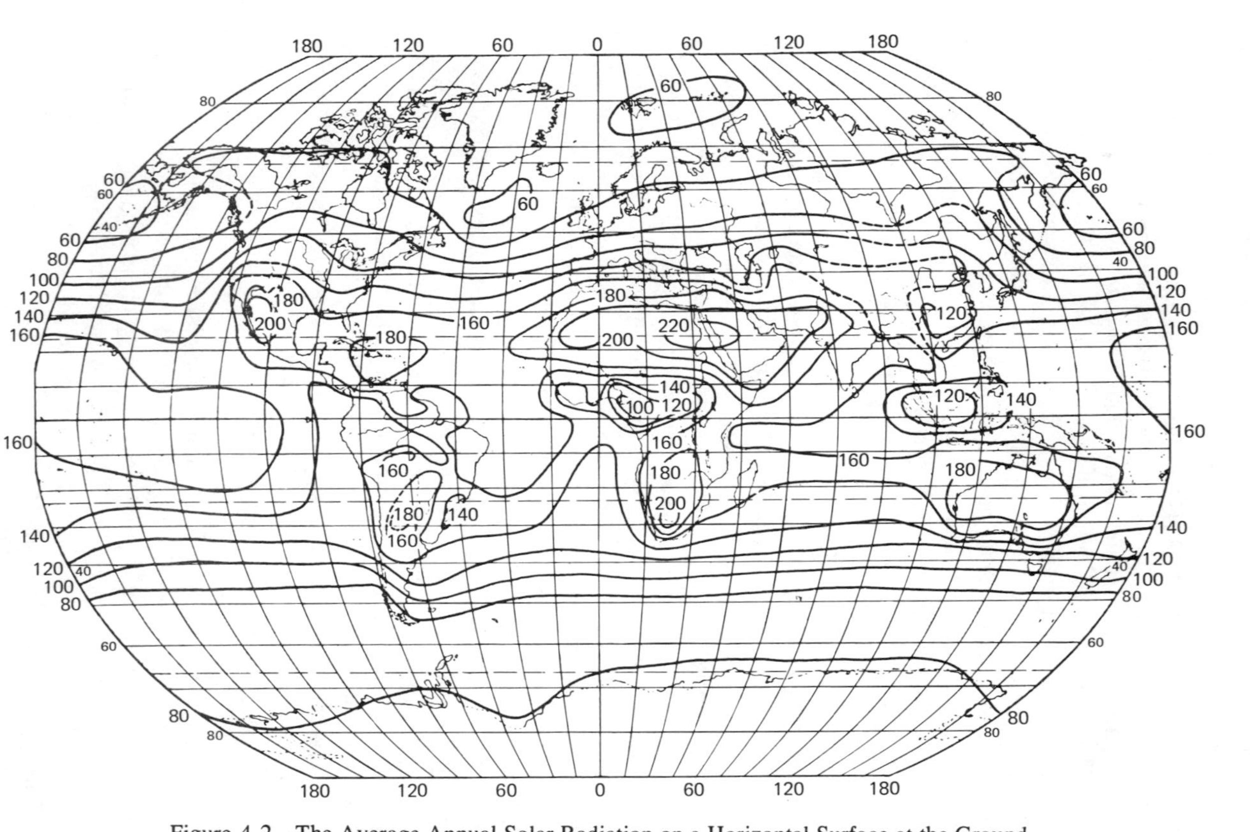

Figure 4.2 The Average Annual Solar Radiation on a Horizontal Surface at the Ground in Kilocalories per cm^2 per Year. SOURCE: Reproduced by permission of the University of Chicago Press.

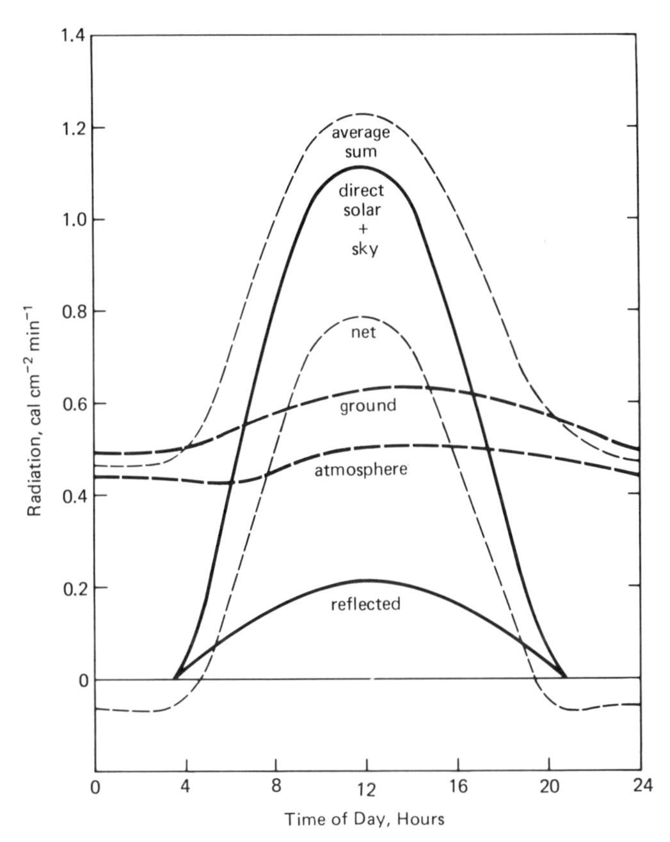

Figure 4.3 The Daily Variation of the Radiation Fluxes Incident on a Horizontal Surface as Measured at Hamburg, Germany, on June 5, 1954, a Day with Cloudless Sky. The Net Radiation Is Given and Also the Average Sum of the Fluxes Incident on an Upward-Facing and a Downward-Facing Horizontal Surface.

We can describe the intensity of radiation from a surface by the following formula:

$$R = \epsilon \sigma T^4 \qquad (4.1)$$

where $\sigma = 8.13 \times 10^{11}$ cal cm^{-2} K^{-4} min^{-1} and T is the surface temperature.[1] This law is known in physics as the Stefan-Boltzmann Law of Radiation. All surfaces emit radiant energy according to this law. Our own surface emits infrared radiation to the extent of 150 to 200 watts. Surfaces at ambient temperatures radiate infrared wavelengths only, but very hot surfaces, such as a

[1]This and the following equations are included for those readers with a mathematical background.

tungsten filament light bulb, become incandescent. Infrared thermal radiation is the single most ubiquitous property of our environment. The climate of the rooms in our houses is primarily and predominantly a radiation environment, of radiant heat emitted by floor, ceiling, walls, and objects. If the walls are at 20°C we receive a stream of infrared radiation of 0.6 cal cm^{-2} min^{-1} and at 25°C 0.64 cal cm^{-2} min^{-1}. We in turn radiate to the walls. A nude person has a skin temperature of about 28° and would radiate 0.67 cal cm^{-2} min^{-1} to the walls of the room. When wearing clothing we radiate a lesser amount since the surface temperature of our clothing is less than the skin temperature.

Out of doors all surfaces, such as soil, rock, trees, buildings, radiate according to their surface temperatures. These streams of radiation which flow to us as we stand out of doors are significant attributes of the climate around us. Without the energy we receive from these surfaces our environment would feel very cold indeed. If it is winter and the snow surface is at −10°C, we receive 0.39 cal cm^{-2} min^{-1} from it, and yet, though small in value, it is significant and better than no energy at all, such as we would have if we were surrounded by surfaces at absolute zero.

The clear sky overhead emits infrared radiation, which seems somewhat surprising since it appears so thin and transparent. Yet there are molecules in the atmosphere that absorb and emit infrared wavelengths of radiation even though they are perfectly transparent to visible light. These are the molecules of carbon dioxide and water vapor, which have very intense absorption bands beyond the visible in the infrared. In fact, this is one of the wonderful, but curious, properties of our atmosphere. The atmosphere is transparent to sunlight, allowing sunlight to reach the ground, but is partially opaque to infrared radiation. By being partially opaque at infrared wavelengths, the atmosphere forms a sort of radiation "blanket" over the Earth. Since the ground surface, at ambient temperatures, emits entirely at infrared wavelengths this radiation is absorbed by the atmosphere and not allowed to flow freely from it to the cold sink of space. The water vapor and carbon dioxide molecules of the atmosphere then emit the terrestrial radiation they have absorbed. But they emit this energy in two directions; half of it outward to space and half of it downward back toward the ground. In this way some of the energy is returned to the surface and to the ground which remains warmer than it would if the sky were completely transparent to infrared radiation. Less radiation is absorbed and returned toward the ground on extremely clear cold nights than on clear, but warm and muggy, nights. Not only does the sky radiate to the ground and to space during the night but also, of course, during the day. These streams of infrared radiation are present all the time and are, in fact, the most persistent feature of our environment. Also the ground radiates skyward a flux of infrared radiation according to the temperature of the surface. This flux is shown in Figure 4.3 on a diurnal basis as well as the downward flux from the sky. The two streams of radiant energy, the downward flux from the atmosphere and the upward flux from the ground, represent a substantial amount of energy within the environment day or night. The solar radiation is only an additional modulation to this ever-present thermal radiation.

An overcast of clouds reduces the amount of sunlight reaching the surface during the daytime, but at night it represents a marvelous blanket of warmth overhead. Clouds radiate energy according to the fourth power of the absolute temperature of their surface which from their underside is their base temperature. In warm summer weather the base temperature of clouds may be 5°C above freezing, but in the winter it may be −5°C or −10°C or lower. Since clouds are opaque they act essentially as blackbody radiators at these temperatures and emit a substantial flux of radiation toward the ground. When the air gets cold and is near the freezing point citrus growers are always happy when the sky is overcast rather than clear. Although there is a flux of radiation from a clear sky to the ground, which is distinctly more energy than for a completely transparent nonradiating sky, a cloudy sky does emit a much greater flux of radiant heat. If one camps out at night the differences in sky cover are very noticeably in terms of body comfort.

We hear about the CO_2 greenhouse effect or the warming of the global climate caused by an increasing concentration of carbon dioxide in the atmosphere. What is this all about? Since the Industrial Revolution there has been a rapid proliferation of industry throughout the world. Fossil fuels (coal, oil, and gas) are a source of power for industry, and products of combustion—carbon dioxide and water vapor—are released into the air. Carbon dioxide is released in vast quantities as a product of respiration by all organisms on Earth, but the amounts released by man's activities have caused the atmospheric concentration of carbon dioxide to increase. In fact it has been increasing steadily since about 1880. The average concentration of carbon dioxide in the atmosphere today is approximately 320 ppm (parts per million) per volume of air, but in 1870 the concentration was about 283 ppm. Hence there has been a 13 percent increase of carbon dioxide concentration in a period of 100 years. This amounts to 500 billion tons of carbon dioxide released by industrial activities. The current rate of increase of carbon dioxide is approximately 0.75 ppm per year, which in another 40 years will amount to 30 ppm or nearly another 10 percent increase. The influence of industrial CO_2 on climate is through the effect of this gas on the infrared transmission of the atmosphere. Increased CO_2 concentration results in increased absorption of terrestrial radiation and an increased amount of radiation emitted back toward the ground. The result is more energy at the ground surface and a slight warming of the surface and of the atmosphere. A 20 percent increase of CO_2 would produce approximately a 1°C rise in the mean temperature of the atmosphere. This may not seem like much of an increase, but an increase of 3°C or so in the mean temperature could result in sufficient warming to melt all of the ice caps of the world and to flood most major coastal urban areas. Whether it is coincidental or not, there was a general warming trend of the global climate beginning about 1880 and continuing through 1940. The increase of mean global temperature was about 0.5°C (0.9°F) during this period.

While we are concerned with the effects produced by carbon dioxide concentrations in the atmosphere, we should remember that plants require carbon dioxide for photosynthesis. Rates of growth or productivity for most plants are limited by the normal atmospheric concentrations of carbon dioxide. An increase of atmospheric CO_2 would produce an increase of plant productivity. During the daytime when great amounts of photosynthesis are occuring in a forest, crop, or grassland, the atmospheric concentrations of CO_2 nearby are pulled down to 240 ppm or less as the plants assimilate the carbon into carbohydrates. During the night, when photosynthesis is absent because of lack of light, the plants and all other living things continue to respire and to release CO_2 into the air and consequently the CO_2 concentrations near the forest, crop, or grassland rise as high as 400 ppm or more. Hence the plants begin the day with high local concentrations of CO_2 for photosynthesis, except when strong winds carry them away and mix them into the vast atmosphere above.

Between about 1940 and 1950 there began an abrupt cooling trend, so that in nearly 30 years the average temperature dropped by about 0.3°C. We can only speculate as to why this has occurred. It is entirely possible that this resulted from purely natural causes and that man's activities had little to do with it. On the other hand it is possible that the massive growth of technology since 1940 has had something to do with the climate change. But just what by-products of technology could be producing such a change? Is it the jet contrails which seed the skies every day and produce thin veils of cirrus clouds overhead to reflect sunlight back to space and cool the Earth's surface? Is it dust and aerosols from our massive polluted cities which are reflecting sunlight to space? We really do not know for sure. We know that the carbon dioxide concentration of the atmosphere is increasing, but we know also that the turbidity, the dustiness, of the atmosphere is increasing as well. The transparency of the air over remote mountain areas in Switzerland and Hawaii is several percent less than it was half a century earlier. The transparency is less, presumably because of the increased dustiness of the atmosphere. Does this result in a cooling of the mean temperature? It is entirely likely that it does. The insidious thing about this whole matter of climate change is that we really do not know what is going on but when we do learn precisely the relation between cause and effect it may be too late to do anything about it. The atmosphere is massive. It can take a lot of insult and perturbation. But once it has begun to go into oscillation, out of control, because of enormous inputs of foreign matter, it may develop abnormalities that are extremely difficult to correct.

Very few surfaces, either of ground, buildings, plants, or animals, are horizontal. It is interesting to consider the amount of direct sunlight which is incident upon vertical walls, hillsides, or various inclined planes. The plant and animal life of a north-facing hillside is often very different from that of a south-facing hillside because of the irradiation differences by sunlight. This phenomenon is particularly striking in mountain canyons in the Rocky Mountains and the Sierras where climate limitations on plant growth are critical. As one drives west from Denver and enters the glacial valleys whose rushing mountain streams drain the continental divide, one notices the vegetation on the south-facing slope on the north side of the valley to one's right to consist of ponderosa pine and juniper while the north-facing slope to one's left, which is the south side of the valley, is covered with spruce and fir. The two vegetation types meet along the stream in the center where all species tend to mix, but where there are also poplar, mountain maple, and many other species. The south-facing slopes are clearly the more sunlit, the warmer, and drier and the north-facing slopes are shadier, cooler, and more moist. However, there are times of the year when a north-facing slope at these northern latitudes may receive much more illumination than one normally would expect.

Figure 4.4 shows the direct illumination of various slopes by the sun for a position on the Earth of 40°N latitude at the time of the summer solstice, June 22nd. One can notice here that sunrise is, of course, earliest for the east-facing slopes, later for the horizontal slope, considerably later for the 45° south-facing slope, and much later in the morning for the vertical south-facing slope. A north-facing 45° slope has a considerably longer day (nearly three hours longer) than a south-facing 45° slope on this date in the summer, but the south-facing slope has a much greater amount of irradiation at noon. However, by the time of the equinox, that is, March 21st or September 21st, the irradiation on the north slopes is almost negligible, and during the winter months all north-facing hillsides with slopes greater than 22.5° have no direct irradiation from the sun. This in itself certainly accounts for the fact that these mountain slopes are cooler, wetter, and with more snow accumulation than are the south-facing hillsides throughout the year.

A north-facing vertical wall, which may have ivy growing on it and certain insects and birds living within it, will receive some sunlight early in the morning on June 22nd but by 0800 it will be in shadow until 1600 when the evening rays will illuminate it again for a few hours. The early morning and late afternoon sunlight is relatively weak since the slant rays of the sun pass through a long atmospheric path before striking the north-facing wall. On the other hand, the climate of a vertical south-facing wall is completely different than is the climate of the north-facing one. When the sun casts a shadow on the north face of the wall at 0800 on June 22nd it is just beginning to illuminate the south face which then has direct sunlight on it until 1600, after which time the sun strikes the north face of the wall again. We should keep in mind that the wall in shadow, whether north or south face, receives diffuse scattered skylight throughout the day. It is interesting to follow the illumination of the vertical south-facing wall by the direct sun at noon through the season. On June 22nd it receives very much less radiation than a horizontal surface, by the time of the equinox it is receiving

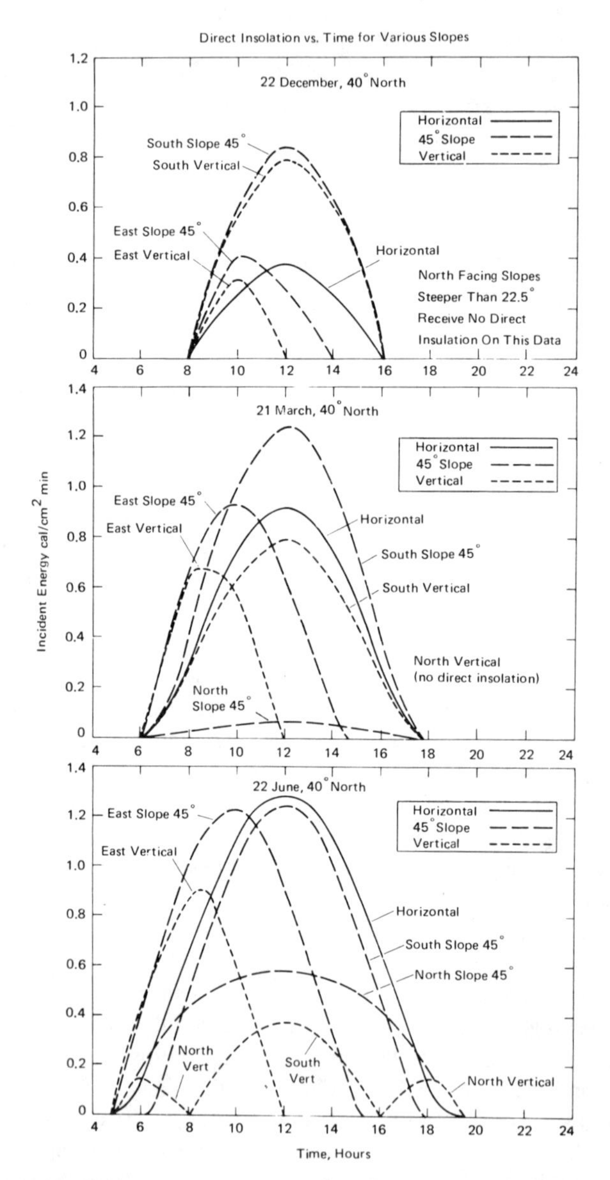

Figure 4.4 The Amount of Direct Sunlight Incident upon Various Slopes at 40° N Latitude as a Function of the Time of Day for the Winter (December 22nd) and Summer (June 22nd) Solstices and for the Spring and Autumn Equinoxes Both the Same as for March 21st.

nearly an equal amount, and by the winter solstice the vertical south wall gets very much more direct radiation than does the horizontal surface. Plant growth and animal activity reflect intimately the solar radiation climates of the various slopes.

The climate of slopes and walls is altered in many other ways than just the direct solar radiation incident upon them. The drainage of water varies from slope to slope, the wind changes enormously among them, and the accumulation of snow is strikingly different depending upon the direction of the prevailing wind. Not only is the radiation incident upon vertical walls different on the two sides but the radiation which reflects off the wall is very different. As a result the total amount of radiation incident on the ground surface at the base of the wall is much greater on the south side than on the north side of the wall and evaporation rates are very different.

The spectral quality of sunlight and skylight is shown in Figure 4.5. Approximately 50 percent of the energy from the sun is in the infrared part of the spectrum and about 50 percent is in the visible and ultraviolet portion. The direct sunlight is filtered by the atmosphere on its way to the ground. As a result of its passage through the atmosphere the infrared part of the solar spectrum is strongly attenuated by absorption bands of water vapor and carbon dioxide. The ultraviolet wavelengths are terminated abruptly at about 2900 Å, or 0.29 microns because of intense absorption by stratospheric ozone. In addition to atmospheric absorption throughout the spectrum, there is scattering of sunlight by the air molecules of the sky and by dust. Some of the light that is scattered out of the direct sunlight beam we see visibly as blue skylight coming toward the ground from the hemisphere of the sky. Small particles scatter most effectively radiation that has a wavelength of about the same size as the diameter of the particle. Since air molecules, O_2, N_2, O_3, and so on, are very small, the primary scattering occurs at the short wavelength end of the spectrum. Rays of ultraviolet and blue light are scattered much more effectively than are red and infrared rays. This is precisely the reason why the sky is blue and the reason why sunsets are red. Sunlight, as seen by an astronaut in orbit outside the Earth's atmosphere, appears brilliantly white. As sunlight penetrates the atmosphere the direct beam is depleted of ultraviolet and blue light which is scattered into the sky to either side. When the sun is near the horizon, as at sunrise or sunset, the loss of short wavelengths is so great that the sun appears red.

The spectral quality of sunlight, skylight, or daylight in general is of major significance to plants and animals. Every animal has its own characteristic coloration. The appearance of an animal in daylight is determined by the amount and quality of the illumination and the spectral reflectivity of the animal surface. Just how conspicuous an animal is in the natural scene may determine its vulnerability to predators. Adaptive coloration among animals is a well-established evolutionary trait which has worked selectively for the survival of all kinds of animals. White kangaroo mice are found on the white sands of New Mexico,

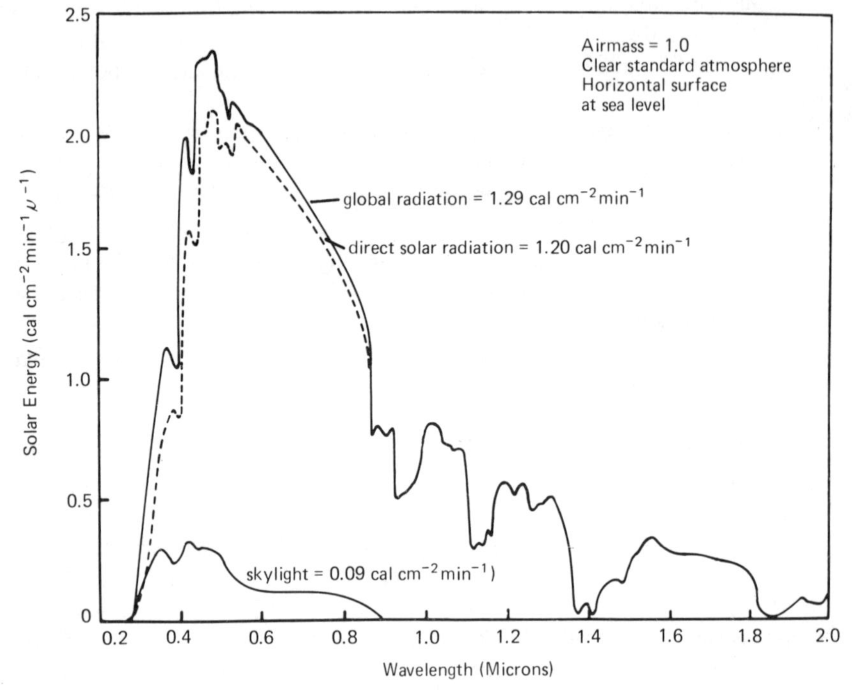

Figure 4.5 Spectral Distribution of Direct Sunlight, Skylight, and the Global Radiation (Sum of Direct Plus Skylight) as a Function of the Wavelength.

reddish colored ones on the red lava beds, and dark brown kangaroo mice on darker soils. Green katydids match closely green leaves and many of the true bugs have an identical appearance to some of the vegetation on which they feed. A famous case is that known as industrial melanism whereby several species of moths that are normally gray or whitish in color have changed to a dark mottled appearance which nearly matches the pollution stained bark of trees within the industrial cities of Europe. These dark colored mutants of the normal population survived predation by birds within the dirty city while their light colored relatives were vulnerable against the soot darkened backgrounds. The arctic hare is white amidst the winter snows as also is the ptarmigan, which makes them more difficult to locate by the hawks and owls circling overhead.

The coloration of an animal affects also its heat budget during the daytime since it absorbs more sunlight if it is dark than if it is light. It is difficult to know just how much evolutionary adaptation has occurred for this reason. But, as we shall see later, the absortive properties of an animal are very important from an energy budget standpoint in sunlight. At night, when there is thermal radiation exchange only, animal coloration has absolutely nothing to do with the energy budget. There is an oft-stated gross misconception in textbooks that the thermal energy budget of animals is affected by coloration and that the snowshoe rabbit is white in order to lose less radiation. This is absolutely not true in any way. I have seen statements made that a dark colored sleeping bag will lose more heat at night to the cold sky overhead than will a light colored sleeping bag. This also is not true. Coloration is the result of pigmentation which interacts with visible parts of the spectrum only and to some extent with the ultraviolet and very short infrared, but not at all with the longwave infrared heat exchange which occurs at night.

Energy Exchange for Plants

A plant in any environment responds to the amount of radiation, air temperature, wind, and moisture by assuming a specific temperature and losing water at a definite rate. The extent to which a plant interacts with these environmental factors depends upon the characteristics of the particular plant. The temperature of plant leaves is very important to the plant just as your own temperature is important to you. A plant is like a cold-blooded animal in that its temperature is near to the environmental temperature, whereas a warm-blooded animal must have a nearly constant body temperature.

Processes of photosynthesis, respiration, growth, and so on, are quite temperature sensitive just as most other chemical reactions are dependent upon temperature. Chemical reactions proceed very slowly when the temperature of the mixture is cold, react more rapidly when warmer, and when too hot the reaction is usually destroyed by the rapid energetic collision of the molecules. A typical chemical reaction with temperature is shown in Figure 4.6. There is clearly an optimum temperature at which the reaction proceeds at its maximum rate. In Figure 4.7 are shown the temperature dependences of photosynthesis and respiration. The temperature optimum is clearly evident for photosynthesis. Respiration increases with temperature and reaches a maximum rate at some fairly high temperature above which even it will slow down and eventually cease. Of particular interest is the fact that plants that have evolved to live in cold climates, such as alpine or arctic regions, have relatively low temperature optima and grow poorly under warm conditions. Plants that have evolved in warm or hot regions of the world have developed enzyme systems which are most responsive to higher temperatures while plants of temperate zones are intermediate in response.

When we attempt to understand the effect of climate on plants we must imagine how the plant temperature affects its physiological response. If photosynthesis cannot proceed at an adequate rate, or if respiration exceeds photosynthesis, then the plant will not grow. Although temperature is an important factor it is not the only one. Plant growth is limited by light intensity and by the availability of carbon dioxide. The light intensity affects the energy budget for a plant, and hence the plant temperature, but also light itself is

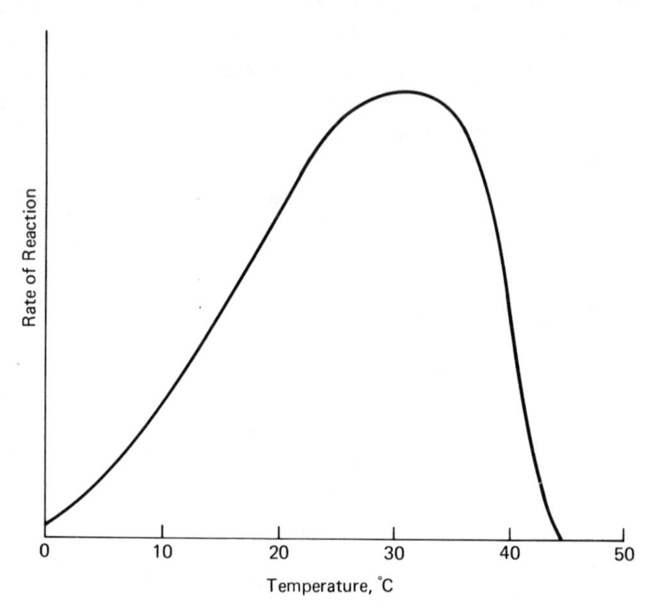

Figure 4.6 A Typical Chemical Reaction Rate as a Function of the Temperature of the Solution.

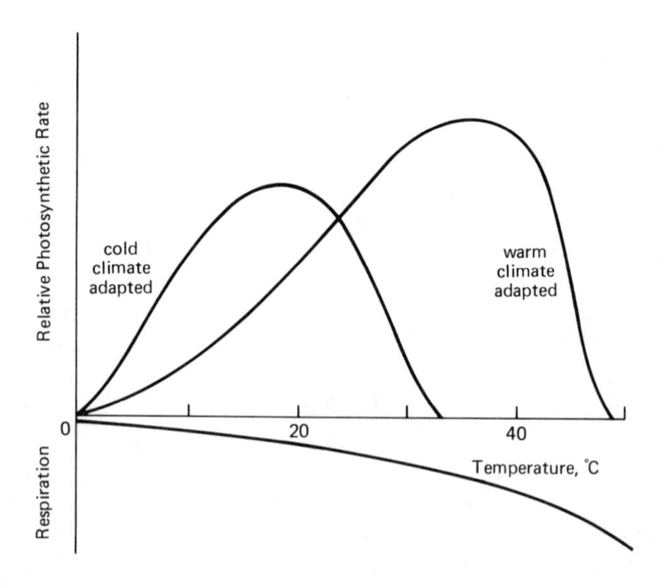

Figure 4.7 The Temperature Dependence of Photosynthesis for a Cold Climate and a Warm Climate Plant and a Typical Respiration Rate Curve as a Function of Temperature.

necessary for the photosynthetic process, which is a photochemical reaction. In habitats where the quantity of light available is low, plant growth is limited. This is the situation beneath the forest canopy where seedlings in the forest will often spring up but growth is extremely limited because of the dim light. Seedlings of maple trees in the woods, after ten years or more of life, may be only a few inches tall. If a large tree falls down, its loss creates an opening in the canopy, light floods the ground cover and the seedlings shoot up at an enormous rate until competition among them for moisture, light, and nutrients will determine eventually which individual will replace the fallen giant in its place in the forest.

Photosynthesis in plants involves chemical reactions which are light and temperature dependent, by which the carbon dioxide molecule is split, the carbon is assimilated to form carbohydrates, and water and oxygen are released in the process. The plant absorbs carbon dioxide through its pores, called stomates, and diffuses water vapor and oxygen. If inadequate amounts of carbon dioxide occur in the air nearby a plant then the rate of photosynthesis, the rate of formation of carbohydrates, and the rate of growth become limited. The response of a plant to climate or to environmental conditions depends upon its temperature, the quantity of light, and the availability of carbon dioxide. All these things must be considered when attempting to understand plant growth. There is no more fundamental question to be answered by man than this one, concerning primary productivity, which has such pervasive importance to all life on Earth. The growth of plants makes food available to the entire food chain of animals, and the release of oxygen to the atmosphere is absolutely critical for animal life.

The several processes which affect the flow of energy between a plant and its environment are radiation, convection, conduction, evaporation, and chemical activity or metabolism. From the thermal or heat budget standpoint the chemical processes of metabolism are small and can be ignored in the energy budget. Averaged over a reasonable length of time a plant leaf must be in energy balance with its environment. If a leaf receives more energy than it loses then it will become hotter and hotter and thermal catastrophe will result. If it loses more energy than it absorbs it will become progressively colder and life will cease. Therefore a plant leaf must remain on the average in essentially energy balance although frequent transient excursions may occur. The primary source of energy entering a plant leaf is radiation. A certain fraction of incident sunlight, skylight, and reflected light, usually between 50 and 60 percent, is absorbed by a green plant leaf and nearly all of the thermal radiation streaming to it from ground or atmosphere is absorbed. If we add up all the streams of radiation which are absorbed by a plant leaf we get a quantity, Q, which will normally have values between 0.6 and 1.4 cal cm^{-2} min^{-1} during the daytime, depending upon the amount of sunshine, and values between 0.3 and 0.8 at night. If the leaf can get rid of this absorbed energy only by the emission of radiation it will emit a quantity σT^4 according to the leaf temperature, T. If there is no other process of energy dissipation then the energy budget of the leaf is written as:

$$\text{Energy In} = \text{Energy Out}$$
$$Q = \sigma T^4 \qquad (4.2)$$

where σ is the Stefan Boltzmann constant described earlier. Shown in Table 4.1 are the temperatures a leaf would assume for various values of Q if there were no other processes of energy loss other than the emission of thermal radiation. Under these circumstances leaves would become very hot indeed with temperatures as high as 89°C when in full sunshine, but when in the shade or at night with low amounts of incident radiation, leaves would be unusually cool.

Table 4.1 Leaf Temperature in C When the Energy Input to the Leaf by Radiation Is *Q* and When the Leaf Is Cooled by Radiation Only, by Radiation plus Convection, and by Radiation, Convection, and Transpiration

Q cal cm^{-2} min^{-1}	Radiation Only	Radiation and Convection V(cm sec^{-1})			Radiation, Convection and Transpiration V(cm sec^{-1})		
		10	100	500	10	100	500
0.6	20	28.4	29.4	29.7	25.2	27.2	28.5
1.0	60	40.9	34.3	32.0	33.4	31.1	30.5
1.4	89	53.0	39.2	34.4	40.5	34.8	32.5

Air temperature 30°
Diffusion resistance 2 sec cm^{-1}
Relative humidity 50%
Leaf dimension 5 x 5 cm

In order that the reader may understand somewhat better the amounts of radiation likely to be absorbed by leaves in various environments, we have worked out a general association between the quantity of radiation in an environment versus the air temperature. This is shown in Figure 4.8. Essentially the warmer the air the greater the total flux of radiation is likely to be. It is not appropriate, nor necessary here, to give the details as to the way Figure 4.8 was derived except to say that it was worked out carefully for real situations.

If a leaf is in air the energy exchange process of convection will take away heat if the leaf is hotter than the air but will deliver heat to the leaf if the air is warmer than the leaf. The effectiveness of the air to cool or warm a leaf depends not only upon the wind speed near the leaf, but also upon the size of the leaf. If the leaf is very small the air can readily flow over its surface and the rate of convective heat exchange will be large. Air flowing over a large leaf, like a banana or tobacco leaf, will exchange heat effectively near the leading edge of the leaf and then the air flow will become detached from the leaf surface as it proceeds downwind across the surface. Near the surface of every object is a boundary layer of adhering air which is in fact a transition zone in temperature, moisture, and air movement between the object surface and the free air beyond. A boundary layer adhering to the surface of an oak leaf in still air is seen in Figure 4.9. The adhering boundary layer is indeed a buffering zone between leaf and air. If air were a perfect insulator then it would isolate the leaf temperature entirely from the air temperature nearby. If, on the other hand, the air were a very good conductor, like a sterling silver spoon, then every small wisp of temperature change of the air would immediately be conducted to the leaf surface. The same phenomenon applies to ourselves, as we shall see later on. As the wind blows and air flows across the leaf surface it wipes away the boundary layer. A very thin layer of air always adheres to the leaf but the thinner the layer becomes the greater is the rate of convective heat transfer. A large leaf has a thick boundary layer adhering to it and a small leaf is enveloped in a very thin layer of stagnant air. This is the reason why the rate of convective heat transfer between a leaf and the air is inversely proportional to the width or length of the leaf in the direction of air movement.

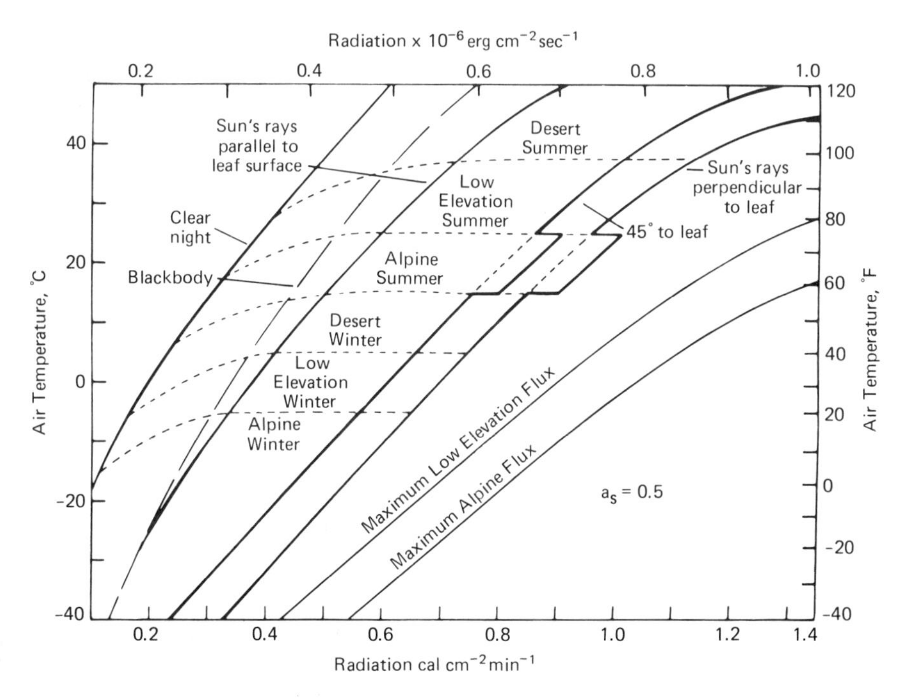

Figure 4.8 The Average Amount of Radiation Flux Incident upon the Two Sides of a Flat Leaf as a Function of the Air Temperature for Various Habitats, and the Average Amount of Radiation Flux Absorbed by a Leaf of Absorptivity to Direct Sunlight, Reflected Light, and Skylight of 0.5. The Absorptivity of the Leaf to Infrared Thermal Radiation from the Ground and Atmosphere Is 1.0.

Many careful laboratory measurements by heat transfer engineers show us that the rate of convective heat transfer between an object and the air is proportional to the square root of the wind speed, V, is inversely proportional to

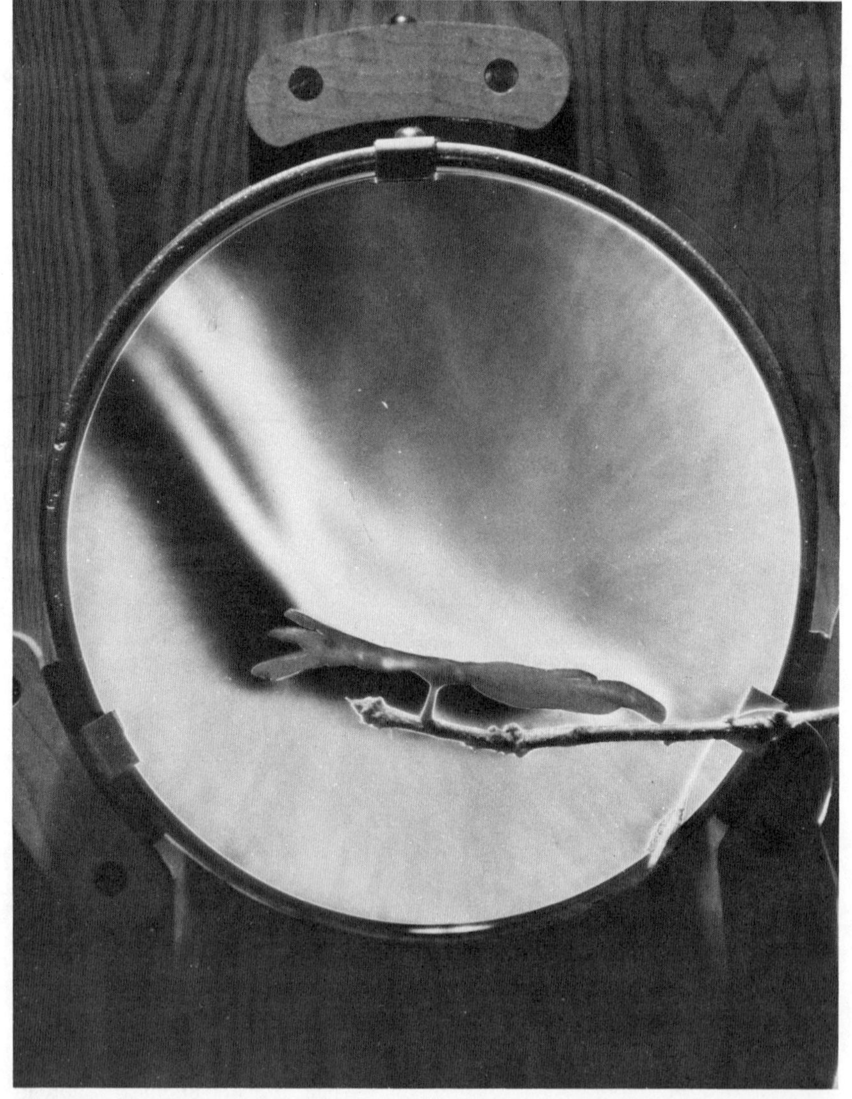

Figure 4.9 The Boundary Layer Adhering to the Surface of a Bur Oak Leaf in Sunshine as Seen in This End on View Taken by Means of Schlieren Photography. The Leaf Was in Still Air and the Only Air Movement Was the Plume of Warm Air Rising from the Leaf by Free Convection.

the width or characteristic dimension of the leaf, D, and is proportional to the temperature difference, $(T - T_a)$, between object (or leaf) and air. Now we can write the energy budget for a nontranspiring leaf in air for which the rate of heat transfer depends upon radiation and convection only. Thus,

$$\text{Energy In} = \text{Energy Out}$$

$$Q = \sigma T^4 + k\sqrt{V/D}\,(T - T_a) \tag{4.3}$$

where $k = 6.0 \times 10^{-3}$ in units such that V is in cm sec^{-1}, D in cm, and $(T - T_a)$ is in °C and the whole convection term is in cal cm^{-2} min^{-1}. This equation states in mathematical symbols that a quantity of radiation, Q, which is absorbed by the leaf is lost by reradiation and by convection by the leaf at a temperature, T. Now we can see in Table 4.1 that the effect of convection on leaf temperature is rather substantial when Q is large. For instance, when $Q = 1.4$ cal cm^{-2} min^{-1}, the leaf temperature was 89°C when radiation only could cool the leaf, but in still air, that is, where $V = 10$ cm sec^{-1}, at 30°C the leaf temperature dropped to 53°C, and in wind at 100 and 500 cm sec^{-1} to 39.2 and 34.4°C, respectively. However, when the amount of radiation absorbed by the leaf is only 0.6 cal cm^{-2} min^{-1}, a condition which would occur at night, the leaf temperature would drop to 20°C with radiational cooling only, but when the air around the leaf is at 30°C convective heating will warm the leaf to 28.4°C in still air ($V = 10$ cm sec^{-1}), to 29.4 and 29.7°C in air flow of 100 and 500 cm sec^{-1}, respectively. In this latter case the convective heat transfer amounts to heat input to the leaf and the term $k\sqrt{V/D}\,(T - T_a)$ becomes negative because $T_a > T$. This is equivalent to moving this term to the left hand side of the equation. For the first time one can understand from Equation 4.3 the way the air temperature and the wind speed affect the leaf temperature. It is only by means of convective heat transfer that the leaf temperature and energy budget are influenced by the air temperature.

A plant in reality uses water, and its leaves transpire through the stomates, which are very small pores or openings connecting the interior of the leaf with the outside air. A plant leaf absorbs carbon dioxide which it assimilates into carbohydrates by means of photosynthesis. The carbon dioxide cannot penetrate the leaf cuticle. The leaf has stomates through which carbon dioxide may diffuse into the leaf. However, when the stomates open, water vapor escapes. Liquid water exists within the cellular tissue of the leaf and then within the intercellular spaces is vaporized to water vapor, a process which requires approximately 580 calories to change a gram of water into vapor at a temperature of 30°C. Two events must occur simultaneously in order for a leaf to transpire, assuming that the stomates are open: energy must be available to convert liquid water to vapor, and gas diffusion must occur from within the leaf out into the free air beyond the adhering boundary layer. For gas diffusion to occur there must be a concentration gradient or difference between the concentration of water vapor inside the leaf, designated as d_l, and the concentration in the air, d_a. The air inside the leaf is considered to be at saturation at the leaf temperature. The concentration of water vapor in the air outside the leaf is the concentration of the air at saturation ${}_sd_a$ multiplied by the relative humidity, h. In fact, the relative humidity is defined as the ratio $d_a/{}_sd_a$. Hence, $d_a = (h_s){}_sd_a$. Water vapor diffusing from inside a leaf out through the stomates encounters a resistance to the flow both within the stomate channel and through the boundary layer. The internal resistance through the stomates we refer to as r_l and the boundary layer resistance as r_a. The gas diffusion equation describing the rate of transpiration E for water vapor diffusion from inside the leaf out through the stomates and across the boundary layer is as follows:

$$E = [d_l - (h_s)\ {}_sd_a]/\bar{r}_l + r_a) \tag{4.4}$$

E is in gm cm^{-2} min^{-1} when d_l and ${}_sd_a$ are in gm cm^{-3}, and r_l and r_a are in min cm^{-1} Actually, the boundary layer resistance varies inversely with the square root of the wind speed and with the square root of the characteristic dimension of the leaf. Hence, $r_a = k'\sqrt{D/V}$ where $k' = 3.0 \times 10^{-2}$ min cm^{-1} sec$^{-1/2}$ when V is expressed in cm sec^{-1} and D in cm. I wish the world were more simple, but it is not and this is the way the loss of water vapor relates to the humidity of the air and to the resistance of the diffusion pathway.

Agricultural plants, that is, crops, and certain other plants have large stomates and as a result have low resistances of the order of 2.0 sec cm^{-1}, while many plants have resistances between 5.0 and 10.0 sec cm^{-1}, and some have even larger resistances. A few plants may have resistances as low as 1.0 sec cm^{-1}.

Now we can write the full energy budget of an actual transpiring plant leaf in air for which energy is transferred between it and the environment by means of radiation, convection, and transpiration.

$$\text{Energy In} = \text{Energy Out}$$

$$Q = \sigma T^4 + k\sqrt{V/D}\,(T - T_a) + 580[d_l - (h_s)\ {}_sd_a]/(r_l + r_a) \tag{4.5}$$

Here we see directly the simultaneous fashion in which the amount of radiation absorbed by the leaf, the air temperature, the wind speed, and the relative humidity affect the leaf temperature and the transpiration rate. There is absolutely no other way in which to realize the exact interaction of the environment, or climate, with a plant leaf. One cannot ask what the effect of the air temperature is upon a plant without specifying at the same time the amount of radiation, the wind speed, and the relative humidity. Likewise, one cannot understand the influence of wind on a leaf outside the context of radiation absorbed, air temperature, and relative humidity.

We shall inspect the results of the analysis represented by Equation 4.5 in several ways. First, consider Table 4.1 once again. We have seen that when a leaf loses energy by means of radiation only, it will get very hot if the amount of radiation absorbed is large, such as a leaf encounters when in full sunshine. We have seen that the air around a leaf can cool it substantially by convection under these same circumstances. Now if the leaf transpires, we see from Table 4.1 that it is cooled much more. The leaf, in effect, turns on its evaporative cooler and very significantly dissipates some of the heat load. For example, if $Q = 1.4$ cal cm^{-2} min^{-1}, whereas with radiational cooling only it was at 89°C, and in air with convective cooling but no transpirational cooling the leaf temperatures were 53.0, 39.2 and 34.4°C at wind speeds of 10, 100, and 500 cm sec^{-1}, respectively, now with transpirational cooling in addition the leaf temperatures become 40.5, 34.8, and 32.5°C, respectively. Even though transpiration appears to be the inadvertent consequence of opening the stomates in order to take in carbon dioxide, the result of releasing water vapor from the leaf and of converting liquid water to vapor in order to do so means a substantial cooling of the leaf. The photosynthetic rate of a leaf is a function of the leaf temperature as shown in Figure 4.7. At very low temperatures photosynthetic rates are slow, then there is an optimum temperature at which rates maximize, and then at higher temperatures photosynthesis stops. Usually leaf temperatures greater than 45°C are damaging to plants. Alpine and arctic plants grow best when temperatures are about 12 or 15°C. Plants of temperate regions have temperature optima at 30°C and some tropical species may reach maximum growth rates at temperatures as high as 40°C.

The two dependent variables, transpiration rate and leaf temperature, are displayed in Figure 4.10 as a function of the air temperature and the internal diffusion resistance of the leaf to water vapor at a constant amount of radiation absorbed by the leaf of 1.0 cal cm^{-2} min^{-1}, in still air with air movement considered as 10 cm sec^{-1}, at a relative humidity of 50 percent for a leaf of dimensions 5 x 5 cm. If the leaf resistance is 2 sec cm^{-1} and the air temperature 30°C, the leaf is at 34°C, and when the air is 40°C, the leaf is 39.5°C. Transpiration rates are 28×10^{-5}, 42×10^{-5}, and 48×10^{-5} gm cm^{-2} min^{-1} at 10, 30, and 40°C air temperatures, respectively. If the stomates of a leaf are partly closed and the internal diffusion resistance goes from 2 to 10 sec cm^{-1}, the transpiration rates diminish enormously to 5×10^{-5}, 9.5×10^{-5}, and 12×10^{-5} gm cm^{-2} min^{-1} at air temperatures of 10, 30, and 40°C, respectively. If the stomates are closed, entirely, perhaps because of water stress in the plant as the result of dryness in the soil, then the transpiration rate is nearly zero and the leaf temperatures are 27, 41, and 47.5°C at air temperatures of 10, 30, and 40°C. Clearly one can change the value of Q, wind speed, or relative humidity and present other graphs of this kind which will give further information concerning transpiration rate and wind speed.

It is of particular interest to find out how much of a difference in transpiration rate and leaf temperature occurs as the width or size of a leaf varies from one plant to another. Figure 4.11 reveals the influence of dimension. It is noticed that if the leaf is small, say less than 1 x 1 cm, its temperature will always be within a few degrees of air temperature for the conditions presented here. By contrast if

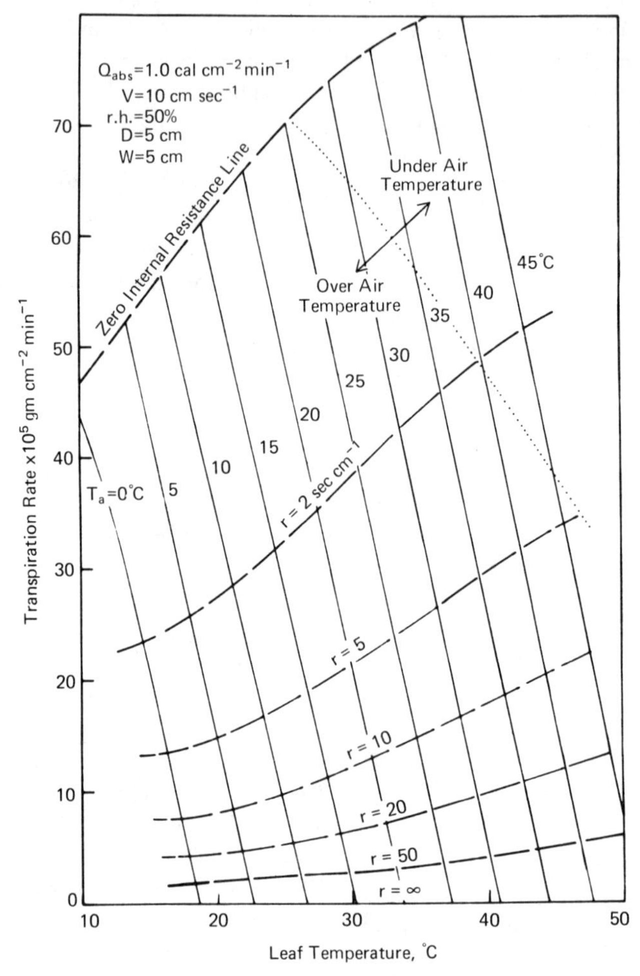

Figure 4.10 Computed Transpiration Rate and Leaf Temperature as a Function of the Air Temperature and the Internal Diffusion Resistance of a Leaf for the Conditions Indicated.

the leaf is large, say 10 x 10 cm or greater, the leaf temperature is often 5°C above air temperature and may be as much as 10°C above or more. When in arid regions, moisture is in short supply, diffusion resistance is large, and air temperatures are high, it is far better for a leaf to have a temperature close to air temperature than one many degrees above the air temperature.

A person can walk through a garden and by touching the leaves of various plants can assess their temperatures and quickly discover that different plants side by side in full sun in the same environment have strikingly different temperatures. Each and every species of plants in the world has its own specific properties and behavior for interacting with its environment. Every species has evolved physical and physiological features that will permit it to function and survive in its particular environment.

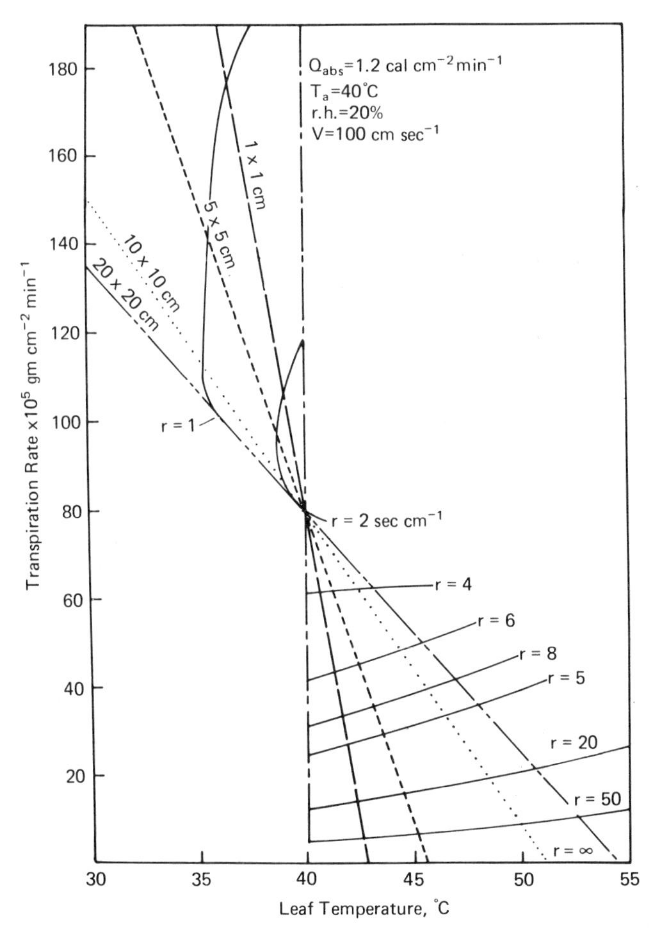

Figure 4.11 Computed Transpiration Rate and Leaf Temperature as a Function of the Air Temperature and the Dimension of a Leaf for the Conditions Indicated.

Energy Exchange in Animals

The interaction of animals with their environment is similar to that of plants in that it is basically an energy flow process except now the metabolic heat is a significant component. A warm-blooded animal, the homeotherm, maintains body temperature nearly constant, by the regulation of metabolic rate, by vascular blood flow to peripheral parts of the body, by water loss through panting or sweating, and by changing the quality of body insulation. A cold-blooded animal, the poikilotherm, has less control over physiological processes but can allow body temperature to vary over a relatively large range of temperatures. The body temperatures of some lizards, for example, may reach as low as 2 or 3°C and as high as 44°C.

An animal exchanges energy by radiation, convection, conduction, and evaporation of water. A very large animal, an elephant, for example, will lose energy primarily by radiation. Most very small animals, such as insects, lose energy primarily by convection. Salamanders and other amphibians as well as some insect larvae lose a great deal of energy by means of a loss of water through a highly permeable skin. Those organisms living in soil or water will lose most of their body heat by direct conduction into the surrounding medium.

The mobility of animals allows them to select microclimates which are more or less agreeable much of the time. Animals can move from sun to shade, change their orientation or the amount of surface area exposed to the sun. Lizards will bask in the sun during early morning or late afternoon but will seek the shade of shrubs during the mid-day heat or even burrow into the desert sand. Locusts will orient for most warmth early in the day and present the least body area to the sun during the hottest time of day. When the heat load becomes too great, some animals, such as the dog and cat, can pant vigorously in order to create rapid evaporation of moisture from the lungs. Birds can activate gullular flutter and in various ways increase respiratory water loss through air sacs attached to the lungs which allow extensive areas of moist surface for evaporative cooling. Man, the horse, the camel, and, to some extent, cattle can sweat and get the benefit of evaporative cooling from the skin when the heat load becomes too great. For each gram of water evaporated from the body, either through respiratory water loss or by means of sweating, 580 calories of heat are consumed. Although the evaporation of water is one of the most effective ways for dissipating heat there is the ever-constant problem of a limited water supply and the threat of body dehydration. Vigorous use of evaporative cooling can occur for only limited periods of time and although rapid drinking of water may be possible it takes considerable time for water to pass from the alimentary tract into the body tissues. It turns out that most people can drink only a limited amount of water irrespective of how much may be available or how thirsty one may be.

The energy budget of an animal is described very much in the same way it is for a plant. Heat gained must equal heat loss for energy balance. The environmental factors which are important to energy flow are radiation, air temperature, wind speed, and humidity. Here we shall consider an evaluation of the first three only. As with plants the reason for writing the energy budget of animals in quantitative analytical form is that it gives us orientation as to just how each of the factors affects the animal. An animal will have heat input from radiation absorbed by its surface, an amount Q, and from metabolic heat, M, which is the result of burning up of food consumed by the animal. The animal has a deep body temperature, T_b, and a surface temperature of its fur, feather, or exposed skin of T_s. The air temperature is T_a. The animal is represented roughly as a cylinder of diameter, D, in cm for purposes of convective heat transfer. Its body is surrounded by a layer of fat which terminates at the skin which in turn may be covered with a thickness of fur or feathers. In order to keep matters reasonably simple we shall consider that the body has an average amount of heat insulation, I. The animal loses respiratory water, E, measured in gm cm^{-2} min^{-1}, which produces evaporative cooling and energy dissipation. The wind speed, V, is measured in cm sec^{-1}.

The energy budget for the surface of an animal is:

$$\text{Energy In} = \text{Energy Out}$$

$$M + Q = \sigma T^4 + k V^{1/3} D^{-2/3} (T_s - T_a) + 580E \qquad (4.6)$$

Here once again we see the simultaneity of the environmental factors. The air temperature and wind speed enter the energy budget through the convective heat transfer team. The radiation absorbed by the animal's surface is on the energy input side as is also the metabolic heat. The evaporative cooling caused by the rate of water loss by respiration and sweating naturally is affected by the relative humidity of the air, but to keep things simple we will ignore the rather complicated way in which humidity enters this term. Equation 4.6 tells us what goes on at the animal's surface in terms of energy flow. Now we must relate this to the internal body temperature, T_b, which is the really critical thing as far as the animal is concerned.

The body temperature is related to the surface temperature as a simple heat flow problem. If there is a slab of insulating material, such as the wall of your home, and on one side the temperature is T_b and on the other side it is T_s, then a quantity of heat q will flow through the wall. The amount of heat passing through is proportional to the temperature difference and inversely proportional to the quality of the insulation, I. Hence

$$q = (T_b - T_s)/I \qquad (4.7)$$

If we identify the quantity of heat q as the metabolic heat M generated within the animal less the energy consumed with evaporative water loss, which is $580E$, we can write

$$q = M - 580E = (T_b - T_s)/I \qquad (4.8)$$

Rewriting Equation 4.8 we get

$$T_b - T_s = (M - 580E)I \qquad (4.9)$$

A warm-blooded animal will vary M, E, or I in order to maintain T_b fixed, but a constant T_b puts specific restrictions on T_s for given values of M, E, and I.

If we solve Equation 4.9 for T_s and substitute into 4.6 we get a heat budget equation for an animal involving T_b and not T_s. Hence,

$$\begin{aligned} \text{Energy In} &= \text{Energy Out} \\ M + Q &= \sigma [T_b - (M - 580E)I]^4 \\ &\quad + kV^{1/3} D^{-2/3} [T_b - (M - 580E)I - T_a] + 580E \qquad (4.10) \end{aligned}$$

The animal automatically solves this equation every moment of its life in order to remain in thermal balance and remain alive.

If we know the physiological and physical characteristics of a particular animal, that is, if we know M, E, I, T_b, and D, for any environmental conditions we can calculate from Equation 4.10 the values of Q, T_a, and V which must prevail in order for the animal to survive energetically when in steady-state conditions. In other words, if we know the physiological and physical properties of a particular animal we can predict the climate in which it must live in order to survive. Although an animal is pretty well locked in by heredity to certain limitations to its metabolic rate, evaporative water loss rate, insulation quality, body temperature, and body size, it nevertheless can change these quantities both deliberately and inadvertently as environmental conditions change. Metabolic rates always increase with heat or cold from an intermediate temperature condition known as thermal neutrality. Evaporative water loss rates can be increased greatly by many animals by means of either sweating or panting when the heat load becomes great. Water loss rates are usually reduced greatly when an animal is in a cold environment and in fact birds can tuck their bills under their wings when very cold in order to reduce further the rate of water loss generated by breathing. The body temperature of warm-blooded animals can and does vary a small amount. It normally increases when the animal is in a very warm environment and decreases somewhat in a very cold environment. An animal can increase the insulation quality of its coat by fluffing its feathers or by pilo-erection of the hairs of fur. Considering all these things we can predict by means of Equation 4.10 the climate in which a given animal must live.

The climate of an animal is comprised of the amount of radiation it absorbs, the air temperature nearby, and the wind speed. In principle, the humidity of the air is also an important climate factor but is not included here.

A climate diagram showing radiation versus air temperature (Q versus T_a) is given in Figure 4.12 for the desert iguana, a cold-blooded animal, and for the cardinal, a warm-blooded animal. These animals must live within the boundaries of the quadrilateral indicated for each. The desert iguana, a lizard, is obviously limited with respect to the lowest acceptable temperatures. This lizard in bright sunshine can survive an air temperature in wind of 100 cm sec^{-1}, as low as -4°C (25°F), but at night under a clear, cold sky it can withstand air only as cold as 7°C (43°F). However, during the daytime when in full sunshine the lizard cannot be at an air temperature greater than 37°C (98°F) but at night could withstand air as warm as 48°C (118°F). If the wind speed is greater than 100 cm sec^{-1} (2.2

mph) then the lines representing the upper and lower boundaries of the quadrilateral become more horizontal. In other words, the body temperature or energy budget of the lizard becomes more tightly coupled to the air temperature as the wind speed increases. The desert iguana absorbs less sunlight than does the cardinal. The right-hand line in Figure 4.12 is the greatest amount of radiation absorbed by the animal when in full sunshine.

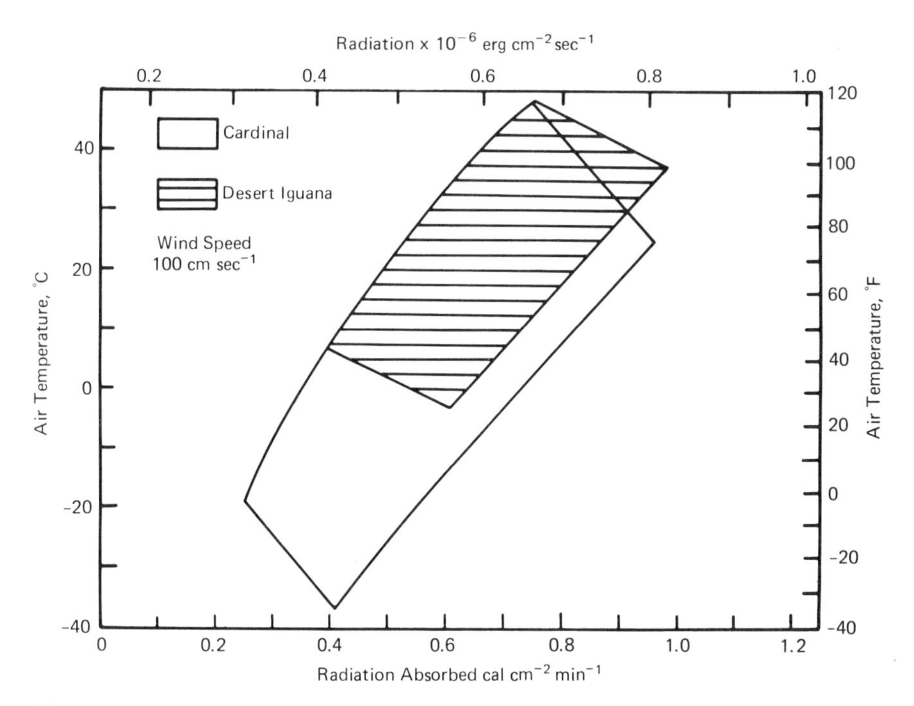

Figure 4.12 Climate Quadrilaterals for a Desert Iguana and a Cardinal in Air at a Wind Speed of 100 cm sec^{-1} (2.2 mph) Showing the Boundaries of Their Climates Expressed in Terms of the Amount of Radition Q Absorbed by the Animal and the Air Temperature T_a. Each Animal, in Order To Survive Energetically, Must Live within the Area Represented by Its Quadrilateral.

The Kentucky cardinal, the red bird, is widely distributed over the southern United States and northern Mexico. In recent years it has extended its range west across the high plains to Colorado and north to Minnesota and Maine. This colorful ubiquitous bird has a very large climate space as seen in Figure 4.12. When in full sunshine the cardinal can withstand air temperatures as low as −36°C (35°F) with air movements of 100 cm sec^{-1} (2.2 mph), In deep shade it can survive temperatures of about −25°C (−12°F) and at night when exposed to the cold sky air temperatures only as low as −18°C (−2°F). One of the reasons the cardinal has been able to extend its winter range to the north is the presence of towns with buildings. These offer warm microhabitats where cardinals can acquire protection from the wind and cold when conditions become too severe by nestling under eaves and other sheltered places. Some of the inner warmth of the building leaks through the walls and affords comfort to birds sitting up close to them. The cardinal can sit in the sun with a slight breeze in air at temperatures up to 25°C (77°F) but if the air temperature gets any hotter then it must get into partial shade. If the air temperature remains at 25°C (77°F) and the wind drops from 100 cm sec^{-1} to still air then the cardinal must also get into partial shade. If on the other hand the wind speed increases from 100 cm sec^{-1}, say, to 200 cm sec^{-1}, the cardinal can remain in full sunshine at temperatures of about 27°C (81°F) or so. If the cardinal is exposed to the cold of the night sky an air temperature maximum of about 48°C (119°F) can be tolerated for a reasonable period of time. The cardinal is able to dump a lot of respiratory water vapor when the conditions are hot, and when the environment is cold it can greatly reduce its water loss in order to conserve heat.

Reference

Gates, David M., *Man and his Environment: Climate.* New York: Harper and Row, Publishers, Inc., 1972.

5

The Regulation of Populations

PETER W. FRANK

University of Oregon

Peter W. Frank was born in Mainz, Germany in 1923. After leaving Nazi Germany in 1936, he eventually entered Earlham College in 1941. After graduating, he spent four years as W.C. Allee's research assistant at the University of Chicago. His doctorate research under the direction of Thomas Park dealt with interspecies competition in water fleas. The year 1951–1952 was spent as a Seessel Postdoctoral Fellow at Yale University under Professor G.E. Hutchinson. After five years' teaching and doing research at the University of Missouri, he moved to the University of Oregon, where he became interested in marine ecology. In the succeeding years, his research has been mainly with intertidal invertebrate populations along the Oregon coast and in the tropics of Australia and the Caribbean.

Some of the problems he has researched are population regulation and the evolution of life history features. Among his papers are "The Biodemography of an Intertidal Snail Population," *Ecology*, 1965; "Life Histories and Community Stability," *Ecology*, 1968; and "Growth Rates and Longevity of Some Gastropod Mollusks on the Coral Reef at Heron Island," *Oecologia*, 1969.

The average organism on earth today has had roughly 10^7 to 10^{12} generations of ancestors. The very persistence of life for such a span is probably a consequence of the negative **feedbacks**[1] collectively termed regulation. Lotka (1925) expressed the phenomenon by analogy of the earth ecosystem with a sun-powered mill wheel:

[1]Terms printed in boldface are defined in the Glossary at the end of this chapter.

> . . . in detail the engine is infinitely complex, and the main cycle contains within itself a maze of subsidiary cycles. And, since the parts of the engine are all interrelated, it may happen that the output of the wheel is limited, or at least hampered by the performance of one or more of the wheels within the wheel. For it must be remembered that the output of each transformer is determined both by its mass and by its rate of revolution. Hence, if the working substance of any of the subsidiary transformers reaches its limits, a limit may at the same time be set for the performance of the great transformer as a whole. Conversely, if any of the subsidiary transformers develops new activity, either by acquiring new resources of working substance, or by accelerating its rate of revolution, the output of the entire system may be reflexly stimulated.
>
> . . . It seems, in a way, a singularly futile engine, which, with a seriousness strangely out of keeping with the absurdity of its performance, carefully and thoroughly churns up all the energy gathered from the source. It spends all its work feeding itself and keeping itself in repair, so that no balance is left over for any residual purpose. Still, it accomplishes one very remarkable thing; it *improves* itself as it goes along, if we may employ this term to describe those progressive changes in its composition and construction which constitute the evolution of the system.

Salient features of regulation can be examined from a more restricted outlook. Focus on a single population provides an admittedly partial view; however, it leads to a better understanding of the various mechanisms by which regulation is ultimately achieved. The quotation above may help us keep in mind the forest while we look at the trees.

In the narrower sense, the problem of regulation may be rephrased as a question: How are the numbers of a population determined or limited? Before an answer can be given, we need a definition of population and of the processes that lead to changes in numbers. To an ecologist a population is a group of similar organisms that occur in a sometimes arbitrary, but always defined, space during a specified time interval. Qualitative restrictions of the degree of similarity are usually implied. Normally we wish to include as members individuals of various ages, of both sexes, and of otherwise different genetic composition, yet to delimit members of other species. Processes that lead to population change (see Figure 5.1) are births, deaths, immigration, and emigration. Sometimes, when considering the effects of a population, numbers alone do not provide a suitable measure of activity. Then **plastic** responses, such as individual growth or metabolic rates, may also have to be determined.

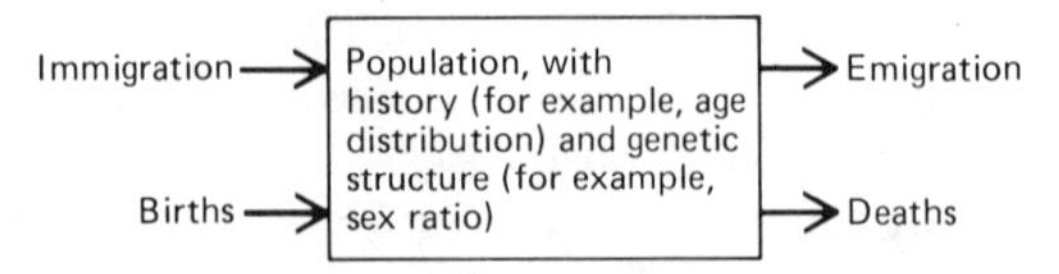

Figure 5.1 Events Responsible for Causing Population Change.

As a result of these processes of population change, areas may be colonized;

in other areas populations may expand, become more dense, contract, decrease, or become extinct. Where extinction is over the entire range of a species, it is clearly irreversible. More commonly, only a small sector of its area of distribution is involved, that is, there is local extinction. Even such processes may not be trivial. The meaning of local extinction to the species is a matter of some controversy regarding the relative importance of **selection** and interbreeding in such units (Ehrlich and Raven, 1969).

Some of the problems with the concept of regulation perhaps may be visualized best by biological models of the extremes of situations encountered in nature. Such extremes are exemplified in the **batch technique** and **continuous culture technique** for growing microorganisms. A bacterial culture in a test tube undergoes a series of predictable changes, during which resources are used up, **metabolites** are produced, and conditions change in such a way that ultimately no further birth processes occur. The population thus typically increases to a peak at which births equal deaths. Thereafter, it declines to ultimate extinction, unless provision is made for emigration, that is, a subculture. The decrease in population may not be seen readily since dead bacteria look like live ones. Furthermore, depending on the kind of bacterium used, a resistant spore may be formed that can persist with little expenditure of energy for long periods of time. However, the organisms clearly respond to their environment and are regulated throughout the process of population change. There may be a period near the beginning of growth when doubling the amount of materials and energy supplied to the culture does not change the rate of increase but merely its duration. Nevertheless, generally the population responses follow from changes induced by the bacteria themselves in the materials provided. The exact nature of the responses need not concern us at present.

At the opposite extreme, consider a population of bacteria in a continuous culture device, the chemostat (see Figure 5.2). Here a population is introduced, and supplied with a constant, continuing input of energy and materials. An output furnishes a continuous sample of the growth chamber's contents. Under normal culture conditions, a steady state is soon established in the growth chamber. Then the number of new individuals produced per time interval equals the number of live bacteria leaving by the overflow. The level to which the population grows at steady state depends on the concentration of some limiting nutrient and on the rate of flow. If the latter is rapid enough, the population may be washed out of the culture; in effect its emigration rate will be higher than its birth rate, and extinction in the growth chamber will ensue. Otherwise, the bacterial poppulation attains a finite steady density, so long as the rate of nutrient supply is high enough to be usable by the organism and the rate of flow remains constant. Genetic change in the population then is the sole source of potential change in its numbers.

The two conditions modeled, batch cultures and continuous cultures, simulate organisms and environments in nature. Continuous culture conditions are supposedly synonymous with ecological **climax**, where materials are

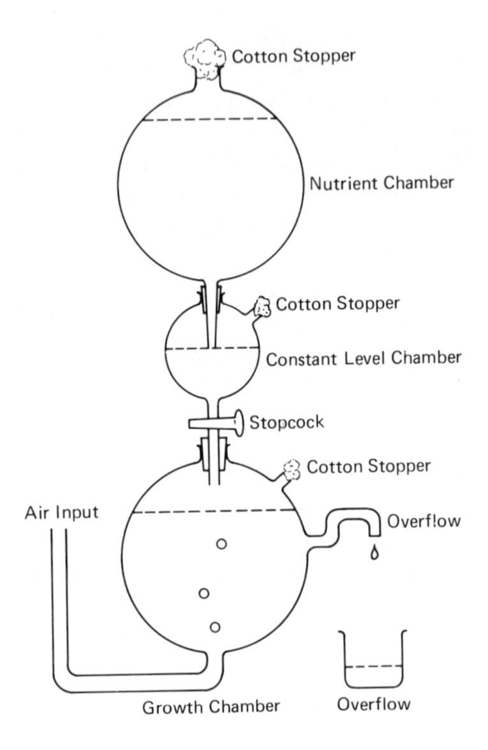

Figure 5.2 Diagram of a Simple Chemostat. The Reservoir (Nutrient Chamber) Should Be Relatively Larger than the Growth Chamber. The Sterile Air Input Keeps the Culture Mixed.

recycled, and a continuous, constant input of energy (disregarding climatic rhythms) leads to populations persisting over many generations. For certain arrested **successional** stages, for example, communities of spring brooks or intertidal rocks, the continuous culture model is roughly appropriate. Even more generally, one can regard species populations in nature as similar to those in a chemostat, providing one allows for frequent and often capricious changes in the rules: the limiting factor, the amount supplied, and the rate of flow. The batch culture may be viewed as the ultimate extension of change in conditions. There is, however, considerable regularity in the various secondary successions for which this model is peculiarly apt. Fallen logs, carcasses, and various other minor and transient habitats do have a predictability that is not implicit in haphazard changes in limiting factors.

One encounters here a problem of scale. Given so large an area that many individual successions in various stages of development are included, one may be able to disregard the many batch cultures involved and consider only the aggregate. This may behave according to the continuous culture model. Conversely, over small areas a climax is not reconstituted generation after generation. Thus, to describe regulation, we must consider population variance in time and patchiness in space; to analyze it, dispersal is equally pertinent as are births and deaths. Unfortunately, because of technical problems, dispersal processes have not been nearly so much studied as have birth and death processes.

Examples of Density Effects on Populations

Given a uniform, favorable setting with an unlimited source of materials and energy, a population of similar individuals will increase exponentially, that is, according to the continuous compound interest equation $N_t = N_0e^{rt}$, where N_t is the population at time t, N_0 the population at time 0, and r the instantaneous population growth rate per individual. This equation has been used much for calculating such fascinating statistics as the number of elephants that could exist 1000 years hence, given an ideal world for elephants, or the savings that you would be able to draw on had one of your ancestors put by a penny for you at the time of Christ. (Even at an annual rate of interest of 1 percent, you would have a tidy sum exceeding three million dollars.) Not surprisingly, such conditions are not encountered. Organisms are exposed to vagaries of various sorts that lead to continuous adjustments of populations. One of the major controversies regarding regulation has been over the extent to which changes in birth, death, and movement rates related to density are actually called into play. It may be instructive to consider a few of the many observations that have been made on effects of changes in density.

Harper (1967) reviews a number of studies of plant populations. After pointing out that **demographic** studies on plants have been neglected because of the plasticity of individuals and their capacity to reproduce asexually, he describes the work of Yoda, *et al.* (1963) on a number of plants as different as trees (*Pinus, Betula*) and herbaceous perennials (*Plantago*). These authors concluded that the chance of a seed producing a mature plant decreased with increasing density; that, beyond some maximum, further increases in numbers of seeds per unit area produced no further increase in the population. Moreover, there was an **asymptotic** density which was correlated with size of individuals, largely regardless of initial conditions.

Among natural populations of animals, the best direct test of the effects of density is Eisenberg's (1966) study of the freshwater snail *Lymnaea.* Eisenberg fenced in some of these snails, but otherwise maintained natural conditions. He then manipulated densities by removing snails from some fenced areas and adding snails to others. The hypothesis Eisenberg wished to test was that the number in high, normal, and low density populations would converge in time. Such convergence might be taken as an indication of regulation by density-related mechanisms. Eisenberg's snails survived equally well regardless of density. However, there was a pronounced effect on the next generation. Clutch size of the parents varied inversely with their density so that major differences in birth rates resulted. Thus, one generation after density manipulation, convergence was achieved, the low density snails making a higher contribution, the high density snails a lower one to the next generation than the snails maintained at their original (control) density. Provision of additional food for some of the experimental snails indicated that the effect of density on birth rate was related to the nature of the food supply.

Unfortunately, although there are many studies of the effects of density on populations in the laboratory, and many observations on birth and death rates in nature, the former can usually be criticized as inapplicable in nature, the latter as being merely examples of correlation without demonstrated causation. The convergence experiment under natural conditions has not been done for species other than *Lymnaea elodes*.

Among the many correlative studies, William Farr's observation more than 100 years ago that human mortality in England increased as the sixth root of density (see Greenwood, 1942) is still noteworthy, even though the meaning of Farr's data is obscured by other variables, such as differences in age structure of city and country populations. More recently, Tanner (1966) examined a variety of data on population growth and density. Of 71 species examined, in 47 there was a significant negative correlation between density and growth rate, and in only 1 was the correlation significantly positive. This latter species was man. Tanner's data are for the world population, presumably over historical time; he does not provide the precise source for the human population estimates. Moreover, several authors (Eberhardt, 1970; Maelzer, 1970; St. Amant, 1970) have pointed out that the statistical devices used by Tanner and by Morris (1965, see below) may attribute negative feedback to data that actually lack it.

Interactions between population density of a prey species and its various predators are of evident interest to the question of population regulation. Among predators or prey, one can distinguish two categories of responses to density: a functional response, that is, one that effects a change in how an individual performs, and a numerical response, that is, an effect on numbers, caused by immigration, births, emigration, or deaths.

In an examination of the effect of bird predators on an insect pest, the larch sawfly, Buckner and Turnock (1965) during one summer studied 43 species of birds thought to be potentially significant as predators. For the majority of these species, stomach contents indicated that sawflies made up a greater proportion of the diet where they were abundant than where they were rare. These birds showed a positive functional response over the range of densities encountered. Moreover, where the density of sawflies was high, the numbers of several of the predator species increased, so that there was likewise a positive numerical response. Nevertheless, when a plot with relatively few sawflies was compared with one where they were 50 times as dense, the percentage of sawflies eaten by birds was higher in the plot with the lower density. The authors suggest that birds may, indeed, be significant in regulating sawflies, but only up to moderate levels of infestation.

Much of the evidence on effects of density on emigration is anecdotal. Mass emigrations of lemmings are proverbial, and similar migrations by Snowy Owls and Siberian Nutcrackers have repeatedly been commented on in relation to density. Quantitative data on emigration rates as correlated with density are not abundant. Waloff (1968) presents observations of Watmough which suggest that the plant lice he studied were probably regulated by emigration (see Table 5.1). Errington (1944) implies an important role for emigration in the regulation of muskrat populations. As marshes dry up and become suitable habitats for decreasing numbers of muskrats, these tend to emigrate in increasing numbers. Since the surrounding areas are generally inhospitable, increased mortality from minks and automobiles causes a decline in numbers made certain by the changed behavior.

Table 5.1 Emigration Rate of a Psyllid (*Arytaina genistae*) as a Function of Adult Density, for the First Generation during the Year (after Waloff, 1968, from Watmough, 1963)

Year	Maximum Adult Density per 100 g plant	Percent of Population Emigrating
1960	21	86
1961	2	65
1962	5	45
1963	0.3	*

*Negligible.

Models of Density Regulation

One of the more useful ways of looking at density regulation graphically is to compare the numbers of successive generations at the same stage in their life history (see Figure 5.3). Then, if the abscissa and ordinate measure on the same scale, a line of slope +1 going through the origin is the locus of all points for which a steady state exists. As Ricker (1954), who suggested this model as a flexible means of achieving various sorts of population growth curves, pointed out, any persistent population must have some range of densities for which points lying above the line of zero change. The precise nature of the function relating the numbers in successive generations will determine the pattern of population growth. If the curve ascends until it reaches the intersection with the steady-state line, successive generations will increase, but normally at a declining rate, until the steady state defined by the particular curve is attained. This is the case for curve *A* in Figure 5.3, which generates curve *A* in Figure 5.4. However, it is possible to draw a curve, such as *B* in Figure 5.3, which reaches a sharp maximum to the left of the steady-state line. Where this happens, fluctuations of considerable magnitude and duration may be generated, as indicated in Figure 5.4, curve *B*, which has been drawn to correspond with curve *B* in Figure 5.3.

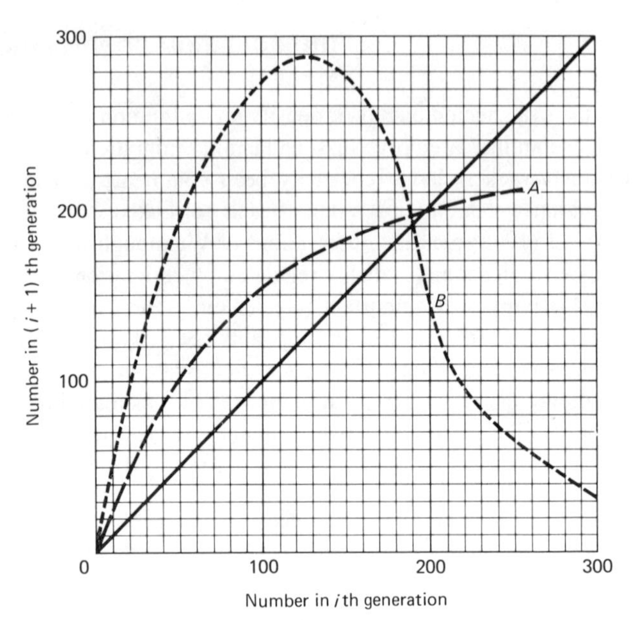

Figure 5.3 Hypothetical Relationships between Densities of Successive Generations in Two Population Types. In A, Numbers of Successive Generations Increase Consistently, but at Decreasing Rates as Density Rises. In B, There Is a Sharp Decrease in Numbers Past an 'Optimum.' The Two Curves Lead to the Population Growth Curves of Figure 5.4. SOURCE: Adapted From Ricker.

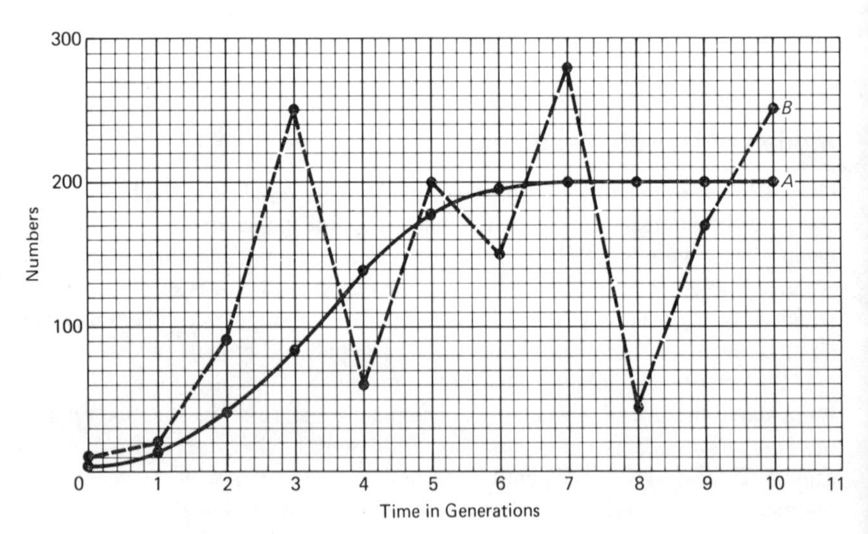

Figure 5.4 Population Growth Curves Generated from the Respective Curves in Figure 5.3. SOURCE: Adapted from Ricker.

How realistic are curves of the sort illustrated? If one were to measure actual numbers of some animal over several successive generations, these would not lie on precisely the same line. Rather, a scatter of points would be observed, from which several alternative models might be derived. One explanation of such scatter is that it is the result of the operation of factors that vary in intensity but that have no relation to density. Another is that several causal mechanisms, each with a different density function, are involved. If these mechanisms operate with differing relative intensities in different generations, there is no reason why a single line drawn in Figure 5.3 should prove adequate to describe and predict population growth. For example, consider successive generations of dragon flies, whose young are aquatic whereas the adults live on land. Oxygen lack one year might cause high mortality among the **nymphs.** Although these nymphs contribute to the oxygen lack through their own respiration, their density is unlikely to be a major factor in the rate of mortality. Another year, lack of suitable prey for the nymphs might become a major decimating factor. This would probably be dependent, functionally and numerically, on dragon fly density in a fairly direct way. Again, predation by birds as the adult dragon flies emerge is conceivably uncommonly important another year, perhaps because an unusually high number of blackbirds has invaded a nearby marsh. This relationship would have yet a different density function from those preceding.

Entomologists, notably Morris (1965) and Varley and Gradwell (1960), have developed methods for sorting out those factors that have increasing effects as density rises. The data required are far more extensive than those needed for Ricker's model. One must have complete fecundity and life tables. These set forth the numbers of eggs produced and give estimates of the magnitude of the various sources of mortality. Furthermore, such data must be sought for populations with different densities. It is then possible, by suitable statistical analysis, to sort out those factors that increase in their effects at greater densities. These are then considered the "key factors" in setting the level at which the population is maintained. Effective analysis depends, of course, on some consistency in mode of regulation.

Returning to Ricker's model, we can suggest possible distinctions between species whose population curves correspond roughly with curve *A* of Figure 5.3, and those with populations more in accord with curve *B*. The latter are characterized by having relatively enormous capacities for increase at intermediate densities, an ability that the former species lack. Such high capacities for increase are typical of the opportunistic species which, earlier, were called species adapted to take advantage of successional communities. The opportunistic species also include virtually all those considered by man to be pests or weeds.

Examples of Regulation of Population Fluctuations

Solomon (1964), in a review of the regulation of insect populations, gives estimates of the range in numbers encountered among various species. For insects with a single generation per year for a period of ten years, fluctuations commonly are over a 3- to 25-fold range. In the red locust *(Nonadacris septemfasciatus)* the range was 750-fold, and *Bupalus* varied in a single ten-year period by a factor of 14,190. MacArthur (1958) assembled some data for several species of warblers. These suggest only 2- to 3-fold variation in numbers over similar periods. Some of the difference in the amplitude of fluctuations may be traced to the birds having longer and overlapping generations. The same individuals may be counted for several years; this will tend to smooth out variations. However, aside from this, there are evidently real distinctions between the precision of regulation among different types of populations. Accordingly, a few examples from rather different situations are needed to illustrate the gamut.

In the intertidal zone, populations of various kinds of invertebrates attain high densities. Thus the barnacle *Balanus balanoides* at times crowds rocks at the upper intertidal levels. Where individuals grow in close proximity, they are stunted and physically may displace others on which they abut. Settlement of larvae can play no role in density regulation. In another genus, Knight-Jones and Stevenson (1950) observed that, on the contrary, settlement occurred preferentially in areas where barnacles had previously existed. In any event, nothing suggests that settling larvae are derived from individuals living in the same area. *Balanus* occupies a rather narrow vertical zone of the intertidal zone. Connell (1961) performed a number of experiments under natural conditions designed to indicate the causes of this zonation. He transplanted rocks with their barnacles to different areas, and traced mortality rates under the different conditions.

The upper limit of the zone of distribution is apparently set by interaction among a number of factors. Fewer barnacles settle there because only the highest tides wash over the area. Desiccation may cause mortality because wave splash may be infrequent. Moreover, conditions for existence are marginal because barnacles are filter feeders and have only minimal opportunities for feeding at the uppermost levels. Thus maintenance becomes increasingly tenuous there, even though competition for space is virtually absent and predators are not abundant. What, then, sets the lower limit to distribution? Predators, especially the intertidal whelk *Thais lapillus* become increasingly significant at successively lower levels. These snails can feed only while under water, and therefore increase their rate of predation with depth. Connell suggests that, generally, the upper limit for intertidal animals is set by physical factors, for example, temperature extremes and desiccation, whereas the lower limit is determined by biotic interactions, especially predation. This is borne out by a study of the limpet *Acmaea digitalis,* which also occurs in the upper intertidal zone (Frank, 1965). Limpets are grazers and move to feed, perhaps to avoid predators such as starfishes. Like barnacles, these limpets have planktonic larvae so that mortality becomes the sole regulating factor in their populations. In summer, the limpets suffered mortality primarily by desiccation, which was most severe at the uppermost levels. At this time, limpets were distributed generally lower than in winter, when storms contributed to heavy mortality, particularly at the lower levels. It was possible to show that limpets moved adaptively, so that in the aggregate they lived higher up in the winter when conditions were most favorable there, but moved lower as summer approached and conditions at the upper levels deteriorated. Moreover, limpets reacted to increased densities by increasing their rates of movement outward from the best areas. Apparently the limpets behaved in such a way that they tended to maximize their protection from desiccation and perhaps from sheer physical damage by molar forces. When density increased, a larger proportion was forced to live in the more hazardous areas of the habitat. As numbers declined, the limpets remaining occupied the optimal areas. Thus behavior, in particular, habitat selection, came to occupy a more significant role in setting the distributional limits than in the sessile barnacles. Ultimately, although one may say that limpets do not occur subtidally because predators are too effective there, this, for *Acmaea digitalis,* is true only in an evolutionary sense. These limpets simply do not remain submerged for long, but actively crawl out of the subtidal area.

The role of habitat selection in determining population numbers is quite possibly often of like significance. A beautiful example is one of the results of a long-term study of titmice by Kluyver (Kluyver and Tinbergen, 1953). The Great Tit *(Parus major)* in Holland breeds in treeholes. It also will readily use nest boxes. The numbers of nesting birds were observed in several woods of various types over a number of years. On one estate, Hulshorst, there is a relatively small area (25.5 ha)[2] of mixed wood of the sort preferred by the titmice. The remainder of the forest is of Scotspine in a pure stand (100 ha). Even when supplied with an excess of nest sites, this woodland supported an average population only about a quarter as dense as that in the mixed wood. Of more interest, however, is the difference between different years (Figure 5.5). The range between the worst (1942) and best (1949) year is from 10–17 breeding males in the mixed forest, but from 0–23 in the pine woods. Kluyver and Tinbergen concluded that social intolerance between tits limited their density in the most attractive habitat, but that in the poorer habitat saturation was not normally reached.

That this sort of territorial behavior exists in invertebrates is perhaps more surprising. Palmer (1968) examined how populations of a scale worm *(Arctonoe vittata)* **commensal** in a number of mollusks are regulated. The preferred host along the Oregon coast is the keyhole limpet *(Diodora aspera).* Characteristical-

[2]1 hectare (ha) = 2.471 acres.

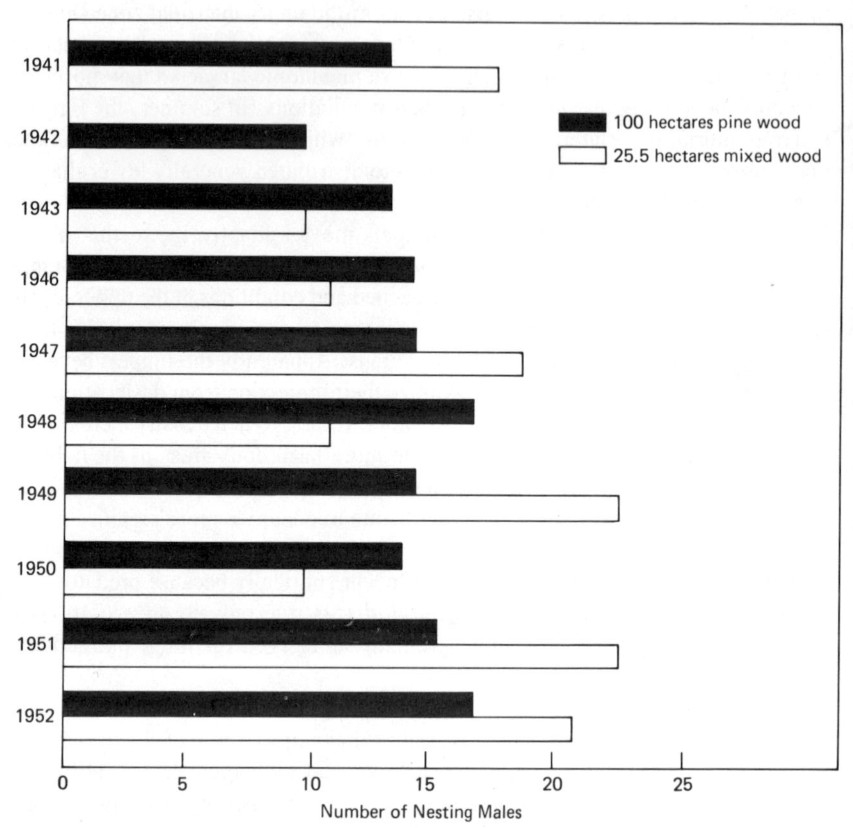

Figure 5.5 Numbers of Males Observed Singing in Mixed and in Pine Wood at Hulshorst during Ten Years. In 1942 No Males Established Nests in the Pine Woods. SOURCE: Data from Tinbergen.

ly, as many as 80 percent of these snails carry a single scale worm. Only very rarely does one encounter a snail that has more than one adult worm in its mantle cavity. Once inside the mantle cavity of its host, the scaleworm evidently remains there for long periods of time, perhaps normally for life. Palmer noted that during the period of settlement of young, from August to October, small scaleworms often occurred on a single host in numbers up to six or seven. However, as the season progressed, these were eliminated, very probably by cannibalism. Scale worms are known to be rather aggressive predators. By the following spring, the pattern of one adult per host was again manifest. In another host, the large gumboot chiton (*Cryptochiton stelleri*), adult scale worms are less commonly seen. Typically only about 20 percent of these chitons bear the commensal. Superficially, this seems like another case of habitat selection, with the chiton a less preferred habitat. However, Palmer's data suggest a somewhat different situation. At the time of settlement, in August–October 1965, only 1 of 64 chitons lacked a commensal worm, and many bore as many as 3 or 4. Evidently, *Cryptochiton* offers a poorer habitat for residence to adulthood and for long-term survival than does the keyhole limpet; however, the newly settling commensal scale worms do not recognize the fact, but seem to settle indiscriminately on either host.

Introduced species offer many natural experiments in population regulation (Elton, 1958). In particular, they yield some of our best evidence of the importance predators and parasites may have in regulating populations. When a new species is introduced, normally it arrives without its original complement of natural enemies. It may, of course, acquire new ones. The conditions for existence are often changed in major ways. Sometimes, as a result, the introduction proves numerically highly successful. In a few cases, as in the introduction of the striped bass to the west coast of the United States, no known changes in the biota have ensued. More often, effects have been deleterious or catastrophic to the existing flora and fauna. Two of the introductions into Australia, that of the prickly pear cactus, *Opuntia*, and of the European rabbit, *Oryctolagus cuniculus*, provide excellent examples of the extent to which predators and disease organisms can effect regulation.

The consequences of the introduction of *Opuntia*, and its eventual control have been well reviewed by Andrewartha and Birch (1954). Following the introduction of the cactus, it became the dominant vegetation on 30 million acres of agricultural land in Queensland. Here prickly pear weights of 500 tons per acre occurred. Studies of natural enemies of the cactus in America eventually focused attention on a moth, *Cactoblastis cactorum*; these moths have a life cycle which depends on the caterpillars' use of cactus as food. Normally, the cactus plants are entirely destroyed. Introduction of the moth in 1925, and continued liberation in the following years of three billion moths resulted in a rapid decline in the numbers of cacti. As early as 1934, no large areas where *Opuntia* was still dominant existed. Because the moths have a relatively high rate of dispersal, major colonies of cactus are quickly found and eliminated. In 1966, occasional cactus plants could be found in Queensland. Evidently, a rough balance has been struck between the powers of dispersal of the cactus and its predator, the moth. Individual plants may still become established. Eventually they are discovered by *Cactoblastis* and killed, but some manage to reproduce successfully before this happens. *Cactoblastis* thus is now less abundant than in its heyday in the thirties. Its effectiveness in controlling prickly pear depended not only on its abilities as a predator on the cactus, but on the fact that its own natural enemies were not introduced along with it, and that its new Australian predators killed less than 25 percent of the caterpillars. Otherwise the levels at which cactus and *Cactoblastis* became regulated might have been rather different.

As a pest species, the rabbit, which is particularly common in the southern half of Australia, may be without peer. Its spread, after an original introduction of wild rabbits in 1859 was remarkably rapid throught the temperate zone. In one case a spread over a distance of 1100 miles in 16 years is recorded (Fenner and Ratcliffe, 1965). Although no good density estimates seem to be available because of the difficulty of sampling a population that spends much of its life in burrows, estimates of the losses sustained by sheep farmers because of competition by rabbits with sheep are above $70 million per year. These estimates derive from the increased yields in wool and lambs following the reduction in the numbers of rabbits by the virus disease myxomatosis. Although no conclusive data seem available, rabbits in the years before the introduction of the disease may often have been limited by a reduction in breeding success concomitant with high density. The reasons for such reduction, whether food shortage or behavioral intolerance are not known. Predators, mainly foxes and cats, seem not to have affected population in a measurable way.

Myxomatosis is a virus disease which, in its native area, South America, causes an endemic but nonfatal affliction of rabbits *(Sylvilagus brasiliensis)*. Late in the nineteenth century a lethal outbreak of disease among a laboratory colony of European rabbits in Uruguay marked the first known appearance of the virulent, fatal, **epizootic** disease that has caused major changes in the Australian and European rabbit populations.

Introductions of myxoma virus into test populations of rabbits in Australia at first were relatively unsuccessful. At best, although local kills were achieved, there were no permanent effects because the disease did not spread. Not until a later series of test introductions was a widespread epizootic initiated. Fenner and Ratcliffe (1966) explain that in Australia the situation for the transmission of myxoma virus differs rather markedly from that elsewhere because the European rabbit was introduced free of fleas. The virus is transmitted by bites of blood-sucking insects. In the early introductions, in dry areas, only where a native flea was abundant was local control achieved. Elsewhere the disease died out without even being transmitted to the local population. The introduction of 1950, which led to an eventual massive epizootic, was along a major river, the Murray, where mosquitoes provided the main vector for spreading the disease. Thereafter, the disease spread so quickly that within three years its area of distribution was virtually as extensive as that of the rabbit, although individual populations in many areas were not yet affected. Man-caused inoculations contributed to the spread of the virus, but were almost certainly not necessary to its eventual success.

Rabbit numbers declined to a low level by 1955 and have since increased. A number of elements entered into this change. Initially, mortality among rabbits exposed to the virus was greater than 90 percent. Immunity, which can be transmitted to offspring, and a decrease in virulence of the virus acted in concert to dampen the initial effect. Conditions were clearly ideal for selecting resistant rabbits and less virulent strains of virus. This was one of the major reasons why

Australian officials attempted to speed the initial spread. The endemic disease is unlikely, however, to permit return of rabbits to their former abundance, although they have again reached destructive numbers locally. One reason is that predators, which are ineffective in controlling rabbits once they have reached high densities, have a significant influence when they are rarer. Moreover, nonfatal cases of myxomatosis nevertheless leave rabbits more vulnerable. Although it is too early to assess the level at which rabbit numbers are going to be maintained, it seems evident that it will be far lower than before the introduction of the disease.

Our knowledge of the ways in which even common animals are regulated is not at all complete. This is probably because the mechanisms are complicated. Lemmings in Alaska, for example, have been studied for a number of years by several groups of investigators. There seems little doubt that they undergo relatively regular fluctuations in numbers, with peaks from three to four years apart, on the average. The fluctuations are not caused by regular climatic fluctuations, nor are fluctuations in the numbers of predators a primary cause. There is general agreement about certain differences in birth and death rates among these animals at different stages in the cycle. Nevertheless controversy still continues over the precise mechanisms that operate in such a way that feedback delays cause repeated fluctuations. Krebs (1966) has summarized the major hypotheses that have been advanced to explain the cycles in recent years (Figure 5.6). Krebs observed an effect on food as lemming densities increased, but there was no indication of quantitative or qualitative food shortage, and, at peak densities, individual weights were high. A hypothesis of exhaustion of adrenal function (the general adaptation syndrome, G.A.S., of Selye) could not be substantiated. Krebs concluded that changes in behavior during different portions of the cycle deserved more emphasis than they have thus far received.

Watts (1969) summarizes extensive work on the woodmouse (*Apodemus sylvaticus*) with a strikingly similar set of conclusions. One of the best demonstrated indications of negative feedback in these mice is a relationship between density in June and the time in the fall when numbers begin to increase. The higher the numbers in June, the later the beginning of the period of increase. Winter survival is also correlated with the acorn crop, but, curiously, even when crops are poor the survivors show no indications of being poorly fed; their weights remain normal. Thus Watts is eminently able to show that density-dependent factors act in such a way that fluctuations are of minor extent, but precise causal mechanisms remain obscure. When one considers what a physiologically complex set of **homeostatic** mechanisms a mammal possesses, it is really not so remarkable to find that behavior is enlisted in the adaptive machinery. The more restricted question of whether different individuals, perhaps with systematic genetic differences, exhibit one sort of behavior at high densities and another at low densities, as suggested by Chitty (Figure 5.6), is as yet unresolved.

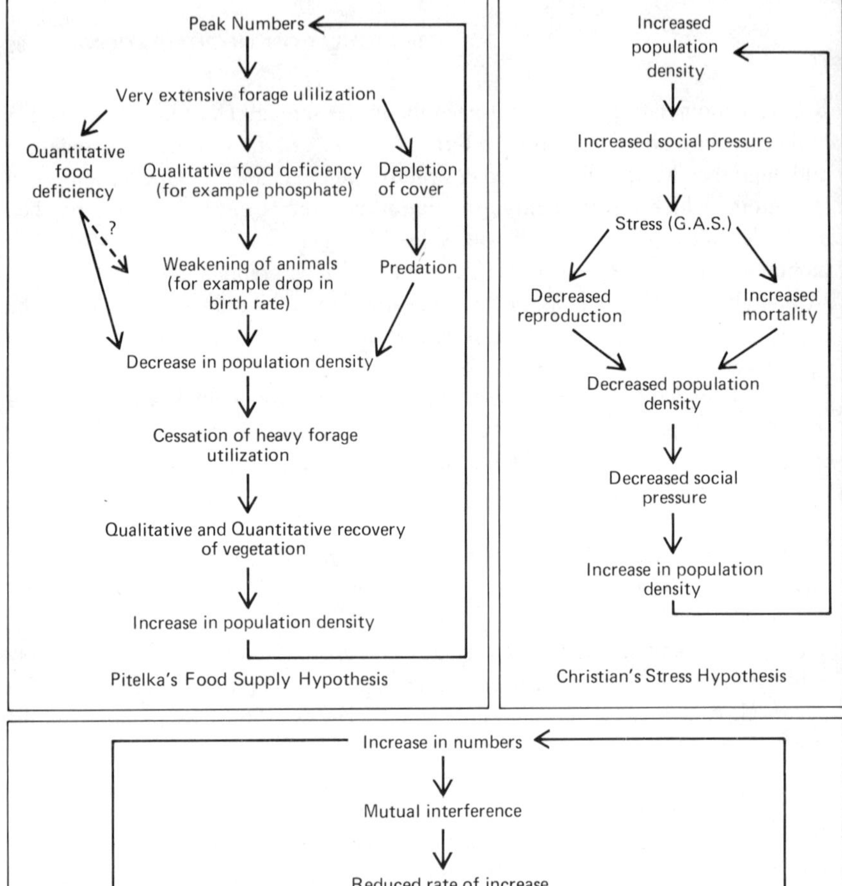

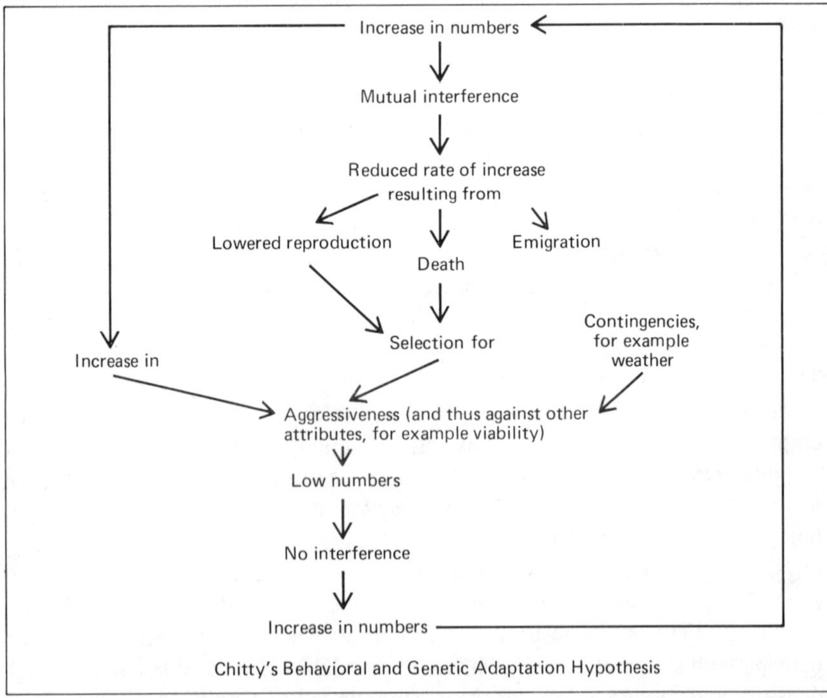

Figure 5.6 Three Models of Regulation of Populations of Small Rodents. SOURCE: Adapted from Krebs, 1966.

As indicated above, certain insects and, in particular, species that infest coniferous forests, such as the pine looper *Bupalus* or the Noctuid moth *Panolis*, undergo local irruptions during which densities become three or four orders of magnitude higher than they are during most years. Massive defoliation, leading at times to destruction of vast areas of forest, then follows. Schwerdtfeger (1968) who has long been a student of such populations, points out that such populations can be modeled, in principle, with two simple assumptions: (1) normally these populations are subject to random, density-independent mortality; and (2) during an outbreak, food shortage, acting as a density-dependent source of increased mortality and decreased natality, triggers the precipitous decline of the population to such a low level that a new set of random fluctuations can follow. Such a model is, unfortunately, not itself predictive. Moreover, it is likely to prove overly simple. Holling (1959) showed that predation by small mammals on pine sawfly cocoons yielded a combined effect of functional and numerical responses whereby at low densities of sawflies, these predators' responses were potentially able to control density. As densities exceeded some moderate value, however, the percentage of predation decreased as prey density increased. The effect was somewhat similar to that Buckner and Turnock (see above) noted in regard to bird predators on the larch sawfly. This observation is perhaps of general significance. Moreover, if several predators and **parasitoids** have increasingly effective responses over the lower range of densities, but fail to act as controlling factors once some high point of density is exceeded, control will normally be assured for considerable periods by the linking together of the several mechanisms. For any system of economic control of such pest species, it is of great significance that one be able to distinguish between these two explanations of the observed changes in density.

From the examples presented, the diversity of responses of different organisms to changes in their numbers may stand out. It is eminently true that even within the same species, different conditions will elicit different mechanisms of control. Thus whatever generalizations can be made will not concern specific mechanisms. This need not preclude several useful generalizations, however.

General Conclusions

The numbers and productivities of various species are evidently under sets of controls that differ considerably in precision. In general, for example, most vertebrates fluctuate less than many insects, and annual plants probably fluctuate considerably more than do trees. Several questions regarding the operation of regulating factors are thus of interest: Are species limited by different factors near the edges of their distribution from at the center? Are opportunistic species regulated in a different manner from others? Are there consistent differences in mode of regulation among different groups of species, for example, among phyla or **trophic levels?**

Watt (1968) argues that climatic factors are most important at the edges of a species' distribution, or where climate fluctuates over a wide range. It is a truism that optimal conditions for it are likely to exist near the center of a species' range, and that at the edges of its distribution stresses increase until the limits of tolerance are exceeded. Moreover, increasing stress by one factor tends to decrease the limits of tolerance to another. Therefore, resources are more likely to become short at the center, and climate and enemies commonly determine where the limits of distribution will lie.

One can consider such peripheral populations, living in fluctuating climates, as occupying the space where they exist only temporarily. They are thus formally similar, in this respect, to opportunists. There is, however, a fundamental difference between opportunists and populations of species that normally rely on a more stable resource base but happen to live near the edge of the species distribution. The former will possess adaptations that provide for emigration or persistence in some altered state when conditions become unfavorable. Organisms of temporary ponds, for example, may have eggs that can withstand drying, or can be blown away when the pond dries up. By contrast, peripheral populations are likely to be wiped out by an unusually large deviation in weather. Here we encounter evolutionary questions: Are peripheral populations genetically distinguishable from central ones, and are such differences adaptive and consistent? Although there is some evidence suggesting the existence of such genetic differences (Carson, 1965), there is as yet no general answer to this question.

The correlation between opportunism and the amplitude of local population fluctuations is undoubtedly real. Arguments that extend this correlation to imply that regulation is necessarily more precise in climax than in successional communities are, however, partly circular. Given suitable conditions all species are opportunistic. Thus we risk failure to distingush between constant and changing conditions, attributing to the species properties of the environments. There is, however, an element of truth in that opportunistic species much of the time encounter situations where no limiting factors exist, so that, for short periods, populations may grow maximally without the imposition of factors with a density function at these levels.

Hairston, Smith, and Slobodkin (1960, 1967) have argued that, on land, herbivores are generally limited in numbers by their predators; the other trophic levels are considered resource-limited. Their concept has been criticized and they have responded to the criticism. To a major extent, the generalization seems a statement about the general nature of the adaptations of land plants. Curiously, their defense mechanisms seem relatively effective in warding off major inroads by their predators, the example of *Opuntia* and *Cactoblastis* notwithstanding. The argument eventually becomes the question of why, when one kind of plant does become limited by its herbivores, another species is available to exploit the resources freed by the first. Since regulation, in the context of this essay, concerns the level of species populations, arguments for entire trophic levels are not really germane.

Differences in mode of regulation among different species certainly do exist. Plants, lacking active movement and nervous systems, have adapted by plastic responses, including the ability to withstand mutilation and to regenerate. Vertebrates with well-developed brains and highly evolved homeostatic mechanisms, in general, involve these in their regulatory responses. Moreover, and importantly, the entire life history of the species is highly adapted to its particular mode of life. An opportunistic species, say, a tapeworm, whose members have a low probability of finding an environment suitable for growth and reproduction, produces many offspring, preferably spread over a long period of time, or makes provisions for persisting over long unfavorable periods. Regulation, in the form of a change in response of the resource or the population to a change in its density, somehow eventually becomes involved in determining abundance and distribution. Where there are time lags in the reaction to density, fluctuations may become rhythmic. Where resources become available suddenly and in varying amounts, populations will fluctuate seemingly without regularity. However, in principle, regulation is never relaxed for long periods and depends on the sort of density-governing mechanisms that Nicholson (1954) has long considered responsible.

Returning to the quotation by Lotka near the beginning of this essay, it is perhaps necessary to ask whether ultimately, when one speaks of regulation, the whole species network need not be considered. The general belief among ecologists that complex, diverse communities are more stable than simpler ones, such as those man institutes by his agricultural processes, is based on several lines of evidence. It seems eminently logical that complex interspersions and food webs should provide for a multiplicity of regulatory mechanisms that can collectively act as more effective limits than could any smaller combination. It is thus possible that regulation in nature is an even more involved process than has been indicated. However, the complexities of **interspecies** interactions have so far defied satisfactory analysis. It thus remains for the future to determine whether the presently shaky theoretical and observational base for the relationship between complexity and stability can be strengthened.

Glossary

asymptotic (*n.* asymptote): approaching a limiting position.

batch: a process similar to that by which a housewife makes dough for bread, by discrete units (see also "continuous culture").

climax (ecological): the last stage in ecological succession (see also). Climax is usually thought of as being self-perpetuating.

clutch size: the number of seeds, eggs or young produced by an organism in a single reproduction period, for example, the number of eggs brooded by a bird in a nest.

commensal (*n.* & *adj.*): literally, eating at the same table; therefore, living with another animal and partaking of its food or other requirements.

continuous culture: a set of processes analogous to those of a factory assembly line, where raw materials are continually fed in at one end and the finished product appears continually at the other.

demographic (*n.* demography): referring to the science of population.

endemic: indigenous to an area.

epizootic: affecting many animals; equivalent to the term epidemic among humans.

feedback: a means of changing a process by having the output affect the input. Negative feedback means that as a process increases its intensity, a signal is fed to the input causing the increase to change downward.

homeostatic: capable of staying the same because of self-regulation, characteristically by negative feedback (see above).

interspecies: between different species.

metabolites: chemical substances produced as a result of life processes.

nymphs: the juvenile stages of several types of insects.

parasitoid: an animal that gains sustenance by living within that of another species at some stage in the life history, but which, as opposed to a parasite, causes the death of the host as a condition of continued survival of the feeding animal. Parasitoids are common among certain insects, which may be used for biological control of other, pest, species.

plastic (*adj.*) (*n.* plasticity): able to be molded, specifically by environmental differences; for example, leaf size in the same plant varies with light intensity.

selection: in the sense of natural selection, the differential survival of certain heritable combinations in the next generation, owing to either differences in the number produced or in the number living to reproduce themselves, in turn.

steady state: a situation in which conditions do not, apparently, change; it is caused by a balance between opposing processes, which seemingly cancel out, although they involve energy expenditure.

successional (*n.* succession): somewhere along the cumulative set of changes that result during the initial exploitation of a habitat by a suite of available species.

trophic level: a grouping of organisms on the basis of the number of steps by which they are removed from sunlight as their energy source. Thus green plants are at one trophic level, herbivores at the next, and so on.

References

ANDREWARTHA, H.G., AND L.C. BIRCH, *The Distribution and Abundance of Animals.* Chicago: University of Chicago Press, 1954, p. 501.

BUCKNER, C.H., AND W.J. TURNOCK, "Avian Predation on the Larch Sawfly, *Pristiphora erichsonii* (Htg.), (Hymenoptera: Tenthredinidae)." *Ecology*, 1965, **46**, pp. 223–236.

CARSON, H.L., "Chromosomal Morphism in Geographically Widespread Species of *Drosophila*," in H.G. Baker and G.L. Stebbins (eds.), *The Genetics of Colonizing Species.* New York: Academic Press, 1965, pp. 503–531.

CONNELL, J.H., "The Influence of Competition, Predation by *Thais lapillus*, and Other Factors on Natural Populations of the Barnacle, *Balanus balanoides*." *Ecol. Monogr.*, 1961, **31**, pp. 61–104.

EBERHARDT, L.L., "Correlation, Regression, and Density Dependence." *Ecology*, 1970, **51**, pp. 306–310.

EHRLICH, P.R., AND P.H. RAVEN, "Differentiation of Populations." *Science*, 1969, **165**, pp. 1228–1232.

EISENBERG, R.M., "The Regulation of Density in a Natural Population of the Pond Snail, *Lymnaea elodes.*" *Ecology*, 1966, **47**, pp. 889–906.

ELTON, C., *The Ecology of Invasions by Animals and Plants.* London; Methuen, 1958, p. 181.

ERRINGTON, P.L., *Ecology of the Muskrat.* Ames: Rep. Iowa Agr. Exp. Sta., 1944.

FENNER, F., AND F.N. RATCLIFFE, *Myxomatosis.* Cambridge: Harvard University Press, 1965, p. 379.

FRANK, P.W., "The Biodemography of an Intertidal Snail Population." *Ecology*, 1965, **46**, pp. 831–844.

GREENWOOD, M., "Medical Statistics from Graunt to Farr." *Biometrika*, 1942, **32**, pp. 101–127; **33**, pp. 1–24.

HAIRSTON, N.G., F.E. SMITH, AND L.B. SLOBODKIN, "Community Structure, Population Control, and Competition." *Amer. Natural.*, 1960, **94**, pp. 421–425.

HARPER, J.L., "A Darwinian Approach to Plant Ecology." *J. Ecol.*, 1967, **55**, pp. 247–270.

HOLLING, C.S., "The Components of Predation as Revealed by a Study of Small-Mammal Predation of the European Pine Sawfly." *Canad. Entomol*, 1959, **91,** pp. 293–320.

KLUYVER, H.N., AND L. TINBERGEN, "Territory and the Regulation of Density in Titmice." *Arch. neerl. zool.*, 1953, **10**, pp 265–289.

KNIGHT-JONES, E.W., AND J.P. STEVENSON, "Gregariousness during Settlement in the Barnacle *Elminius modestus* Darwin." *J. Mar. Biol. Assoc. U.K.*, 1950, **29**, pp. 281–297.

KREBS, C., "The Lemming Cycle at Baker Lake, Northwest Territories, during 1959–62." Arct. Instit. North Amer. Tech. Paper 15, 1966, p. 103.

LOTKA, A.J., *Elements of Physical Biology.* Baltimore: Williams & Wilkins, 1925.

MAELZER, D.A., "The Regression of log N_{n+1} *on log* N_n as a Test of Density Dependence: An Exercise with Computer-Constructed Density-Independent Populations." *Ecology*, 1970, **51,** pp. 810–822.

MacArthur, R.H., "Population Ecology of Some Warblers of Northeastern Coniferous Forests." *Ecology*, 1958, **39,** pp. 599–619.

Morris, R.F., "Contemporaneous Mortality Factors in Population Dynamics." *Can. Entomol.*, 1965, **97**, pp. 1173–1184.

Nicholson, A.J., "An Outline of the Dynamics of Animal Populations." *Austral. J. Zool.*, 1954, **2**, pp. 9–65.

Palmer, J.B., "An Analysis of the Distribution of a Commensal Polynoid on Its Hosts." Ph.D. Dissertation, Univ. of Oregon, 1968, p. 110.

Ricker, W.E., "Stock and Recruitment." *J. Fish. Res. Bd. Can.*, 1954, **11**, pp. 559–623.

Schwerdtfeger, F., *Ökologie der Tiere. II Demokologie.* Hamburg: Paul Parey, 1968, p. 448.

Slobodkin, L.B., F.E. Smith, and N.G. Hairston, "Regulation in Terrestrial Ecosystems, and the Implied Balance of Nature." *Amer. Natural.*, 1967, **101**, pp. 109–124.

Solomon, M.E., "Analysis of Processes Involved in the Natural Control of Insects." J.B. Cragg, (ed.), *Adv. Ecol. Res.*, 1964, **2**, pp. 1–58.

St. Amant, J.L.S., "The Detection of Regulation in Animal Populations." *Ecology*, 1970, **51**, pp. 823–828.

Tanner, J.T., "Effect of Population Density on Growth Rates of Animal Populations." *Ecology*, 1966, **47**, pp. 733–745.

Varley, G.C., and G.R. Gradwell, "Key Factors in Population Studies." *J. Anim. Ecol.*, 1960, **29**, pp. 399–401.

Waloff, N., "Studies on the Insect Fauna on Scotch Broom." J.B. Cragg (ed.), *Adv. Ecol. Res.*, 1968, **5**, p. 193.

Watt, K.E.F., *Ecology and Resource Management.* New York: McGraw-Hill, 1968, p. 450.

Watts, C.H.S., "The Regulation of Woodmouse (*Apodemus sylvaticus)* Numbers in Wytham Woods, Berkshire." *J. Anim. Ecol.*, 1969, **38**, pp. 285–304.

Yoda, K., T. Kira, H. Ogawa, and K. Hozumi, "Self-thinning in Overcrowded Pure Stands under Cultivated and Natural Conditions." *J. Biol.*, Osaka Cy Univ., 1963, **14,** pp. 107–129.

6

Can We Control Human Population?

GARRETT HARDIN

University of California, Santa Barbara

Garrett Hardin was born in Dallas, Texas, in 1915, and grew up in many cities in the Midwest. His undergraduate years were spent at the University of Chicago, where he studied ecology under W. C. Allee. His Ph.D. was earned at Stanford University in 1941 for investigations of microbial ecology carried out under Willis H. Johnson. For four years after receiving his doctorate he was a staff member of the Division of Plant Biology, Carnegie Institution of Washington, perfecting culture methods for algae and studying the feasibility of using them as food. In 1946 he left this institution for the Santa Barbara campus of the University of California, where he has been ever since, except for brief periods as Visiting Professor at the Berkeley and Los Angeles campuses of the same institution, and at Stanford and the University of Chicago.

After some ten years in the laboratory he shifted his work wholly to writing and teaching. His elementary textbook, *Biology: Its Principles and Implications,* has passed through four editions since its first publication in 1949. In 1959 he published a popularization of evolution, *Nature and Man's Fate,* which was later translated into German and Portuguese. His "collage of controversial readings," *Population, Evolution and Birth Control,* was first published in 1964. In 1970 he pioneered a small text, *Birth Control,* designed for high school students. His best known paper is "The Tragedy of the Commons," published in 1968. The argument of this essay was expanded into a full-length book, *Exploring New Ethics for Survival: The Voyage of the Spaceship Beagle,* published in 1972.

He has been a leader in the abortion reform movement since 1963, and served on the board of directors of Zero Population Growth from its inception. His deep interest in ethical problems has led to service as a member of the Board of Directors of the Starr King School for the Ministry (Unitarian-Universalist), as well as to his being honored as Nieuwland Lecturer at the University of Notre Dame in 1970.

Many thoughtful men agree that the two most important problems of our day are thermonuclear war and population. Thermonuclear war is placed first because if we don't solve that problem there will be no others to worry about. But if we succeed in avoiding destruction by war we still face the almost equally dangerous (though not so catastrophic) effects of overpopulation. This second problem is immensely complex. We will, no doubt, be a long time in solving it, for it is interrelated with so many other difficult issues. We must work at this complex rapidly, but we must not be childishly impatient. "Is it too late?" many people ask. It is *always* later than we would like—but *too* late? *No:* better part of a disaster than the whole of it. Let's "keep our cool." Let us deliberately study the interrelated problems in all their aspects while we work our way through to reasonably acceptable solutions.

How Malthus Started a Controversy

The "population problem" was spawned in controversy, from which it has never fully escaped. Though some of the principles governing population growth had been more or less clearly seen long before Thomas Robert Malthus (1766–1834), there was no sustained or general discussion of the problem until the publication of his *Essay on Population* in 1798. Since that time there have been only occasional lulls in the rhetorical storm. Several automatic psychological defenses were called into action by Malthus' essay. The full title even suggests an invitation to controversy: *An Essay on the Principle of Population, as It Affects the Future Improvement of Society. With Remarks on the Speculations of Mr. Godwin, M. Condorcet, and Other Writers.*

The essay is no bare theoretical discussion of the mathematical laws of population growth; it quite explicitly spells out implications in the field of social and political affairs as well. The chain of reasoning is long and contains many hidden assumptions which Malthus' contemporaries were not always willing to accept. We need not spend a great deal of time on the political and social aspects of Malthus' arguments, but a brief résumé of them will reveal some of the emotions invested in population questions from 1798 to the present day.

The preface to Malthus' essay begins with this statement: "The following Essay owes its origins to a conversation with a friend. . . ." The friend was his father, who was a great admirer of Condorcet and Godwin. The Marquis de Condorcet (1743–1794) was one of the many intellectual parents of the French Revolution, which, like all revolutions, was given to parricide. The French Revolution was scarcely under way before Condorcet found himself on the wrong side of politics; he went into hiding, but was ultimately caught and killed (whether by his own hand or others' is not known, and hardly matters). While in hiding, however, he wrote the remarkable document that is one of the principal sources of the "idea of progress" which so dominated the thought of the western world during the subsequent century and a half. The title of this seminal work was *Sketch of a Historical Picture of the Progress of the Human Mind.* In introducing his work, Condorcet says that its object "will be to show, through reasoning and through facts, that nature has assigned no limit to the perfecting of the human faculties, that the perfectibility of man is truly indefinite; that the progress of this perfectibility, henceforth independent of any power that might wish to arrest it, has no other limit than the duration of the globe on which nature has placed us."

Condorcet presents his argument at some length. For the historian of ideas there is a wealth of fascinating pathways in the *Sketch*, but we will resolutely pass them all by to get to the concluding paragraph:

> And how admirably calculated is this picture of the human race, freed from all these chains, secure from the domination of chance, as from that of the enemies of its progress, and advancing with firm and sure steps towards the attainment of truth, virtue, and happiness, to present to the philosopher a spectacle which shall console him for the errors, the crimes, the injustice, with which the earth is still polluted, and whose victim he often is! It is in the contemplation of this picture that he receives the reward of his efforts towards the progress of reason and the defense of liberty. He dares then to link these with the eternal chain of human destiny; and thereby he finds virtue's true recompense, the joy of having performed a lasting service, which no fatality can ever destroy by restoring the evils of prejudice and slavery. This contemplation is for him a place of refuge, whither the memory of his persecutors cannot follow him, where, living in imagination with man restored to his rights and his natural dignity, he forgets him whom greed, fear, or envy torment and corrupt; there it is that he exists in truth with his kin, in an Elysium which his reason has been able to create for him, and which his love for humanity enhances with the purest enjoyments.

Whatever one may think about the particular ideas of the author (most of which we have not mentioned), one cannot but be impressed by his courage in penning such a paragraph while living in the shadow of the guillotine.

At about the same time, on the other side of the Channel, a work of similar emotional tone was written by William Godwin (1756–1836), amateur philosopher, biographer, historian, novelist, journalist; friend of Leigh Hunt and his circle; husband of the early suffragette Mary Wollstonecraft; and father of Mary Shelley (who wrote *Frankenstein*). The subject of this too-involved sentence (and equally involved life) published in 1793 *An Inquiry Concerning Political Justice, and Its Influence on General Virtue and Happiness.* It is a large work with a simple Rousseauesque conclusion: man is innocent by nature—it is his institutions that corrupt him. Destroy his institutions and evil will disappear.

Malthus, Senior, like many other intellectuals of the day, found the views of Condorcet and Godwin intoxicating. His thirty-two-year-old son took a soberer view of the situation: to his mind it appeared that the steady pressure of population upon resources would keep people in a more or less steady state of misery which would preclude the rosy Elysium envisaged by Condorcet and

Godwin. He developed his argument at some length. The father (evidently a remarkable man) was so impressed by his son's reasoning that he urged him to publish it, which he did. The *Essay* was an immediate sensation, calling forth extravagant praise at first (by Godwin, among others), but this was soon followed by equally extravagant abuse by the same Godwin, as well as by Byron, Shelley, Hazlitt, and the literary crowd generally, who made Malthus their favorite whipping-boy. He was the "best-abused man of his age," as his biographer, James Bonar, subsequently remarked. Malthus, remarkably resistant to abuse and to changing his ideas, devoted much of the remainder of his life to revising his essay on population, which swelled from a modest 50,000 words in the first edition to the nearly 200,000 words of the second and subsequent editions. In substance, though not in title, the work changed from an essay to a treatise.

Call it what you will, Malthus' work is a complex of many ideas, theses, arguments, gratuitous assumptions, and ambushing enthymemes. A few technical mathematical ideas are used to derive a host of practical political consequences, the validity of which is far from established. To the second edition of his essay Malthus added a passage which grated on the nerves of the literary establishment like the scraping of a fingernail along a blackboard. This is the famous "passage of the feast," as it has come to be known:

> A man who is born into a world already possessed, if he cannot get subsistence from his parents on whom he has a just demand, and if the society do not want his labour, has no claim of *right* to the smallest portion of food, and, in fact, has no business to be where he is. At nature's mighty feast there is no vacant cover for him. She tells him to be gone, and will quickly execute her own orders, if he does not work upon the compassion of some of her guests. If these guests get up and make room for him, other intruders immediately appear demanding the same favour. The report of a provision for all that come, fills the hall with numerous claimants. The order and harmony of the feast is disturbed, the plenty that before reigned is changed into scarcity; and the happiness of the guests is destroyed by the spectacle of misery and dependence in every part of the hall, and by the clamorous importunity of those, who are justly enraged at not finding the provision which they had been taught to expect. The guests learn too late their error, in counter-acting those strict orders to all intruders, issued by the great mistress of the feast, who, wishing that all guests should have plenty, and knowing she could not provide for unlimited numbers, humanely refused to admit fresh comers when her table was already full.

Such sentiments drove essayists and poets into a positive fury. "Why," asked William Hazlitt, "does Mr. Malthus practise his demonstrations on the poor only? . . . I do not see why they alone should be forced . . . to do the will of God. Mr. Malthus' gospel is preached only to the poor." Shelley spoke contemptuously of "sophisms like those of Mr. Malthus, calculated to lull the oppressors of mankind into a security of everlasting triumph." Later, Karl Marx (1818–1883) referred to him as "the contemptible Malthus" and "this libel on the human race." In *Capital* Marx wrote: "The people were right . . . in sensing instinctively that they were confronted not with a man of *science* but with a *bought advocate,* a pleader on behalf of their enemies, a shameless sycophant of the ruling classes."

Such widespread and passionate rejection of Malthus springs, probably, from two sources. On the one hand, Malthus' ideas seem to negate the idea of "progress," which became an article of faith to the people of the nineteenth century who clung to it with religious fervor. Whether the negation is really true or necessary is an open question. But Malthus' critics—as well as Malthus himself—believed in the negation, and that was all that mattered. No man can attack a living religion with impunity—which is what Malthus had attempted. (There is irony in this, for the author of the *Essay* was an ordained minister—but his religion was not the new religion of Progress, but, rather, the old religion of the preacher in *Ecclesiastes.* It was the religion of eternal stability, of the acceptance of irreducible evils. Such a religion was completely unacceptable to the prophets of Progress.)

But the power of the idea of progress, great as it was, resided almost entirely at the subconscious level. None of Malthus' critics explicitly charged him with sacrilege against the new religion. The overt indictment was against his antihumanitarianism. During the latter part of the eighteenth and all of the nineteenth century a great complex of interrelated social reforms was set in motion: woman suffrage, child-labor laws, antislavery movements, unionization of labor, workmen's compensation acts, penal reform, and various primitive social security measures under the general heading of "Poor Laws." To put the matter simply: To Cain's question, "Am I my brother's keeper?" socially sensitive men of the nineteenth century responded with a resounding "Aye!"

What did Malthus say? If a man is to be judged by his words, the passage of the feast surely says "No, I am not my brother's keeper." In fact, in action, Malthus was quite a different person. In personal relations he was noted for his kindness and gentleness. In civic matters he advocated free universal education, free medical care for the poor, state assistance to emigrants, and welfare payments to casual laborers who had six children or more. Though Malthus praised nature who "humanely refused to admit fresh comers when her table was already full," he himself was too tender-hearted to follow in nature's footsteps. Hazlitt, Byron, Shelley, Marx, and countless others did not know this. They took his words at face value; from these they built an image of a heartless, pessimistic ogre, an ogre who is still passionately attacked, particularly in Russia and China. To use the terminology of the ethologists, the name "Malthus" is still, in some quarters, a *releaser* of fantastic potency.

Exponential Growth

Malthus was graduated from Cambridge University as "Ninth Wrangler." This quaintly named honor designated a student taking the course of study called the "mathematical tripos." We may assume, then, that young Malthus had some mathematical ability—but not very much, since he placed only ninth in a class that was surely not large. This bit of biography helps explain some important characteristics of Malthus' later work. He naturally saw problems in a mathematical light, but the focus was not always as sharp as one might wish.

The principal argument of the first chapter of the *Essay* recalls the method of Euclid's *Geometry:*

> I think I may fairly make two postulata.
> First, That food is necessary to the existence of man.
> Secondly, That the passion between the sexes is necessary and will remain nearly in its present state.

These seem to be about as indubitable propositions as one could want. So far so good. Malthus goes on:

> Population, when unchecked, increases in a geometrical ratio. Subsistence increases only in an arithmetical ratio. A slight acquaintance with numbers will show the immensity of the first power in comparison of the second. . . . Taking the population of the world at any number, a thousand millions, for instance, the human species would increase in the ratio of—1, 2, 4, 8, 16, 32, 64, 128, 256, 512, [etc.] and subsistence as—1, 2, 3, 4, 5, 6, 7, 8, 9, 10, [etc.]. In two centuries and a quarter, the population would be to the means of subsistence as 512 to 10: in three centuries as 4096 to 13, and in two thousand years the difference would be almost incalculable. . . .

With this, the trouble begins. The mathematics is all right, but one of the assumptions is shaky—that subsistence should increase arithmetically. This was rightly questioned by Malthus' critics. The reality behind the assumption we will try to discover later. The other assumption, however, the assumption that an unchecked population grows "geometrically," rests on the soundest of biology.

Consider: What meaning shall we assign to the phrase, "Population, when unchecked . . ."? Surely this: That any species of plant or animal has a maximum rate of population growth which will be realized whenever environmental conditions are optimal (by definition), and which will be constant (otherwise we would suspect some variable, some hidden "check"). Explicitly, in the absence of all constraints

"The amount of increase per unit time per unit population is a constant," (6.1)

which can be stated in mathematical language thus:

$$\frac{dy}{dt} = by \qquad (6.2)$$

where:

y = size of population
t = time

$\frac{dy}{dt}$ = rate of change of y with respect to t

b = a constant ("biotic potential") characteristic of the species

Equation 6.2 gives us the rate of change of the population. If we want to find the size of the population at any particular time we must integrate Equation 6.2, obtaining:

$$y = y_0 eb^t \tag{6.3}$$

where:

y_0 = the "initial" size of the population, at "time zero"
e = the base of natural (Napierian) logarithms (2.718 . . .)

Because e has an exponent in the growth equation, we say that "growth is exponential," that is, that *uninhibited* growth is exponential. Using Equation 6.3 presents no problems to those who are reasonably well versed in algebra and who have a table of exponentials at hand. Others may find it more convenient to use the compound interest equation to deal with exponential growth:

$$y = y_0 (1 + i)^t \tag{6.4}$$

where i = the increment ("interest") per year, and the other symbols have the same meaning as before. Equation 6.3 assumes "instantaneous" growth, that is, the population is growing slightly at all times, at a constant rate per unit time, whatever the unit. Equation 6.4 analogizes population growth to the growth of a bank account to which the interest is added only once a year, at rate i. This is not quite realistic for the biological situation, but it is exact enough for our purposes here. This equation is also, it will be noted, an exponential one.

Calculations using the compound interest formula are usually made easier by taking logarithms of both sides:

$$\log y = \log y_0 + \log (1 + i) \tag{6.5}$$

Using this equation and a table of common (Briggsian) logarithms it is easy to verify the figures given in Table 6.1.

To illustrate his points, Malthus often used the rate of increase shown by the United States of his day, "where the means of subsistence have been more than ample," as a consequence of which the population had been doubling every 25 years. (No attempt was made to correct for the immigration component, which was then probably minor.) It will be noted from Table 6.1 that an "interest rate" of only 2.8 percent per annum, when compounded yearly, will cause a doubling in 25 years. For money at interest, this would be considered a low rate. For population growth it is a high rate.

Consider, for example, the average growth rate of world population from 1650 to 1950 (Table 6.2). To grow from 545 million to 2,406 million in three

Table 6.1 Population Doubling Times

Rate of Population Increase (percent per year)	Time Taken to Double Population (number of years)
0.016	4333
0.1	693
0.5	139
1.0	70
1.5	47
2.0	35
2.5	28
2.8	25
3.0	23
3.5	20
4.0	18

Table 6.2 Population, in Millions, of the World and Several Regions (Carr-Saunders' estimates to 1900; United Nations data thereafter)

Date	Europe and Asiatic U.S.S.R.	Asia, Excluding U.S.S.R.	Latin America	World Total
1650	103	327	12	545
1750	144	475	11	728
1800	192	597	19	906
1850	274	741	33	1171
1900	423	915	63	1608
1920	485	997	92	1834
1930	530	1069	110	2008
1940	579	1173	132	2216
1950	594	1272	162	2406
1960	641	1665	208	2972

centuries required an annual interest rate of only 0.5 percent. Actually, this is the *average* rate. We know that the rate of increase grew during the latter part of this period to around 2 percent per year, therefore the rate of increase must have been much less during the early part of the period (as inspection of Table 6.2 shows). But even in the earliest part of the seventeenth century the rate of increase was considerably greater than it had been during most of man's existence on earth—a point we shall return to later.

The graph of a population growing exponentially is shown in Figure 6.1. The curve rises ever more steeply. If the subject is money earning compound interest, the interest becomes principal and earns more interest. In a human population, the children become parents and produce more children. Both cases are instances of what engineers call "positive feedback"—the output (interest-money, or children) feeds back into the system to produce more of the same. As a result, the

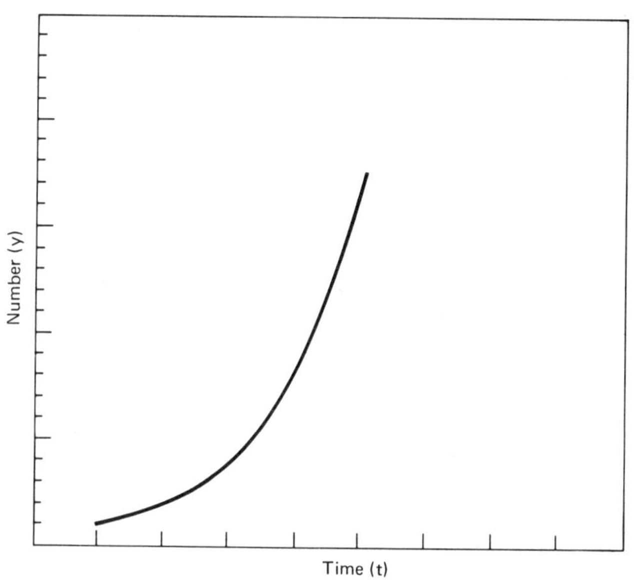

Figure 6.1 Curve of Population Growth, When the Rate of Growth Is Constant, Plotted on Ordinary Arithmetic Graph Paper

curve of money or population soars off toward infinity and it does not matter very much how low the growth rate is, so long as it is positive. On the time scale of geology and evolution even the slowest known rates of biological reproduction rapidly push the curve of potential population growth toward infinity. Darwin, meditating on the ideas of Malthus, was much struck with this fact:

> The elephant is reckoned the slowest breeder of all known animals, and I have taken some pains to estimate its probable minimum of rate of natural increase; it will be safest to assume that it begins breeding when thirty years old, bringing forth six young in the interval, and surviving till one hundred years old; if this be so, after a period of from 740 to 750 years there would be nearly nineteen million elephants alive descended from a single pair.

It is a commonplace to point out how rapidly the descendants of a single bacterial cell, dividing once every twenty minutes, would soon produce a population equal to the weight of the earth. But slow-breeding elephants, in the scale of geological time—or even of historical time—are almost equally impressive. Whatever the organism, it is clear that in a finite world—the only kind of world available to worldly organisms for billions of years—exponential growth can never be continued for more than a tiny segment of time. *Most of the time the average growth rate of any population is, and must be, zero.* Most of the time two parents leave, on the average, just two adult descendants in the next generation, whether they be salmon that lay millions of eggs or elephants that have less than a dozen babies.

Prediction versus Projection

During the past three centuries man has been living in an exceptional period in that the population growth rate has been distinctly greater than zero. Why this had been true we will see later. For the moment we note that it took more than a century of sustained population growth for thoughtful men to become aware of the fact. It was not clear that it *was* a fact until the first periodic censuses were taken (beginning in 1790 in the United States and in 1801 in England). Once it was clear that the population was growing it was natural for men to try to predict its size at some future date. The basic method used in arriving at such predictions can be shown most easily by introducing semilog graph paper. Although many of the older population prophets were unfamiliar with this kind of paper, the method they used was, in principle, the one described below.

Semilog paper is graph paper that has an arithmetic scale along one axis and a logarithmic scale along the other. Time is graphed on the arithmetic axis (abscissa) and population size on the logarithmic axis (ordinate). If the population is growing exponentially, the "curve" of population growth will be a straight line. Figure 6.2 shows such a graphing of Malthus' example of "geometric" growth series, 1, 2, 4, 8, 16 . . . and so on. "Log graphing" has a number of advantages over ordinary graphing:

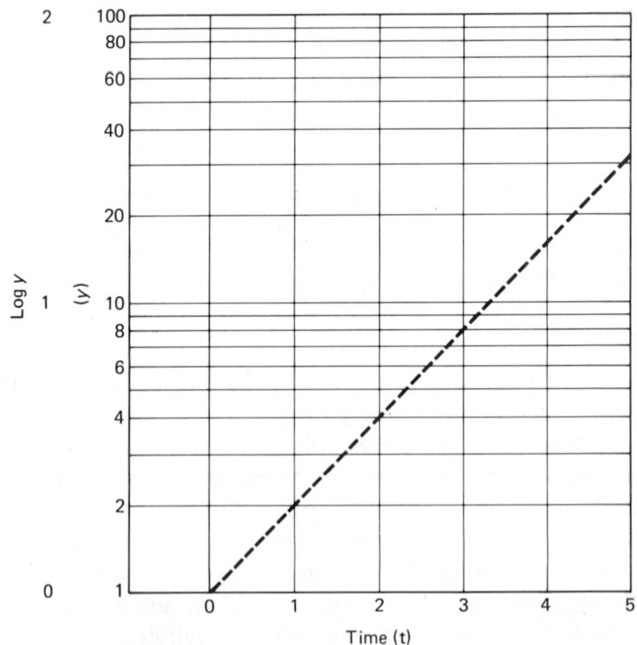

Figure 6.2 Curve of Population Growth, When the Rate of Growth Is Constant, Plotted on Semilog Graph Paper

1. The compression of the scale makes it possible to graph a wide range of values without loss of information.

2. Since exponential curves become straight lines on semilog paper, a "curve" can be fitted with greater confidence to the observational data points, about which there is always some uncertainty. Whenever the curve is a straight line on semilog paper, the growth rate is said to be "logarithmic," which is merely another term for "exponential."

3. Slight differences in the growth rates of different populations are more easily established by comparing two straight lines on semilog paper than they would be by comparing the corresponding curved lines on ordinary arithmetic paper.

4. When growth ceases to be logarithmic (exponential), that fact will become immediately apparent by the departure of the growth curve from a straight line on semilog paper. On arithmetic paper it is difficult to tell when the curvature changes.

The graph in Figure 6.3 of the population data given in Table 6.3 illustrates some of these advantages. Note that from 1790 to 1880 the growth rate of the United States was almost constant, as indicated by the closeness with which a straight line "fits" the census points. Then around 1890 the growth rate slackened; this correlated with the closing of the American frontier to the west. The growth rate continued to fall slightly decade by decade until after World War II, when to everyone's surprise it began a new upward spurt.

Graphing population on semilog paper is a convenient way of summarizing the past. Can it also be used to predict the future? It can; but let us see the results.

Table 6.3 Population, in Millions, of the United States, Exclusive of Alaska and Hawaii

Year	Population
1800	5.3
1810	7.2
1820	9.6
1830	12.9
1840	17.1
1850	23.2
1860	31.4
1870	38.6
1880	50.2
1890	62.9
1900	76.0
1910	92.0
1920	105.7
1930	122.7
1940	131.7
1950	150.7
1960	178.5

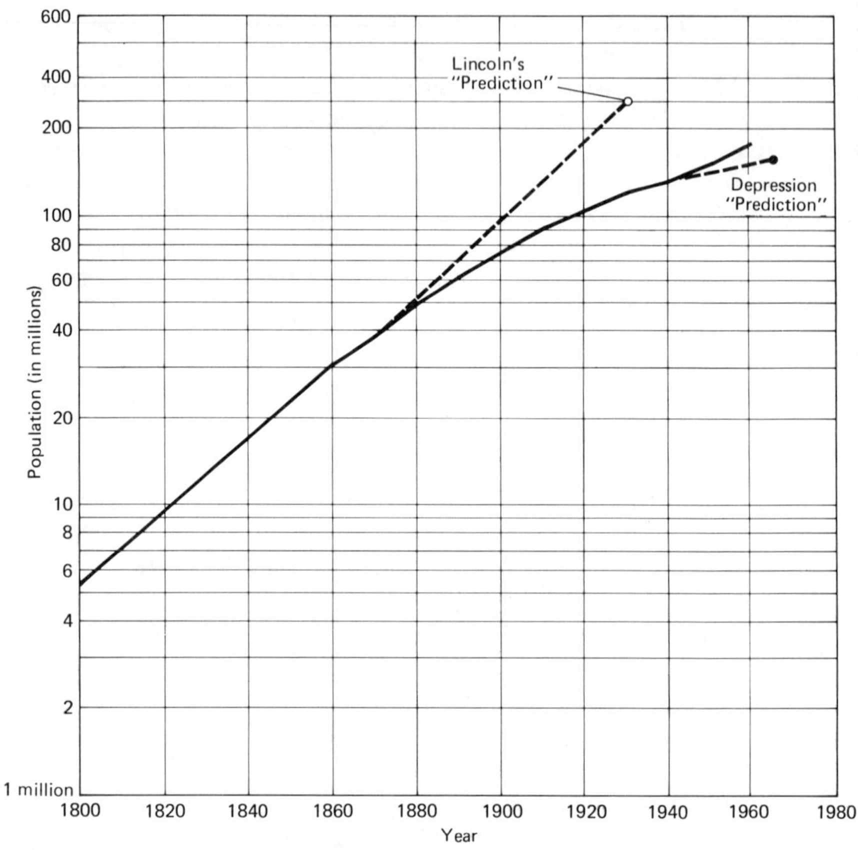

Figure 6.3 Population Growth of the United States, Plotted on Semilog Graph Paper. Note How Extrapolation of Past Trends Led to Two Mistaken "Predictions" (really, *projections*).

Darrell Huff (1954), in his amusing and enlightening book *How to Lie with Statistics*, pointed out that Abraham Lincoln (in his Second Inaugural Message to Congress) predicted that the U.S. population would reach 251,689,914 in 1930. (The actual population turned out to be less than half that much.) On the other hand, the very considerable fall in birth rate that took place during the Great Depression of the 1930s (not closely mirrored in the decennial censuses) led many population experts to predict that the United States population would never surpass 150 million. Textbooks with this statement were still in use in colleges in 1953, according to Huff, though the supposed ceiling was violated by 1950, as shown in Table 6.3. By now, the limit has been ultra-violated.

How were these erroneous predictions arrived at? Without trying to unearth the actual reasoning, we can say that, whatever method was used, it amounted to projecting a straight line through some past points on a semilog plot of the population curve into the future. To illustrate: On Figure 6.3 lay a straightedge along the population points for the years 1850 and 1860; the straightedge intercepts the 1930 line very nearly at the figure given by President Lincoln, way above the level actually achieved. On the other hand, the same procedure used with data from the Depression years results in a population projection that was falsified in less than two decades.

A significant change in terminology was made in the paragraph above, which began with a question about *predictions* and ended up with a discussion of *projections*. No demographer worth his salt will claim to make predictions any more; all he can get out of the data is projections. He can show only what the present trend of population growth will produce *assuming no change in the trend takes place*.

Estimating the trend may be done as simply as we did it here, that is, by laying a straightedge along points on a semilog graph and seeing what the projection of the line is. This amounts to allowing for the "velocity" of the population. In a more sophisticated way, a demographer may try to estimate the "acceleration," that is, the change in "velocity"; this also produces a projection. Asserting that the presently recognized trends, and only these, will operate to produce the future is to commit the error the contemporary philosopher Karl Popper has called *historicism*. We can explicitly state the rules for making projections on the basis of recognized trends. But no one in the 1920s foresaw the Great Depression and its effect on family formation; and in the 1930s no one foresaw the remarkable change in the temper of young people in the late forties and fifties that led them to desire larger families. If we insist on saying that the future is determined by the present we must admit that we are unable to predict the future because of the importance of unrecognized elements of the present. It is not predictions that we make, but projections only—and these must be taken with more than a grain of salt.

What's the Point of Projections?

"If the present growth trend continues, there will be standing room only for man on the earth in less than 600 years."

"By 3550 A.D. the earth will be one solid mass of human protoplasm all the way to the center, if humanity keeps on breeding the way it is now."

These are true statements—in their hypothetical form. But since we know the pictures they paint—an S.R.O. world or an earth of solid human flesh—will never develop, why make such ridiculous statements? What's the point?

"The future," said the British engineer Dennis Gabor (1963), "cannot be predicted, but futures can be invented." Inventing it is the business of the social engineer, and for this, "the engineering of human consent is the most essential and the most difficult step." The making of population projections of the sort just quoted is part of the process of engineering human consent. Such projections must be evaluated in terms of their effectiveness in producing such consent, and not fruitlessly criticized as failing to *predict* the future.

Consent to what? *To planning for a world in which the average rate of population growth is zero.* This is the revolutionary message that human ecologists are trying to get across. Ironically, the message during all but the last 300 years of man's existence would have been regarded as a needless truism rather than a disturbing hypothesis. "Everyone" knew (until recently) that the population stayed always the same, if one ignored temporary fluctuations caused by disease and warfare.

It has been estimated that the world population was about one million in the year 50,000 B.C. (The figure is a very unreliable one, but the uncertainty of the figure makes little difference in the conclusions we will soon reach.) Today there are over 3800 million people. What rate of exponential growth would be required to produce such an increase? Just 0.016 percent per year. Would people living in a world where population increase was taking place at that rate be aware of growth? Hardly—especially considering that epidemics frequently wiped out as much as 10 percent of the population in one year, and occasionally as much as 25 percent. Fluctuations far outweighed the long-term trend in impressing themselves on the consciousness of men living in such a world.

Consider, for example, the psychological world of a man born into a village of 100 people at the time, say, of Alfred the Great. *On the average*, by the time he died 80 years later (if he was so lucky as to live that long) such a village would now have 101 inhabitants. But during his lifetime the village population would certainly have fluctuated between 90 and 110 and perhaps between 75 and 125. Under the circumstances our hypothetical participant-observer could be excused if he failed to see the long-term trend. Life appeared to follow cycles: now better, now worse, but never really different. Mostly, for thousands of years, thoughtful observers could detect only fluctuations. The few who thought they could detect long-term effects perceived only downward trends. As late as 1761 the wise and worldly Montesquieu (1689–1755) quite reasonably remarked, "I dwelt for more than a year in Italy where I saw nothing but the ruins of that ancient Italy, so famous in former times. . . . Why is the world so thinly peopled in comparison with what it was once? How has nature lost the wonderful fruitfulness of the first ages? Can it be that she is already old and fallen into decline?" In the absence of careful censuses, Montesquieu's impression of a population decline was forgivable.

By the time Montesquieu died, population (we now know) was growing more rapidly than ever before. Soon, aided by Condorcet and a host of others, the idea of progress engulfed the world. As concerns population, this idea implied that *growth is normal*, and can be expected to continue forever (so long as we don't do something horrible, like dropping hydrogen bombs). *What most people regard as the wisdom of the ages is anything that has been believed for three generations.* Belief in the normality of population growth falls into that category.

Yet the most elementary reasoning shows that, in a finite world, population increase must be a most rare and exceptional event in the life history of a species, which is typically measured in many millions of years. If the world available to humanity is infinite, that is another matter—but that remains to be demonstrated. In the meantime, as we set about creating the future, prudence dictates that we assume and plan for a static population. In fact, if thorough analysis indicates that we have already overshot the optimum point—as we probably have—then we should embark on a program of negative population growth, until we correct for past excesses.

For two hundred years our intoxication with the idea of progress has prevented us from seeing the reality of population. It is time now to sober up.

A Demographic View of History

It is very hard to break with the past. It is hard to view the world markedly differently from the way our immediate ancestors viewed it. To free ourselves from the preconceptions of our immediate ancestors we must take a long view of history.

If we were to make a graph of the human population of the world for the last million years, allowing one millimeter to the century, we would have a graph 10 meters long, or about 33 feet. Neglecting fluctuations, the slope of this population curve would be very nearly zero for most of its duration. There would, however, be three eras in which the curve would move up sharply to a higher level. By introducing large gaps in the curve we can get it all on one sheet of paper (see Figure 6.4).

The first shift in the "set point" of the population came with the

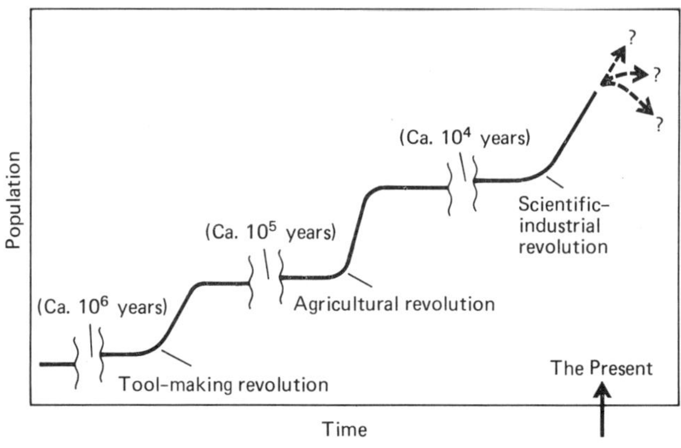

Figure 6.4 Demographic History of the World, Compressed by Judicious Omission. Note Great Change in Set Point of Population at Each Great "Revolution," with Almost No Growth between Revolutions. SOURCE: Adapted from G. Hardin, *Biology: Its Principles and Implications*, 2nd ed. (Chicago: W.H. Freeman and Co., 1966).

"tool-making revolution." Equipped with flints for skinning animals, then bludgeons and bows and arrows for killing them, man was able to feed a far larger population on the same area of land. When men—or more likely *women*—learned to grow crops from seed, the resulting "agricultural revolution" moved the set point up yet another notch, following which further increases took place at only a slow pace for nearly 10,000 years. Then with the coming of the scientific-industrial revolution, in about 1600 A.D., the set point again moved upward.

It is still moving upward. The question is: Will population increase forever? Or will it level off—and if so, at what level?

Is "Space" the Solution?

From time to time someone suggests that we may be able to avoid the necessity of population control by shipping our excess production of people off into space, to colonize other celestial bodies. What are the prospects of such a solution?

First of all, it is clear by now that no other planet of our solar system is even remotely suitable for continuous human habitation. We once had great hopes for Venus; now we know that its surface temperature is greater than 500°F. Mars is far less hospitable than the South Pole or the top of Mount Everest; until we colonize those rugged areas to the density of Manhattan island it would be ridiculous to think of going to Mars. The more distant planets are, for a variety of reasons, even less suitable.

The only possibility remaining is that of planets of stars other than our sun. The nearest star is Alpha Centauri, 4.3 light years away. Because it is a triple star it probably has no planets in orbits suitable for life, but let's assume that it has—and that such a planet has no humanoid life already occupying it; or that, if it has, we can exterminate this life to make room for our burgeoning millions. How difficult would it be to pursue this solution?

In 1959 I made an attempt to calculate the cost of interplanetary travel, on the basis of technology foreseeable in the next century. Deliberately and grossly underestimating the cost at every point I estimated that it would cost $3 million per man to ship off our excess population. A population of more than 200 million (the 1970 figure for the United States), increasing at one percent per annum, produces an excess of 2 million people, which would cost some $6000 billion per year. The Gross National Product of the United States is about $1000 billion per year. So, on economic grounds alone, the solution of the population problem by interstellar migration is out of the question.

There is an even more decisive reason for rejecting the proposal. Recall the basic motivation for proposing this solution: to escape the necessity of population control. Would a program of interstellar migration enable us to escape this hard necessity?

At the speeds of our present rockets it would take about 100,000 years to get to Alpha Centauri. Assuming fantasic improvements in technology it might be possible some day to cut this time to 350 years—a period of time roughly equal to the entire time of existence of the American colonies and the American nation.

Consider the position of the emigrants on board the spaceship, which could not be much more spacious than a fine submarine, and in any case, would be non-expansible. Reproduction would have to be rigidly controlled to make sure that the population on board did not expand. Thus a paradox is introduced: the people who would propose such a solution to the population problem because they dislike population control would be unfit themselves to be passengers on the spaceship. On the other hand, suitable candidates for the long space journey would be those who were willing to control their reproduction on earth and who therefore would have no need to resort to such an outrageous solution to the population problem.

The paradox is decisive; there is no need to pursue further the technological aspects of the problem. Space travel, while conceivably permitting the colonization of other celestial bodies, will not enable us to escape the necessity of controlling population growth here on earth.

The Maximum Possible Population

Malthus saw that the size of a population was dependent on the amount of "subsistence," but what did he mean by this word? Apparently, he meant food (though he may have had housing and other necessities in mind). From Malthus' day to the present students have been fascinated by the question "What is the maximum population possible?"

In all parts of the world until less than two centuries ago, and in two-thirds of the world even now, food has been a limiting factor. It was natural therefore for Malthusian thinkers to focus on food; even today, some people praise the "Green Revolution" in India, resulting from the introduction of improved strains of wheat and rice, because it will permit more people to live—as though eating were all there is to living. It is hard to escape the Malthusian emphasis on food.

The land area of the world is fixed, but the efficacy with which land is used to produce food greatly improved during the nineteenth and twentieth centuries. Inproved harvesting machines, improved motive power (tractors, unlike horses, "eat" only when they are working), improved strains of plants, greater use of irrigation and fertilizers—all these greatly increased the amount of food produced per acre and led people to wonder what the upper limit was. The answer is not clear even now.

The development of organic chemistry in the nineteenth century led to the following important generalization: *in principle* any carbon-containing compound can be converted into any other. In practice, the yield of the wanted compound may be ridiculously low; but we can always hope for practical improvements. In principle, we can make food out of air and water, given a supply of energy.

Plants themselves are not very efficient in this synthesis. Under highly specialized laboratory conditions the efficiency of photosynthesis in capturing radiant energy may approach 3 percent (if we make no allowance for the capital cost of the laboratory apparatus). In the field, under ideal conditions, the energetic efficiency is closer to 0.3 percent. If we include the entire day, and all twelve months of the year in our reckoning, the efficiency drops to 0.03 percent. (These broad generalizations mask a great deal of variability.)

Such low levels of efficiency have led scientists to hope that they might be able to beat nature at the game of capturing solar energy. To date, this dream has not been realized, in spite of millions of man-hours of investigative effort. Some photoelectric cells and wafers have a fair efficiency, but the total reckoning must include the cost of making the sophisticated apparatus itself, which is high.

With energy, we can make food from air and water. The practical limit of population would seem to be not food as such—"subsistence," in Malthus' language—but the capturable supply of energy. Various calculations have been made of the limiting population level supportable by solar energy, assuming various efficiencies in the capture process. The significance of these calculations, all hypothetical, was abruptly diminished by the development of nuclear energy.

Although only a small fraction of the energy theoretically present in uranium-235 is capturable as electrical energy in an atomic energy installation, a plant "burning" 1.8 lbs of U^{235} produces the electrical equivalent of the solar energy falling on one square mile of the earth's surface in an average day. Similar amounts of energy are available in the fusion of heavy water, of which the oceans have a supply that is more than sufficient for any foreseeable human needs. (At the moment we can extract this energy only explosively—from a hydrogen bomb—but much research is being carried out to learn how to extract it non-explosively, in a controlled fashion.)

The fantastic quantities of energy available from nuclear reactions led the physicist J.H. Fremlin to look at the population problem in a radically new way. With *practically* unlimited quantities of energy it makes little sense to calculate the population limit as a function of the *supply* of energy. Another problem replaces this one.

The earth receives energy from the sun every day. *Why does it not get hotter and hotter?* Because it radiates away an equal amount of energy "at night," that is, from the side of the earth that is not facing the sun. Absorption and reradiation of energy are in balance, and keep the surface at a temperature that is (fortunately for us) suitable for life as we know it, the only kind we are interested in.

What would happen if the sun became much brighter? The earth's temperature would rise to a new equilibrium level. The equilibrium level is fortunately not very sensitive to increases of energy inputs, because radiation increases as the fourth power of the surface temperature (stated in degrees

Kelvin). Nevertheless, an increase in energy input would raise the equilibrium temperature.

If we augment solar energy with atomic energy, all that extra energy ultimately must be radiated out into space. The equilibrium temperature of the earth would rise. More people would require more energy input, resulting in a higher equilibrium temperature. The earth cannot sweat; the way it can lose heat is by radiation, and this is subject to physical limits. Fremlin calculated that the limit of human existence would be reached at a population of 10^{16} to 10^{18} people (compared with the present 3.8×10^{9}). Beyond that point the temperature of the earth would be too high for human existence.

We say that a moderately active person "consumes" 3000 kilocalories of energy a day: we often forget that he just as surely *releases* 3000 kilocalories a day, which must be radiated out into space. Beyond a population limit of about 10^{18} people—a billion billion—the world would become one great Black Hole of Calcutta, unfit for man. 10^{18} is the absolute limit.

Quantity or Quality?

What would human life be like near "Fremlin's Limit"? Let the author himself describe it. Consider, he says,

> the housing problem for 120 persons per square meter. We can safely assume . . . the construction of continuous 2000-storey buildings over land and sea alike. . . . Very little horizontal circulation of persons, heat or supplies could be tolerated and each area of a few kilometers square, with a population about equal to the present world population, would have to be nearly self-sufficient. Food would all be piped in liquid form and, of course, clothes would be unnecessary. . . . Occasional vertical and random horizontal low speed vehicular or moving-belt travel over a few hundred meters would be permissible, however, so that each individual could choose his friends out of some ten million people, giving adequate social variety, and of course communication by video-phone would be possible with anyone on the planet. . . . Little heat-producing exercise could be tolerated.

At this point the obvious question intrudes: Is this the kind of life we want? Granted that technology is capable of supporting a population of 10^{18} people, is this what we want? Should we tolerate public policies that push us in this direction?

Surely we can agree that it is not the *quantity* of human life that we want to maximize, but its *quality*. "Quality of life" is difficult to define. Many people have many different opinions about this. Reconciling these opinions may be a disputatious, perhaps even a bloody, business. The solution we reach with great travail may be far from the optimum—but it surely will not be as near the *pessimum* as is the world described by Fremlin.

The idea of an optimum is simple; it is illustrated by the curve in Figure 6.5. We understand the problem only in principle, but that is something. The units in

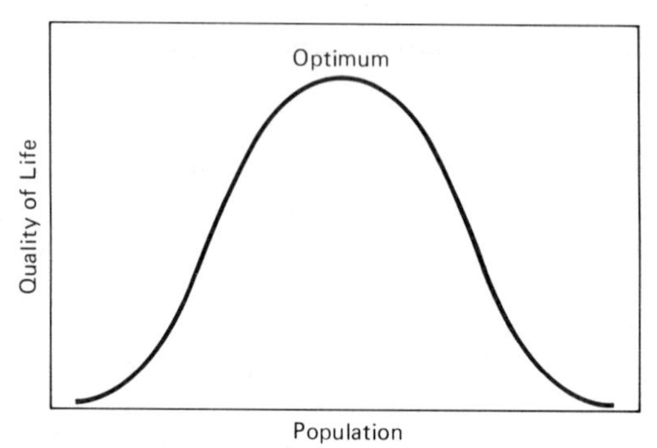

Figure 6.5 Graphical Representation of the Meaning of "Optimum Population Size"

which "quality of life" are measured are yet to be worked out. We don't know exactly where the optimum point is, but surely we can agree that the optimum point must be closer to the world of Thoreau, with a population density of 0.000002 persons per square meter, than it is to Fremlin's limit of 120 per square meter.

"Popollution" Problems

Progress in any field is dependent upon asking the right questions. As we have seen, the question "What is the maximum possible population?" is not the best question to ask, if we are concerned with the future quality of human existence. "What is the optimum population?" is a better question, but it may be too difficult. It may take too long to determine the highest point of the curve in Figure 6.5. While we are at work on this interesting theoretical problem the quality of life may deteriorate rapidly, perhaps beyond the "point of no return."

All we really need to know is this: *Will a further increase in population increase or decrease the quality of life?* That is, are we moving toward the peak of the curve in Figure 6.5, or away from it? If we can answer this, we don't need to know exactly where the peak is.

It is a matter of considerable urgency that we tackle this problem with resolution. We don't have much time. Exponential growth is a runaway process. Notice, in Table 6.4, how each new billion people is added to the world's population in less time than the last. This means that the need for a speedy decision is greater by the hour. Even a suboptimal decision is preferable to drifting. The longer we delay the decision to bring population growth to a halt, the greater will be the suffering when we do. No matter what we decide, what we do, we may be in error; but bringing population to a halt prematurely will surely be a less serious error than waiting too long to do so. Underpopulation can be easily and quickly corrected; overpopulation, with its consequent destruction of the environment, may be fatal to man's prospects.

Two of the dogmas of the religion of Progress are (1) *the bigger the better*

Table 6.4 Time Required to Add the Next Billion People

World Population in Billions	Year	Interval, in Years
1	1833	
		96
2	1929	
		32
3	1961	
		15
4	1976	
		12
5	1988	
		7
6	1995	
		5
7	2000	

NOTE: The first two intervals are derived from demographic data. Subsequent intervals from demographic projections. (SOURCE: R.E. Lapp, unpublished.)

and (2) *the more the better*. Are these dogmas true? It depends on the level of the population.

A *very* small population is severely limited in what it can do. The principality of Monaco, with a population of 23,000, does not manufacture automobiles, and for good reasons. The making of complex machinery requires many specialized tools, and is economical only on a large scale. The manufacturing process enjoys *economies of scale*. Monaco is large enough to run a gambling casino but not large enough to run an auto factory. A highly technological society requires a rather large population.

How large? No precise answer is possible, but it is worth pointing out that Sweden, with only 8 million inhabitants, manufactures two different brands of automobiles, as well as other complicated machinery.

What is not so generally recognized is that *in principle* each manufacturing process has its own optimum size; beyond a certain point, *dis*economies of scale occur. For a nation as a whole, such diseconomies unquestionably show up eventually. *The more the better*?—Let's take a critical look at some U.S. statistics (Table 6.5).

Notice in Table 6.5 that as population increased at an average rate of 1.7 percent per year during the years 1950–1965, the increase in the number of hunting licenses sold yearly amounted to 2.9 percent per year, while the number of fishing licenses went up 3.4 percent per year. That is, the number of licenses per capita increased. Is this good? Does this indicate a richer life for Americans? It seems most unlikely. It is a matter of universal knowledge that hunting and fishing have been getting steadily worse every year, for many decades. Increased sale of licenses does not indicate better sport; it merely indicates more disposable income, and a boundless fantasy life among our citizens.

Table 6.5 Average Annual Rates of Growth of Various Activities in the United States, 1950-1965

Statistic	Average Percentage Increase per Year
Population	1.7
*Hunting licenses sold	2.9
*Fishing licenses sold	3.4
Motor vehicles registered	4.1
Average daily number of phone conversations	4.9
Annual number of visitors to national parks	8.2
*Annual number of campers in national forests	14.7

*Terminal year, 1964.
SOURCE: Calculated from data in the *Pocket Data Book, USA, 1967.*

During the same period, visits to the national parks increased at 8.2 percent per year, while camping in the national forests increased by nearly 15 percent per annum. The amount of park and forest land remained virtually static, therefore the degree of crowding in these facilities increased fantastically over this period—a 680 percent increase in camping, for example. Dreaming of the wilderness of John Muir, a prosperous and too numerous population turned Yosemite Valley into an outdoor slum not too different from Coney Island. Progress? The more the better?

Increase in size has an interesting effect on communications. The average daily number of phone conversations in the United States increased at a rate of nearly 5 percent per annum. Increasing prosperity no doubt had something to do with this; but is the increase all progress? The per capita amount of talking on the telephone almost exactly doubled in fifteen years. Is this good or bad? To put the problem in perspective we need a bit of mathematical analysis.

Communication is a matter of relationships. Suppose there are n persons in a group and each person wants to communicate with all the others—how many lines of relationships are there to keep open and functioning? If there are 3 people, there are 3 relationships; with 4, the number increases to 6; with 5 to 10. A set of 10 people can have 45 relationships; 100 people, 4950. In general, for n people, the number of relationships, r, is given by:

$$r=\frac{n(n-1)}{2}=\frac{n^2-n}{2} \qquad (6.6)$$

Variables (like n) are more important than constants (like 2), so we will not be too far off if we say: *the number of relationships increases approximately as the square of the number of individuals to be related.* Given that *time* is a "resource" that cannot be increased, this means that as the number of individuals in a group increases we must choose among these alternatives: (1) either we must have relationships with an ever smaller fraction of the population; or (2) the duration of each relationship must be shortened.

In any case, we are face to face with a condition of *information overload* which increases with the size of the population. Information (and communication) suffer from a diseconomy of scale. Frictions, selective blindness, arbitrary decisions, and misunderstandings must all increase with increasing size of the population unit. The probability of serious social disorganization (or totalitarianism) must also increase. In an unconscious effort to diminish social losses we undoubtedly increase the amount of time we spend communicating with each other. This no doubt accounts for part of the increase in the number of phone conversations per capita with increase in population. The resource we call time is eroded by population growth.

Automobiles increased 2.4 times as fast as the population during the 1950–1965 period. Why? In large part, no doubt, because of increasing prosperity. What were the consequences of this exuberant multiplication of cars? *Pollution.* The amount of air pollution increased greatly in that period. (The number of miles driven per car per year also increased, adding further to the pollution.) Though we are without adequate statistics it is probable that the number of traffic jams, the amount of time spent in queues, and the amount of time required to commute to and from work also increased. In a word, the quality of life was decreased by increasing automobilization.

As a first approximation we may say:

$$\text{Population} \times \text{Prosperity} = \text{Pollution} \tag{6.7}$$

For example, we have more air pollution because we have more people to drive cars (population) and more disposable income (prosperity) to spend on them. Keeping population constant we can diminish the amount of pollution from this source by equipping automobiles with expensive smog-control devices—which must be paid for by diminishing the individual's prosperity, the amount of income he has to dispose of for things that are intrinsically enjoyable.

In a larger context, the word "pollution" should be understood to include any sort of *environmental disruption* (E.D.)—air pollution, water pollution, soil erosion, aesthetic deterioration, extinction of species, and so on. The Japanese word *kogai* covers all these, but English has no equally suitable word. "Pollution," or "E.D.," must do. Social disorganization, which occurs at high levels of information overload, also may be regarded as a form of pollution.

Because of diseconomies of scale, population and pollution problems are essentially connected. For this reason some people speak of "popollution" problems. Improved technology ameliorates pollution (at a cost), but the relationship given in Equation 6.7 implies that ultimately the solution to pollution is not to be found in technology but in population control. How can we achieve that?

Political Systems and Population Control

Birth control is not population control; it is a necessary, but not a sufficient, condition. In every society, according to Kingsley Davis, the number of children

Table 6.6 Political Systems of Environmental Exploitation

	Rules of the Game						Results of the Game		
	Exploitation of Environment by:		Profits Go to:		Gain from Stressing the System:				
Case	Individual (1a)	Group (1b)	Individual (2a)	Group (2b)	Overall Gain (3)	Gain to the Decision Maker (4)	Intrinsic Responsibility (5)	Temptation to Sabotage Information (6)	Name of the Game
I	✓		✓		–	–	+	0	Private Enterprise
II		✓		✓	–	0	0	+	Socialism
III		✓	✓		–	+	–	(0)	System of the Commons
IV	✓			✓	–	+	0	+	?

desired per family is, on the average, greater than the number required for population equilibrium. Technological control must, then, be accompanied by political change. Of what sort?

The population problem should be put into a general framework that includes all the principal systems of exploiting the environment for man's use. For such exploitation there are four possible political systems. (Each of these has many variants, but we need consider only their ''pure'' forms.) These systems and their characteristics are displayed in Table 6.6, which needs to be carefully studied. One of the four possible systems (Case IV) seems to be a class without any real members, and has no name (unless it is ''private philanthropy''). It is included in the table only for logical completeness, and will not be discussed further.

Let's see how the three major systems function in the exploitation of the environment. As a way of visualizing the problem let us imagine that we are concerned with raising cattle in a pasture, and that we want to so adjust the number of animals that we obtain the maximum sustainable yield, year after year, indefinitely. Every pasture land has its inherent ''carrying capacity.'' We will assume that the number of cattle is initially exactly at this carrying capacity. To add one more animal would be to damage the pasture and produce less profit, year after year. We indicate this fact by a negative gain in the whole of column 3.

The pasture (environment) may be managed under a system we call ''private enterprise'' (Case I). Each man fences in his own portion of the pasture, on which he runs all of his cattle, and only his cattle. The owner is the decision maker. Under these conditions, the overall gain from his decision accrues entirely to him; as a result, the signs are identical in columns 3 and 4. We may say that the owner is *intrinsically responsible* for his decisions (column 5). If he makes a wrong decision he is not tempted to cover up this information, for if he does he will suffer (column 6). The system works admirably *for those who are in it;* those who are left out of the system of private enterprise may seek to destroy it, a fact that should never be forgotten by the ''in-group.''

A substitution commonly sought by the ''out-group'' is ''socialism'' (Case II). It is believed by many to be more just. Perhaps it is; but it has some significant operational disadvantages. Someone has to make the decisions. What happens to the decision maker when he makes a bad decision? Not much, intrinsically; the negative gain resulting from a bad decision is shared by all of society, and the portion devolving upon the decision maker is, practically speaking, zero (column 4). Intrinsically, he is not motivated to exert himself very hard to reach the right decision. He is likely to become a time-server, a ''bureaucrat''—in the invidious sense. To motivate him, society may embroider the system with adventitous rewards for good performance. Threatened with such *contrived responsibilities,* the decision maker who makes an error will try to sabotage the information system. He will try to see to it that unfavorable reports are classified as national defense secrets, or repressed as a matter of ''executive privilege.''

No great nation today is operating wholly under one system or another. Russia is a mixed economy; so is the United States. Each basic system is subject to myriads of modifications in the various areas of management. This is well known. What is not so well known—and what we must thoroughly understand if we are to make sense of environment and population problems—is that *there is a third system,* a system first described in 1833 by William Forster Lloyd.

Suppose that each herdsman brands his cattle, which are then all run together on a common pasture, without fences. When it comes time to sell the cattle each man obtains all the proceeds from all the cattle bearing his brand. Under this system (Case III), when a man adds one animal too many to his herd he actually benefits at sale time (column 4). It is distinctly to his interest to overload the environment. The herdsman who does not do so is at a competitive disadvantage vis-à-vis the others. Each man's intrinsic responsibility is less than zero; it is, in fact, negative (column 5). Consequently, each one adds more and more animals to his herd, even though the end result is ruin for all. They are all trapped by the system. *Inevitable* ruin is the very essence of what we call tragedy. This is why I call the result of this political system the "tragedy of the commons."

Private enterprise has its disadvantages but it can be made to work, more or less. Socialism has *its* disadvantages, but it also can be made to work, more or less. But, in a crowded world—and we will know no other for our entire lifetime—the system of the commons should be regarded as intolerable.

I am confident that it will be so regarded *when we recognize it.* But men are very clever at hiding the truth under a mass of distorting verbiage. What are we to say, for example, of an advertisement that speaks of the glories of American technology in the following terms? Beneath a painting of fantastic new fishing boats it says: "The hungry of the world are being fed, by the ingenuity of American free enterprise harvesting the inexhaustible wealth of the seas." There are three falsehoods in this statement, and the worst of them is the implication that the oceans are being exploited by a free enterprise system. This is tragically not true. The oceans of the world are being treated as a commons, and maritime nations are exploiting them to ruination. No pretty words, no nice intentions, no shiny technology can alter this brutal truth. Only by changing the political system of exploitation can we hope to save any of the "inexhaustible" wealth of the unfenced seas for our grandchildren.

This kind of political analysis also throws light on pollution problems. Steel mills and power plants belch thousands of tons of industrial garbage into the air every day, because we unthinkingly allow the air to be used as a commons. In a competitive market, the manufacturer has not enough to gain from the commons to pay for the financial loss incurred by installing expensive control equipment. Every decision maker is caught in the same bind. Consequently the pollution continues indefinitely.

Decision makers in Russia behave in the same way. Managers of Russia's pulp-mills are now ruining Lake Baikal with their wastes in the name of "socialism." American managers of paper mills dump mercury into the Great

Lakes, making fish too poisonous to eat, and call their work ''private enterprise.''

Everywhere in the world the environment is polluted because we are trapped in the tragedy of the commons. It is an intrinsically irresponsible system. We will not get rid of pollution until we get rid of this iniquitous third political system. In a crowded world we all suffer when any man is allowed to pollute the common media of air, water, and the ''ether'' of radio and light waves.

This political analysis also illuminates the problem of population control. A few centuries ago family size was determined largely by a private enterprise system. Parents who were so unwise or so unlucky as to produce too many children had to suffer for it: the stress of too many weakened the whole family and often some of the members starved to death. The outside world did little to interfere with the intrinsic responsibility of this cruel system.

The last two centuries have seen a great increase in the philanthropic activities of the state. It is regarded as a disgrace to the community when a child starves to death. More and more of the costs of education are assumed by the community. Few will argue against the desirability of this change, but we should notice its political significance.

As we approach ever closer to the ''welfare state'' we produce an ever-purer example of the tragedy of the commons in population matters. The Marxian ideal ''to each according to his needs'' in fact defines a commons. If parents are free to have as many children as they want because they know that all can dip into the commons, then population control becomes difficult if not impossible. Contrived responsibility—scoldings and the like—must be completely effective or they will fail utterly, in the long run. Even if the great majority of the people are sensitive to the larger needs of the community, the few who are not will cause the system to break down. Unrestricted freedom to breed will, in the long run, produce ruin for all. This is what Malthus was trying to tell us in his celebrated ''parable of the feast.'' The message was ignored when he first wrote it in 1803; but we cannot ignore it much longer.

There is a cliché that says that ''freedom is indivisible.'' Properly interpreted, this saying has some wisdom in it, but there is also a sense in which it is false. *Freedom is divisible*—and we must find how to divide it if we are to survive in dignity.

There are many identifiable freedoms, among which are freedom of speech, freedom of assembly, freedom of association, freedom in the choice of residence, freedom in work, and freedom to travel. You can make the list as long as you like. After you have finished, ask yourself this question: Is there one freedom on the list that would *increase* if our population became twice as great as it is now?

Freedom *is* divisible. If we want to keep the rest of our freedoms we must restrict the freedom to breed. *How* we can accomplish this is not at this moment clear; but it is surely subject to rational study. We had better begin our investigations now. We have not long to find acceptable answers.

References

Committee on Resources and Man, National Academy of Sciences, *Resources and Man.* San Francisco: W.H. Freeman, 1969.

*DAVIS, KINGSLEY, "Population Policy: Will Current Programs Succeed?" *Science,* 1967, **158**, pp. 730–739.

EHRLICH, PAUL R., and ANNE H. EHRLICH, *Population, Resources, Environment.* San Francisco: W.H. Freeman, 1970.

*FREMLIN, J.H., "How Many People Can the World Support?" *New Scientist,* 1964, No. 415, pp. 285–287.

GOLDFIELD, EDWIN D. (ed.), *Pocket Data Book, USA 1967.* Washington, D.C.: U.S. Bureau of the Census, 1966 (revised biennially).

*HARDIN, GARRETT, "Interstellar Migration and the Population Problem." *Journal of Heredity,* 1959, **50**, pp. 68–70.

HARDIN, GARRETT (ed.), *Population, Evolution, and Birth Control,* 2d. ed. San Francisco: W.H. Freeman, 1969.

*HARDIN, GARRETT, "The Tragedy of the Commons." *Science,* 1968, **162**, pp. 1243–1248.

LLOYD, WILLIAM FORSTER, *Two Lectures on the Checks to Population.* Oxford, 1833.

MALTHUS, THOMAS ROBERT, *An Essay on the Principle of Population.* 1798. (There are several reprints of the first edition, as well as subsequent editions.)

*Reprinted in the seventh reference.

7

Air Pollution

JAMES A. OLIVER

Director,
The New York Aquarium

James A. Oliver was born in Caruthersville, Missouri, January 1, 1914, the son of Arthur L. and Mary E. (Roberts) Oliver. He studied at the University of Texas and went on to earn the B.A. (1936), M.A. (1937), and Ph.D. (1941), from the University of Michigan, where he was also appointed a University Fellow (1938–1940) and a Hinsdale Scholar (1940–1941). After serving as Instructor at Northern Michigan (1941–1942) he went to The American Museum of Natural History in 1942, as Assistant Curator. After wartime service as an officer in the U.S. Navy he returned to the Museum in 1946, leaving in 1948 to become Assistant Professor of Zoology at the University of Florida. He was appointed Curator of Reptiles at the New York Zoological Society in 1951, Assistant Director of the New York Zoological Park in 1958, and Director the same year. In 1959 he was appointed Director of The American Museum of Natural History, a post he held for ten years, becoming Coordinator of Scientific and Environmental Programs in 1969. In 1970 he became Director of the New York Aquarium.

He has authored numerous scientific and popular works, including two books, *The Natural History of North American Amphibians and Reptiles* and *Snakes in Fact and Fiction*.

Air pollution is not a new phenomenon. Man-made pollution of local human environments has been taking place ever since man learned to play with fire and brought it into his shelter. Even men dwelling in caves knew air pollution. Cave walls smudged with smoke tell us that early man knew the discomfort of smoke-filled eyes and the choking cough that results from incompletely burned fuels. We know that from early times the gaseous contaminants man has dumped into the air have caused widespread irritation. For example, as far back as the twelfth century, shortly after the discovery of coal, the city of London recorded annoyingly heavy palls of smoke from thousands of soft-coal fired hearths. These records are among the earliest ones we have of any community experiencing prolonged exposure to a form of atmospheric pollution caused by man.

Since the days of the industrial revolution such episodes have been occurring at increasingly frequent intervals. Beginning in the early 1900s major disasters have been recorded in such widespread industrial areas as the Meuse Valley of Belgium in 1930, Donora, Pennsylvania, in 1948, London in 1952 and 1962, New York in 1953, 1962, and 1966, and Tokyo in 1967. As long ago as 1881 the need for smoke control was evident in Chicago and Cincinnati, both of which instituted restrictive legislation at that time. By 1912 similar laws had been passed by 23 of the 28 American cities with populations of more than 200,000. Since that time in the United States, local governments and state legislatures have passed additional laws, but the enforcement of these statutes has been only partially successful at best. In 1955 the first federal legislation was passed to start analytical procedures and to offer advisory assistance to states and local governments. Since 1963, when the United States Clean Air Act became law, a whole body of federal legislation has been produced, designed to improve air quality in the United States. International regulations are being discussed in the United Nations, and pacts among European nations are being effected. People have become increasingly aware of the problems of air pollution, and governments are responding to public pressure.

Air pollution must be considered from many points of view: the harmful effects on human health and vitality, including the loss of animal and plant life; damage to property and materials, and the attendant economic loss; the destruction of natural beauty; pervasive deterioration of the quality of life in a heavily affected area; and changes in climatic and weather patterns, which may ultimately prove of even greater significance to the viability of the environment than the effects of land or water pollutants.

Today, the problems of atmospheric pollution are so vast, varied, and complex that they call for analysis and control on a global scale. For example, the world's weather patterns are undergoing subtle but real modifications, which man's own activities are bringing about. Changes in local weather as a result of air pollutants are well-established phenomena, and their detailed effects are being evaluated.

Sources of Pollution

Some of the problems of air contamination are caused by natural phenomena. These include volcanic eruptions, earthquakes, dust and sand storms, naturally occurring forest and grassland fires, pollens, air-borne viruses and bacteria, hurricanes and thunderstorms. We know that these natural phenomena can cause, and have caused, loss of human life as well as the destruction of property, including entire cities. We are beginning to learn to control these phenomena, or to lessen the extent of the devastation they cause. However, they frequently interact with, exacerbate the effects of, or are in turn worsened by man-made pollutants. For example, the most serious episodes of air pollution have all occurred at periods of air stagnation when the normal patterns of air movement were blocked by meteorological conditions associated with temperature inversions. Rates of dispersion of air pollutants are all dependent on local rates of air flow. In valleys or basins, such as the Los Angeles Basin, the flow of air may often be retarded by hills or mountains.

In this review we are concerned primarily with the kinds and effects of man-caused gaseous or air-borne particulate forms of pollution. There are hundreds of different types of contaminants that are dumped into the atmosphere as waste products of technology. Most of these, fortunately, are localized, occurring over short distances downwind of the sources. However, a number of the more common pollutants are produced in such large quantity as by-products of so many industrial processes that they are beginning to accumulate in the atmosphere in amounts that cause global pollution.

In the United States the most widespread substances involved in urban air pollution and their primary sources are shown in Table 7.1. These pollutants are carbon monoxide, sulfur oxides, hydrocarbons, particulate matter,[1] and nitrogen oxides.

The data given in Table 7.1 are from a study by the National Academy of Sciences–National Research Council for the year 1966. For 1968 the United States Department of the Interior calculated that the total amount of air pollution over the continental United States had increased to 143 million tons. A more recent estimate places this figure at 164 million metric tons annually (R.E. Newall, 1970). Thus it can be seen that in the United States the total tonnage of air pollutants being dumped into the atmosphere is increasing. However, abatement efforts have begun to show results and the relative quantities of each kind have changed. The Air Pollution Control Administration estimated that in 1970 a total of 214 million tons of pollutants were emitted over the continental United States and the percentages of the major pollutants as well as their primary sources have changed since the 1966 estimates.

From 1958 to the present the Air Resources Department of New York City has conducted a detailed daily analysis of sulfur dioxide pollution, the most precise compilation of data yet made by any municipality. Its findings indicate that daily and seasonal fluctuations are closely related to temperature. Because virtually all sulfur dioxide in the atmosphere comes from the burning of fossil fuels, more fuel is burned and more sulfur dioxide is dumped into the air when the temperatures are low. Daily fluctuations are most noticeable in the colder months when the lowest levels of sulfur dioxide tend to occur late at night and early in the morning, between 11 P.M. and 4 A.M. In the yearly cycle it has also

Table 7.1 Air Pollution in the United States

Commonest Pollutants	Million Tons per Year	Percent
Carbon monoxide	65	52
Sulfur oxides	23	18
Hydrocarbons	15	12
Particulate matter	12	10
Nitrogen oxides	8	6
Others	2	2
Total	125	100
Sources of Pollutants	**Million Tons per Year**	**Percent**
Transportation	74.8	60
Industry	23.4	19
Generation of electricity	15.7	12
Space heating	7.8	6
Refuse disposal	3.3	3
Total	125.0	100

[1]Under the term "particulate matter" or particulates are included a wide variety of usually solid, inert substances that vary from microscopic materials up to the clearly visible pieces of fly ash. The microscopic particles may be intimately associated with tiny liquid droplets of vapors in what are termed "aerosols." The particulates that we see are aesthetically disturbing, and make up the dust, dirt, and smoke of annoying air pollution. However, it is the unseen minute particulates, less than one thousandth of a millimeter in diameter, that are the health hazards in this type of pollutant.

been observed that, as would be expected, the highest amounts occur in the late fall and early winter months. In the period July 1, 1969 to June 30, 1970, for example, the annual monthly low of .06 parts per million was recorded in July, whereas the high of .17 ppm was registered in January. Similar seasonal averages have been recorded for most of the other major pollutants.

These fluctuations may also reflect daily and seasonal changes in human activity, as indicated in the daily fluctuations in the levels of sulfur dioxide cited above.

The gasoline-powered internal combustion engine is one of the chief sources of the more common forms of air pollution, accounting for more than half the measured contaminants and more than 75 percent of the largest single pollutant, carbon monoxide. The more than 80 million automobiles registered in the United States in 1967 produced 61 million tons of carbon monoxide, 16 million tons of hydrocarbons, 6 million tons of nitrogen oxides, and 210,000 tons of lead; and the number of registered automobiles is increasing at a rate faster than that of the population.

The United States Department of Transportation estimated in 1970 that approximately 1 percent of the total amount of air pollution in the United States came from aircraft engine emissions. Despite this relatively small amount,

Table 7.2 Types and Sources of Air Pollutants in 1970

Primary Types of Pollutants	Percent
Carbon monoxide	47
Sulfur oxides	15
Hydrocarbons	15
Particulate matter	13
Nitrogen oxides	10
Main Sources of Pollutants	**Percent**
Transportation	42
Stationary sources of fuel combustion	21
Industry	14
Forest fires	8
Solid waste disposal	5

several communities with busy commercial airports have initiated pollution regulations for aircraft and have notified airlines of planned schedules for tighter controls in the future.

The important sulfur oxides and particulates, which in 1966 amounted to 23 million tons and 12 million tons, respectively, are produced chiefly by industrial plants, including electrical power plants. Fuel burning industries also discharge large quantities of carbon dioxide and water (steam) into the atmosphere.

In addition to producing these widely distributed and common pollutants, industrial plants also dump, spray, or release into the atmosphere a variety of particulate contaminants. In some areas these special pollutants may cause more serious problems than the substances mentioned earlier.

Effects of Air Pollution

It is difficult to determine the precise effects caused by specific pollutants, because there are so many pollutants and so many variables related to their occurrence. The pollutants rarely, if ever, occur in isolation, so that it is virtually impossible to measure the extent of an individual's exposure to a single pollutant. Furthermore, pollutants interact, and it is difficult to determine whether an effect is direct, or synergistic. Experiments on animals and plants cannot be accepted conclusively because of marked differences in sensitivity to different substances. Experiments with volunteer subjects have been made on such a small scale that the results are inconclusive.

Nevertheless, we have circumstantial evidence that the presence of certain pollutants is harmful, and among the hazards the effects on human health are of the highest concern. Many scientists have concluded that air pollution can kill, or shorten life.

In the major air pollution disasters noted at the beginning of this chapter the number of excess deaths attributed to the pollution of the air varied from 17 in Donora, Pa. to as high as 4000 in London. The striking feature of the Donora occurrence was that nearly half of the population reported symptoms of discomfort. These varied from minor psychosomatic to near fatal conditions.

Information being accumulated supports a positive association between polluted air and throat irritation, increased deep coughing, presence of phlegm, bronchial asthmas, bronchitis, emphysema, lung cancer, and other respiratory problems. However, to date we do not know which specific pollutants cause specific ailments. The data have not yielded conclusive answers.

For example, in a detailed series of studies on a large sample of white males and females in the Erie County (Buffalo-Niagara Falls) area of New York, researchers found no conclusive evidence of association between oxides of sulfur and total mortality or respiratory system mortality in white males 50 years of age and over. They did, however, find statistically significant positive association between mortality from chronic respiratory diseases and sulfation in the lowest two economic levels of a five-level classification. They also found a positive association between cough with phlegm and particulate air pollution among nonsmoking women of 45 years of age and older. The researchers doubted the existence of any synergistic effect between particulates and sulfation. They cite not only their own researches but the fact that during recent fogs when particulate air pollution in London was dramatically reduced the mortality rate remained near average even though sulfates remained high.

The results of all attempts to evaluate the effects of individual pollutants are influenced both by smoking and by factors in early environment, including economic status. In the Erie County studies the investigators found an interesting interaction among white women smokers with residential mobility and respiratory symptomatology. Those with stable residential histories showed that cough with phlegm was positively associated with air pollution at place of current residence; in those smokers who had moved within the previous five years, it was inverse. These observations are suggestive but not conclusive. Two rather similar surveys were conducted in the 1950s in Soviet Russia and in Maryland on selected "twin communities" in each country where one community had a noticeably heavier air pollution condition as measured by a simple dust-fall index. In both instances it was found that the communities with noticeably heavier air pollution had significantly greater instances of respiratory ailments.

Carbon monoxide, as already indicated, is one of the most voluminous pollutants emitted in the United States, and is especially prevalent in metropolitan areas that rely heavily on motor vehicular transportation. Carbon monoxide is a colorless, odorless, highly toxic substance. It can cause headaches, impair mental functions, reduce the amount of blood-oxygen, and is associated with blood diseases that may result in heart attacks. In addition, the inhalation of carbon monoxide is a well-known cause of death by accident or suicide. Human sensitivity and resistance to this substance appear to be highly variable. The United States government has stated that 8 to 14 parts per million can be dangerous to health. California and Pennsylvania consider any amount of carbon monoxide above 25-30 ppm hazardous if it persists for 8 hours or more. According to some studies it produces headache at 100 ppm, yet other studies have indicated a higher threshold. In a traffic jam carbon monoxide levels inside an automobile may reach 370 ppm!

Carbon monoxide is known to have an affinity for blood hemoglobin that is 210 times greater than that of oxygen. When the carbon monoxide unites with the blood it forms carboxyhemoglobin, which provides a convenient quantitative indicator of the amount of exposure to carbon monoxide.

Smokers invariably have a greater concentration of carbon monoxide than nonsmokers. In a study made in Sweden in 1969 of the hazard to traffic policemen in urban areas, Gōthe et al. found no significant increase in carboxyhemoglobin in the blood of nonsmokers after rush-hour duty, but the level in nonsmokers was three to five times greater. They suggest more detailed studies on nonsmokers. This study indicates the difficulty in pinpointing precise effects of specific pollutants in human beings because the traffic policemen were also exposed to large concentrations of other toxicants in exhaust fumes, such as acrolein, other aldehydes, alcohols, phenols, acetone, uncombusted hydrocarbons, nitrous oxides, sulfur dioxide, and lead compounds.

The photochemical oxidant, ozone, which occurs naturally in small amounts in the atmosphere, is becoming an increasingly serious problem in some areas. Levels of 0.10 parts per million cause severe eye irritations. Prolonged exposure at even lower levels has been reported to increase the frequency of severe attacks in asthmatics. It has affected the performance of cross-country runners at levels of between 0.03 and 0.30 ppm. California, where the emissions from large numbers of automobiles are exposed to bright sunlight, established some of the earliest regulations dealing with this pollutant. For example, in Los Angeles, when the level reaches 0.35 ppm, all students from the elementary and secondary school grades are excused from all strenuous activity. Since this system of regulation was established in 1955, alerts of this kind have been announced 71 times.

Particulates have been widely connected with the hazards of air pollution. Some researchers have maintained that they are most frequently a hazard to health when they occur in association with sulfur dioxide. Recent results suggest that whereas this may sometimes be true this multifaceted form of pollutant has many diverse effects. Particulates include solid particles, liquid droplets, and gaseous particles that vary in size from 0.001 of a micron up to 1000 microns (1 micron equals 1/1000 of a millimeter or 1/25,000 of an inch). Particulates of 1 micron or less in diamater can be taken into the human body through the nasal passages and carried to the lungs. It is estimated that more than half the particulate matter in the atmosphere is of this size or smaller. Further, particulates of this small size remain in the atmosphere for long periods of time, depending on their size. Smoke, mists, and aerosols belong in the smaller-size range, while fly ash, dusts, and soot belong in the large-size range of this group. Inhalation of heavily dust-laden air or dense smoke is immediately harmful and irritating to the respiratory tract; the smaller, finer particulates take longer to produce their unhealthy results. In 1969 the U.S. Public Health Service reported that particulate levels of between 100 and 130 micrograms per cubic meter can induce respiratory diseases in children and cause death in people over 50.

Cities employing a daily air pollution index generally rate any level of particulate pollution above 100 micrograms per cubic meter as unsatisfactory. One study, carried out in Erie County, suggests that a rise in the death rate will occur in regions with an annual average concentration of from 80 to 100 micrograms per cubic meter. This study also suggests an association between these levels of particulates and gastric cancer in men 50 to 69 years old.

A common particulate associated with the building industry, particularly in industrial or urban areas, is asbestos. The brake linings of automobiles also contribute asbestos particles to the air. This substance has long been recognized as an occupational hazard, but has become increasingly present in the urban atmosphere. A number of municipalities have passed specific ordinances to regulate the use of this highly toxic substance.

The effect on human health of the nitrogen oxides and the hydrocarbons is not well known, although nitrogen oxide is said to produce effects similar to those of carbon monoxide in reducing the oxygen capacity of the blood, and certain of the hydrocarbon chemicals are associated with the production of cancer.

Emphysema is the fastest growing cause of death in the United States. Since World War II it has doubled in frequency every five years. Its cause has been intimately associated with smoking as well as with general air pollution. Lung cancer is similarly related, but the precise relationship is still unknown. In England, the rate of air pollution is generally higher than that in the United States, and the death rate in males from emphysema and lung cancer is about twice that of the United States. In efforts to remove the effects of smoking and smoke-related occupations from studies of air pollution, epidemiological surveys of young children have been conducted in Japan, Great Britain, and British Columbia. All showed that children living and going to school in polluted areas had lower pulmonary function and suffered more frequent and more severe respiratory tract infections than children living in relatively unpolluted areas.

Dr. Glenn Seaborg, the Nobel Laureate and former Chairman of the Atomic Energy Commission, has said, "Studies indicate that if air pollution were reduced 50 percent in our major cities, a newborn baby would have an additional three to five years life expectancy; that the same reduction in air pollution would cut death from lung diseases by 25 percent and death and disease from heart and circulatory disorders by 10 to 15 percent, and that all death and disease would be reduced annually by about 4.5 percent with a saving to the nation of at least $2 billion a year." Other scientists question this view and maintain that proof is lacking.

Dr. Henry A. Schroeder, Professor of Physiology at the Trace Element Laboratory at the Dartmouth Medical School, has asked for immediate controls on three trace elements which come from industrial contaminants in the air. One is cadmium, another is lead, and the third is nickel carbonyl.

Medical World News, reporting Dr. Schroeder's concern, stated that "at tissue concentrations found in human adults, cadmium causes high blood pressure in rats. Death rates from cardiovascular diseases are highly correlated with airborne cadmium and cadmium in milk. Lead has been found in vegetation growing along highways in concentrations high enough to abort cows subsisting on the greenery. Nickel carbonyl, which comes from diesel oil, causes lung cancer in animals and exposed workers."

Air pollution produces similar or comparable effects on animals and plants, except that they often show a greater sensitivity to the pollutants than humans do. Dairy cattle in the downwind areas of aluminum factories in New York State have aged and stopped producing milk prematurely as a result of heavy fluorine fumes. Citrus groves have withered around the periphery of the Los Angeles Basin as the pall of pollution has spread outward. Ponderosa Pines in the San Bernardino National Forest 50 miles east of Los Angeles have been killed by the smog of that city. The vegetation covering a 30,000 acre area downwind from a copper smelter in Tennessee has been wiped out.

In January, 1971 the Governor of Maryland appealed to the Environmental Protection Agency to hold hearings on charges that a large stand of Christmas trees in his state had been damaged by air pollution from an electric power generating plant in a neighboring state. In New Jersey, pollution has damaged 36 commercial crops and harmful effects of air pollution have been observed in every county. Rats and mice exposed to laboratory experiments involving sulfur dioxide and ozone have produced results along the lines expected from the results of epidemiological studies. In one interesting investigation laboratory animals developed cancer when they were infected with influenza virus and later exposed to smog containing ozonized gasoline, whereas the animals exposed to either smog or influenza, alone, did not develop lung cancer.

Air pollutants damage a wide variety of materials from synthetic hose to limestone statuary. They corrode metals and erode concrete, discolor or peel paint, cause glass to become brittle, stain and disintegrate cloth, crack leather and rubber. In short, pollutants cause considerable damage not only to human beings but also the things used or valued by humans. The relationship of pollution to morbidity and mortality is the most important concern. As Surgeon General W. H. Stewart (1966) has said, "While cleanliness, comfort, beauty and dignity are among our goals, the safety of people is an absolutely minimum requirement in pollution control."

Effects on Weather and Climate

As indicated in Chapter 4, local weather patterns and regional climates are determined by the interworking of several factors, all largely determined by the energy cycle, of which the primary source is sunlight. The sun's energy reaches the earth in the form of visible and invisible light. Some of this light is absorbed as it passes through the atmosphere, but most reaches the earth's surface where it is absorbed to produce the various phases of the planet's energy cycle. Some of the heat absorbed by the earth is radiated back into the atmosphere where it is trapped by the atmosphere. One of the phenomena that makes the earth habitable for life is the retention of heat by the atmosphere.

The many forms of pollution that man is pumping into the air have had a marked effect on local weather conditions and are beginning to modify global climate. The contrails of high-flying jet aircraft remain as visible clouds denoting their contributions of water vapor to the atmosphere. Observations made in Colorado and Utah in the area traversed by many transcontinental flights showed that from 1950 to 1958, before the jets came into wide use, only an average of 8 percent of the sky was covered with high clouds. From 1965 to 1969 about twice that amount was recorded. In the summer of 1970 a group of scientists, gathered at Williams College at a conference sponsored by the Massachusetts Institute of Technology, concluded that a large fleet of supersonic jets might cause increased stratospheric cloudiness. It was reported that the SSTs might cause water vapor in the atmosphere to increase by 10 percent globally, or by as much as 60 percent over the North Atlantic, where the traffic would probably be heaviest. Moreover, these scientists were not sure about the effects on the climate of the discharge by the SSTs of large quantities of particulate matter that could form an atmospheric smog. Such particulates might occlude sunlight from the sun, or might absorb the radiant heat from the earth.

It is known that above the stratosphere, where the supersonic jets would fly, water vapor from the contrails may remain for as long as 18 months before it mixes with the lower atmosphere. A rise of the stratospheric water vapor content from 2 to 6 parts per million would increase the temperature on the surface of the earth by about 0.9°F. The water vapor in the atmosphere, like the carbon dioxide discharged into it, absorbs the heat radiating back into the atmosphere from the earth. Both the increasing water vapor and the increasing carbon dioxide content of the atmosphere serve as heat traps.

Scientists have been concerned about the role of carbon dioxide in the thermal changes in the atmosphere. This substance occurs naturally in large quantities. The normal cycle is for green plants and trees to take up sizeable quantities of carbon dioxide in the spring and summer. When the leaves drop and vegetable matter decays, the plant cycle is reversed. All animals give off carbon dioxide in the process of respiration. To this natural cycle man has added vast quantities as a result of the burning of fossil fuels—all plant derivatives. The increase in carbon dioxide expected from the burning of fossil duels is 1.8 parts per million per year. However, the increase in the atmosphere is only about 0.7

parts per million per year. The rest must go into the biosphere or, more likely, the largest part goes into the oceans, which contain about 60 times as much carbon dioxide as the atmosphere.

In the air, carbon dioxide absorbs incoming solar radiation, but in small amounts. However, it does trap the long-wave heat radiations back from the earth. This is the frequently termed "greenhouse effect."

Among the perplexing questions are what quantity of carbon dioxide remains in the atmosphere, how long does it stay, and what happens to the remainder. One study conducted from 1958 to 1960 estimates that only about half the carbon dioxide produced by the burning of the fossil fuels in that period remained in the atmosphere. Another report states that only 40 percent or less of the carbon dioxide produced by man between 1964 and 1969 has stayed in the atmosphere. Yet there has been a gradual build up of the total amount of carbon dioxide in the atmosphere. Studies conducted in both Sweden and the United States indicate that carbon dioxide concentrations in the atmosphere increased from 312 to 320 parts per million between 1958 and 1970. This increase of an annual amount of 0.7 parts per million would double the amount of man-made carbon dioxide in the atmosphere in about 23 years. It is estimated that if the trend toward increasing carbon dioxide is not halted before it reaches a quantity of 400 parts per million, there will be an increase in the earth's surface temperature of about 1.4°F. This would be enough to cause considerable change in global climates. One view holds that this trend would cause the melting of polar icecaps with the consequent flooding of many coastal cities. Others believe that we need more knowledge before we can predict, with any certainty, what the future will bring.

Large metropolitan areas, with their great concentration of people, dwellings, and industry, have formed climatic islands. This has produced increased heat radiation, which causes cloud covers and fogs to occur more frequently over and downwind of the cities than in the surrounding countryside. Temperature differences found in cities may be 1° to 3°F. higher than in the surrounding areas. Cloud cover and fog may be as much as 50 to 100 percent greater over the city, and precipitation often is 5 to 10 percent greater. On the other hand, wind movements are generally retarded over urban areas. The increases in cloudiness, fog, and precipitation over and downwind of cities are caused by a variety of factors. The heat radiating upward pushes up the surrounding air to colder levels, producing precipitation. Also, the particulate matter forms clouds and condensation nuclei around which water vapor collects. Detailed studies of 22 urban weather stations during a cold season indicate a trend, in the eastern United States, for precipitation to increase by several percentage points on weekdays. This appears to reflect the changing nature of man's activity from weekdays to weekends.

Other Pollutants in the Air

Two relatively new forms of man-produced pollution have been increasing to the point where almost everyone is aware of them. These are radioactivity and noise pollution. Neither is properly considered a form of air pollution because both are forms of energy. But both may be transmitted through the air and both may cause human health hazards. (Noise pollution is discussed in Chapter 11.)

Radioactive materials have become a great concern since World War II. Since that time a variety of radioactive materials have been released into the atmosphere by tests of defensive devices, accidental release of nuclear materials, or from nuclear power plants. These include strontium 90, cesium 137, iodine 131, tungsten 185, manganese 54, iron 55, rhodium 102, cadmium 109, and krypton 85, to mention only a few of the several hundred radioactive substances known to have been released. Strontium 90 is one of the earliest man-caused radioactive materials to have been analyzed as an increasing hazard in the environment. It has been identified in human bone structures and in both cow and human milk. It has been reported to produce cancer.

One of the difficulties associated with radioactive materials is their long duration. For example, krypton 85 has a half-life of about 10 years.

Laboratory experiments on mammals and circumstantial evidence from human beings indicate that the germ cells and embryos are more sensitive to radiation than other body cells. The Atomic Bomb Casualty Committee established by the United Nations reported on studies of 1613 children whose mothers were pregnant at the time of the bombings of Hiroshima and Nagasaki. The incidence of mental retardation found in the children increased proportionately with the dose of radiation to which the mother had been exposed. The degrees of exposure are measured in rads. Thirty-six percent of the children born to mothers who had received doses of 200 rads or more were mentally retarded. The group whose mothers had received 100 to 200 rads had a rate of 9.3 percent of mental retardation. Because of widespread fear by the public of hazards from radioactive materials, no one wants to have a nuclear-powered electrical plant located near him. This creates serious problems for our expanding economy and ways are being sought to meet the growing demands for power with a minimum disruption of the environment.

Air Pollution Control: Problems and Progress

Criteria

With the present state of our knowledge it is difficult to establish air quality criteria, ambient air standards, and emission standards that have a significant relevance to health hazards and cost/benefit factors for a given locality over a given period of time. Only enlightened estimates can be provided at this time, but it is only on these estimates that air quality criteria can now be based.

The American Conference of Governmental Hygienists has recommended a "Threshold Limit Value" for the major air pollutants. (See Table 7.3.) These are guidelines of conditions considered safe for eight hours. A few states have refined further these criteria pending adoption of national standards.

Table 7.3 Threshold Limit Values for Major Air Pollutants

	TLV*	State†
Carbon monoxide	50 ppm	25 ppm (Pa.‡) 30 ppm (Calif.)
Sulfur oxides	5 ppm	0.1 ppm (N.Y.‡) 0.25 ppm (Calif.) (Mont.)
Hydrocarbons	500 ppm	none
Particulate matter	15,000μGm/m‡	500μGm/m‡ (Pa.) 125μGm/m‡ (Tex.)
Nitrogen oxides	5 ppm	0.25 ppm (Calif.§) 0.1 ppm (Colo.§)
Oxidant	0.1 ppm	0.15 ppm (Calif.§) 0.1 ppm (Colo.§)

*TLV—Threshold Limit Value for eight hours.
†Standards adopted by states.
‡Considered safe for 24-hour average.
§Considered safe for one hour.

These recommended standards are examples of the criteria that are being developed. The Air Quality Act of 1967 permits the National Air Pollution Control Administration of the Department of Health, Education, and Welfare to establish air quality control regions throughout the United States. By 1968, 32 regions had been so designated, and an additional 25 regions were proposed in 1969. The formal criteria designations of the United States government have been revised several times and are still under review.

The Air Quality Act of 1967 also provides for a series of advisory panels to aid in the establishment, adoption, and enforcement of the criteria. At present, wide variation exists among the criteria of local, state, and federal agencies, as well as from region to region. In the instances where no local criteria are adopted, the Department of Health, Education, and Welfare has the authority to impose standards. This has resulted in confusing and complex bureaucratic regulations. An industry might have to meet ten local, four state, and two federal air pollution abatement requirements before formal approval is given to build a new plant. The establishment of the Environmental Protection Agency has greatly improved this situation. This is discussed in Chapter 13.

Regulation and Enforcement of Antipollution Measures

The State of California, with far more automobiles registered than any other state, has pioneered in programs to analyze and regulate the pollutant emissions of its motor vehicles. The year 1940 is sometimes chosen as a landmark year for Los Angeles as far as clean air goes. In that year gasoline consumption in the Los Angeles area was 1.9 million gallons per day. By 1970 the consumption was up to 8.5 million gallons. It is estimated that at the present rate of increase in automobile usage the consumption will be 11 million gallons per day by 1980 and will reach 15 million gallons per day by the year 2000. This consumption of gasoline provides an indirect indication of the increase in pollutants emitted by automobiles. The voters have established the strictest regulations of any state. In 1969, 60 faculty members of the University of California at Los Angeles Medical School passed a series of recommendations urging residents to move out of certain areas in the Los Angeles region because of the dangerous air pollution conditions. In several California communities both private and municipal organizations have promoted experimental use of vehicles powered by engines using nonpetroleum fuels, such as liquid natural gas, propane gas, and electricity.

A number of municipalities have instituted strict regulations on the sulfur content of fossil fuels burned within their political boundaries, dramatically reducing the sulfur oxides released into the air. In 1969, for example, New York City reduced the sulfur dioxide content of the air by half. The same city reported general improvement in most major forms of pollutants, enabling the Air Resources Department to report that in 1970 there were *only* 80 days in which the air quality was rated "unhealthy." This represents about a 30-percent reduction from the 144 "unhealthy" days in 1969, which indicates it is an effective program although it still leaves a great deal—nearly three months—of unhealthy air. London and many other large metropolitan areas have made dramatic and substantial improvements in cleaning up a number of pollutants. Formerly 80 percent of London's smoke came from domestic chimneys. The British Clean Air Act of 1956 was so simple and modest a measure that few thought it would produce the desired result. It merely declared certain areas as smokeless zones, banning the burning of smoky fuels and subsidizing 70 percent of the cost of changing to smokeless fuels. The results have been astounding. The remarkable decrease in particulate pollution has given London many more clear days in the winter season and has reduced the number of deaths above average during recent severe fogs.

Improvements such as these are not without complications. Generally when the smoky, dirty particulates are removed, giving clearer air and allowing more sunlight to reach the surface of the earth, the photochemical pollutants produced by sunlight reacting with auto-exhaust chemicals increase under these conditions. Similarly, when many communities have made drastic changes (reduction) in one source of pollutant another becomes more prolific than before. Average conditions vary from locality to locality, depending on weather and the sources of pollutants. At the end of 1970 the U.S. Congress passed the 1970 National Air Quality Standards Act which represents the stiffest antipollution law ever passed. Enforcement of the letter of the law, which now seems reasonably assured by the establishment of an effective Environmental Protection Agency, should go far toward providing dramatically cleaner air in the 1970s. The Congress noted that "many facilities will require major investments in new technology and new processes. Some facilities will need altered operating procedures or a change of fuels. Some facilities may be closed."

Thus, the Congress has given a firm mandate to the automobile industry to produce a pollution-free internal combustion engine; and a signal to the electric power producers, the paper and steel industries, and all others who have been employing inadequate technology to make urgently needed changes.

The growing environmental conscience and understanding of ecological principles has resulted in a growing determination to improve the quality of our atmosphere. Thus, private citizens, government officials at virtually every level, and representatives in the United Nations are all working to bring about cleaner air for better living. There are many difficulties to be overcome, but substantial programs have been started.

The magnitude of the job is impressive. The cost may seem staggering. The manpower requirements are, at the least, difficult to fulfill. However, the important fact is that the 1970s are being thought of as the decade of the environment and we have made a start.

References

American Association for the Advancement of Science, *Air Conservation.* Washington, D.C., 1965.

American Chemical Society, *Cleaning Our Environment. The Chemical Basis for Action.* Washington, D.C., 1969.

BLADE, ELLIS, and EDWARD F. FERRAND, "Sulfur Dioxide Air Pollution in New York City: Statistical Analysis of Twelve Years." *Jour. Air Pollution Control Association,* 1969, pp. 873–878.

BORGSTROM, GEORG, *Too Many: The Biological Limitations of Our Earth.* New York: Macmillan, 1969.

BROWN, HARRISON, *The Challenge of Man's Future.* New York: Viking Compass Edition, 1954.

COLE, LAMONT C., "Can the World be Saved?" *Bioscience,* 1968, **18,** pp. 679–684.

DARLING, FRANK F., *et al.,* "Human-engineering the Planet." *Impact,* 1969, **19**, pp. 105–219.

EISENBUD, MERRIL, "Environmental Protection in the City of New York." *Science,* 1970, **170**, pp. 706–712.

Environmental Pollution Panel, *Restoring the Quality of Our Environment.* Washington, D.C., 1965.

GÖTHE, CARL-JOHON, *et al.,* "Carbon Monoxide Hazard in City Traffic." *Arch. Environ. Health,* 1969, **19,** pp. 310–314.

HICKEY, RICHARD J., *et al.,* "Relationship between Air Pollution and Certain Chronic Disease Death Rates." *Arch. Environ. Health,* 1967, **15,** pp. 728–738.

LANDSBURG, HELMUT E., "Man-Made Climatic Changes." *Science,* 1970, **170,** pp. 1265–1274.

National Academy of Sciences, *Waste Management and Control.* Washington, D.C., 1966.

National Academy of Sciences, *Resources and Man.* San Francisco: W.H. Freeman, 1969.

NEWALL, REGINALD E., "The Global Circulation of Atmospheric Pollutants." *Scientific American,* 1971, **224,** pp. 32–42.

U.S. Public Health Service, *Sulfur Dioxide and Other Sulfur Compounds. A Bibliography with Abstracts.* Washington, D.C., 1965.

U.S. Public Health Service, *Carbon Dioxide. A Bibliography with Abstracts.* Washington, D.C., 1966.

WEAVER, NEILL K., "Atmospheric Contaminants and Standards." *Jour. Occup. Med.,* 1969, **11,** pp. 455–461.

WINKELSTEIN, WARREN, Jr., *et al.,* "The Relationship of Air Pollution and Economic Status to Total Mortality and Selected Respiratory System Mortality in Man." *Arch. Environ. Health,* 1968, **15,** pp. 728–738.

WINKELSTEIN, WARREN, JR., and SEYMOUR KANTOR, "Respiratory Symptoms and Air Pollution in an Urban Population." *Arch. Environ. Health,* 1969, **18,** pp. 760–767.

8

Water: Life Blood of the Land

MARION T. JACKSON
Indiana State University

Dr. Marion T. Jackson is Professor of Life Sciences at Indiana State University where he covers such topics as general ecology, plant ecology, biogeography, and man and environment. He received his Ph.D. in plant ecology from Purdue University in 1964.

His research interests include forest ecology, phenology, microclimatology, and natural area preservation. He has authored several scientific articles and has contributed chapters to two books.

The earth holds a silver treasure,
cupped between ocean bed and tenting sky.
Forever the heavens spend it,
in the showers that refresh our temperate lands,
the torrents that sluice the tropics . . .
Yet none is lost;
in vast convection our water is returned,
from soil to sky, and sky to soil,
and back again to fall as pure as blessing.
There was never less; there could never be more.
A mighty mercy on which life depends,
for all its glittering shifts
water is constant.

DONALD CULROSE PEATTIE AND NOEL PEATTIE
A *Cup of Sky*

Throughout his brief history on earth, man has depended on three basic natural resources for his survival: air, land, and water. Of these three, none has been more abused and neglected by man than water. Yet without water, the land produces nothing; without water, life vanishes from the face of the earth. Civilizations arose and have prospered for millennia where water was abundant. The Nile Valley is a prime example. When water disappeared, civilizations withered and died. Mesopotamia is a classic example.

Despite his complete dependence on water, man's record in managing and conserving this precious resource has been dismal indeed. His drainage, lumbering, plowing, and paving have accelerated the flow of rainfall from the land to the sea, ravaging his floodplain cities and farms in the process. He has contaminated the streams and lakes with wastes of every description to a point that water endangers his own survival. He has used the vast oceans as the ultimate sump to such an extent that the very survival of the biosphere may be in question. In short, he has violated almost every precept of sound water stewardship.

In earlier times, this indifference mattered little. The atmosphere was thought inexhaustible; water was plentiful, and soil was considered expendable. There was always clean water upstream and new land just beyond, in the wilderness ahead. Floods were accepted as seasonal nuisances since there was relatively little of value for them to destroy. Today, however, as we approach the end of the twentieth century, this indifference has changed to serious concern. There is scarcely a country or a section of the biosphere that is not faced with critical problems with respect to the conservation of water and related land resources. Indeed, the solution to the problems of water quantity and quality is likely a key to future survival.

The Nature of Water

> The fall of dropping water wears away the stone.
>
> LUCRETIUS,
> *De rerum natura I*
> 1st Century B.C.

Helen Keller presumably learned her first word—water—by the simple acts of her teacher pumping the cool liquid over the child's hand, allowing her to taste its refreshing goodness, then tracing the word "water" with the child's finger. Her sight, hearing, and speech missing, Helen Keller's instruction began through the senses of touch and taste. Although the feel of a warm shower, the taste of cool water, the sight of a clean lake, and the sound of a rushing faucet are all routine sensations for most Americans, relatively few persons really understand the chemical and physical properties of this familiar fluid. What is this magic liquid called water?

As early as 1781, Lavoisier, a Frenchman, and Cavendish, an Englishman, separately discovered the chemical composition of water. Both scientists revealed that two atoms of hydrogen are combined with one atom of oxgyen to make this transparent fluid, as is expressed by the best-known of all chemical formulas—H_2O. In reality, the formula is not that simple. More recently it has been discovered that hydrogen and oxygen each have two additional isotopes which differ in molecular weight from the parent elements. These six isotopes plus various ions, which result from the addition or removal of an electron, make 33 possible combinations or kinds of the substances called water. One isotope of hydrogen, deuterium (discovered by Harold Urey in 1934), has twice the atomic weight of hydrogen. It combines with oxygen into D_2O or "heavy water" which was essential in the development of the early atomic bombs and is widely employed as a tracer in biological and chemical research. It is of interest that the ratio of D_2O to H_2O has apparently remained the same for long periods of time. If water stored in glacial ice for thousands of years is compared with that in the cell sap of a corn plant today, the proportion of the two substances is essentially equal.

No other substance behaves as does water in response to temperature changes. Although it is present on earth in all three physical states and moves from liquid to gas or from liquid to solid with amazing ease, it decreases in density as it changes from liquid to solid. Fresh water is the only known substance to reach its greatest density at a few degrees above the freezing point (4° C). Upon freezing at 0° C, water increases about one-ninth in volume, thereby lowering the density. The volume increase apparently results from the ability of water molecules to form clusters which do not mesh perfectly when crystallization into ice occurs. Spaces between clusters of molecules serve both to reduce density and to increase volume. Because of this phenomenon ice floats on the water surface, thereby serving as an insulating layer which prevents cooling of the underlying water. If it were not for these peculiar density-volume relationships, bodies of water would freeze from the bottom up and destroy the habitat for aquatic life. The same physical property causes the death of living cells when the water contained within freezes and expands, rupturing the cell wall or membrane. Sea water both freezes and reaches its greatest density at about –2° C. As a result, when sea water cools, it sinks and causes a mixing action known as upwelling—a process that serves to recharge nutrients in surface layers, which, in turn, increases biological productivity.

The amount of heat energy and time necessary to heat water is understood in a general way by any housewife when she observes that "the watched pot never boils." This resistance of water to temperature change obtains from its high specific heat, which is listed as unity for comparative purposes. Most solids and other liquids require much less heat energy per gram to elevate their temperatures. Consequently, they have specific heats below unity. By designating the specific heat of water as one, the amount of heat energy needed to raise the temperature of 1 gram of water in the liquid state by 1° C automatically becomes 1 calorie. Although this relationship is not linear throughout the temperature range, it is closely approximate.

The changes of physical states of water require vastly greater energy inputs. For example, it takes 79 calories to change 1 gram of ice at 0° C to water at the same temperature; whereas, to convert 1 gram of liquid water at 100° C to steam at the same temperature requires 539 calories of heat energy. Many other simple compounds containing hydrogen (for example, ammonia and methane) have low specific heats and vaporize at lower temperatures. Because water boils at such a high temperature (100° C or 212° F) and requires so much energy to change to vapor, water is retained on earth. Otherwise, all the liquid water on earth would have evaporated long ago. With the possible exception of mercury, water is the only known substance that remains liquid throughout the usual temperature range occurring on earth.

The heat-absorbing capacity of water without an appreciable temperature change enables the oceans of the world to serve as the global thermostat. Solid materials generally require several times less heat energy to raise their temperatures. This difference in the ability of a scorching concrete sidewalk and a cool puddle of water to absorb and retain heat is readily detected by any barefoot boy on a hot summer afternoon. Continents and oceans behave in similar fashion. Land masses both warm up faster and cool more rapidly than do ocean areas; consequently, they experience greater temperature ranges. In addition, as one part of the ocean is heated differentially to another, currents are established which redistribute the heat energy much as warmer water in the center of a tea kettle moves throughout the vessel. Corresponding differences in heating of air masses over land and sea create barometric pressure differences which drive our weather systems throughout the world. This free circulation of heat energy in the interconnected seas and in the atmosphere keeps the entire biosphere within a temperature range compatible to living organisms. Without the temperature modification afforded by water, the planet earth would be just another dead cosmic body.

One of the most remarkable properties of water is its ability to dissolve almost all other substances. Water is the nearest approximation to a universal solvent; it can dissolve, to some degree, all ordinary substances found in soil and rocks. Beginning students frequently fail to understand why the sea is salty. Almost all elements have accumulated in the oceans because of the tendency of water to erode all land masses to base level, dissolving some materials in the process and eventually carrying all to the sea. Chlorine and sodium, the constituents of ordinary table salt, are more generally available and more readily dissolved than other elements of the earth's crust. As such, they are the most abundant elements in sea water, except oxygen and hydrogen which comprise the water itself. Other elements are present in lesser quantities, their abundance reflecting primarily their relative content within the land masses and their relative solubility and susceptibility to water transport. Gold, for example, is present at the average concentration of only 37 pounds per cubic mile of sea water; whereas, the same water volume contains nearly 2 million tons of calcium.

The average salinity of the oceans is about 3.5 percent, which does not approach the salinity of inland saline waters, such as the Dead Sea and Great Salt Lake. The latter is the remnant of a once larger inland sea that largely evaporated in post-Pleistocene times, leaving a lake averaging 28 percent salinity and bordered by the extensive Bonneville Salt Flats, which are part of the barren Great Salt Lake Desert.

Water and its attendant organisms also have an enormous capacity to dissolve and decompose organic matter. For this reason, waters are used everywhere to carry wastes. However, water is seldom changed chemically by all the materials that are emptied into it. When it evaporates, almost all impurities, such as microorganisms, sediment, and salt, are left behind. Except for this remarkable ability to cleanse itself, water would become permanently contaminated and, aquatic, as well as human, life would soon become impossible. Water surfaces are also active interfaces for the absorption of atmospheric gases. Nitrogen gas (N_2) becomes dissolved in water but serves no important function in the water. Both oxygen and carbon dioxide are readily exchanged between air and water to become the raw materials for respiration and photosynthesis in aquatic systems.

The high degree of light transmissivity in clear water increases the depth of the photosynthetic or euphotic zone. Deep, cold, oligotrophic lakes, such as Crater Lake in Oregon, have aquatic mosses growing at more than 400 feet below the surface. In more fertile waters, an abundance of phytoplankton in the surface layers reduces light penetration. Transmission of different wavelengths of light also influences water color. Short blue and violet wavelengths penetrate most deeply in waters with few dissolved minerals, giving the indigo color to youthful lakes, such as Crater Lake.

Water has the greatest surface tension of all ordinary liquids. Surface tension is the ability of the molecules of a liquid to cohere to each other or to adhere to a contrasting material. Cohesion permits rain to fall as drops rather than as a spray of individual molecules, and provides water surfaces with a thin film that is the center of considerable biological activity. Adhesion permits water to cling to solid surfaces and to move by capillarity and by filling the interstices of solid materials such as rocks and soil. Adhesion is a property of vital importance in the formation of ground water.

The chemical and physical properties of water seem to have been designed purposefully to make life on earth possible. But it obviously was not water that was shaped to meet the needs of life. Rather, life originated in the sea and was shaped by millions of years of evolution to become almost perfectly adapted to the nature of water.

Water in Nature

> All the rivers run into the sea;
> yet the sea is not full;
> unto the place from whence the rivers come,
> thither they return again.
>
> *Ecclesiastes*, 1:7

During the Middle Ages it was commonly believed that water flowed from the center of the earth in a massive, never-ending stream. Today, the pathways followed by water in its circuit from sea to atmosphere to land to sea have been charted by scientists and are basic to an understanding of contemporary water management problems. The ocean is the main reservoir; the land is a major recipient; and man is now the principal beneficiary of these global water exchanges of incredible proportions.

These collective movements of water, as it circulates throughout the planetary ecosystem or biosphere, are known as the *hydrologic cycle.* The hydrologic cycle is, in effect, a massive heat exchange system powered by the sun, since about one-sixth of the total solar energy reaching the earth is dissipated in evaporating water. This latent solar energy is stored in the atmosphere, and therewithin released as the water vapor condenses and is precipitated.

In its never-ending journey, water is evaporated from the earth's surfaces and carried over land by the winds (Figure 8.1). About one-sixth (17 percent) of the total water evaporated by the sun's action is obtained from the land, fresh water areas, soils, and transpiring plants (see Table 8.1). The remaining five-sixths (83 percent) is evaporated from the oceans which cover nearly 71 percent of the earth's surface. When atmospheric conditions are suitable, the water vapor condenses and falls to the earth as rain, snow, sleet, or hail. Only about one-fourth of the approximately 265 cubic miles of water evaporated every day falls on land surfaces, the remainder is precipitated back to sea. Average annual rainfall over oceanic areas is about 6 inches greater than the average of 30 inches that falls on the continents. If all precipitation on land areas were uniform—which obviously is not the case, either in time or quantity—and if rainfall on land (25 percent of the earth's total) did not exceed evaporation from land surfaces (17 percent of the world total), the entire land surface of the earth would be in desert or near-desert condition. On the other hand, evaporation from the world's oceans (83 percent of the global total) exceeds the amount received at sea from precipitation (75 percent of the total). This 8 percent differential is what flows from land to sea as runoff, and which is used enroute by man as his major fresh water source of supply.

About two-thirds of land precipitation is held as ice, snow, in reservoirs, as soil moisture, or it serves to recharge groundwater supplies. The remaining one-third returns to the oceans via the streams and rivers in a relatively short

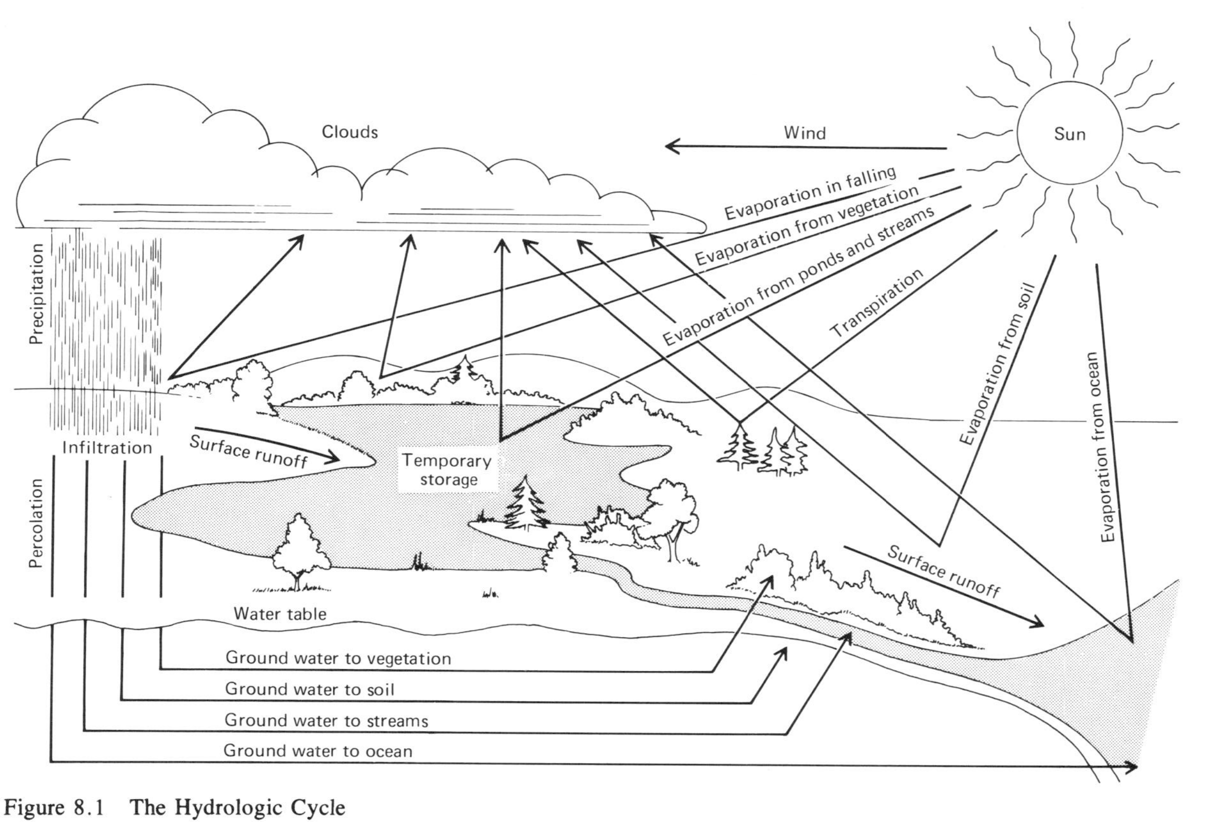

Figure 8.1 The Hydrologic Cycle
SOURCE: Adapted from Raymond F. Dasmann, *Environmental Conservation* (New York: John Wiley & Sons, Inc., 1968).

Table 8.1 The World's Water Budget on an Annual Basis (all quantities in cubic miles)

Partition	World Water Volume	Percentage of Total	U.S.A. Water Volume
Surface water			
Fresh-water lakes	30,000	0.009	4,800
Saline lakes and inland seas	25,000	0.008	2,400
Rivers and streams	300	0.0001	17
Total	55,300	0.017	7,217
Subsurface water			
Soil moisture	16,000	0.005	1,080
Ground water	2,000,000	0.62	144,000
Total	2,016,000	0.625	145,080
Ice caps amd glaciers	7,000,000	2.15	9,600
Atmosphere	3,100	0.001	72
Oceans	317,000,000	97.2	—
Totals (approx.)	326,000,000	100.	9,672
Annual evaporation			
From land	13,400	0.004	960
From oceans	82,600	0.025	—
Total	96,000	0.029	960
Annual precipitation			
On land	23,000	0.007	1,440
On oceans	73,000	0.022	—
Total	96,000	0.29	1,440
Runoff of all rivers			
Water yield per year	9,600	0.003	480

time. (It is of interest that about 25 percent of all runoff of the world is carried by just two rivers, the Amazon and the Congo!) This excess influx from streams (8 percent of the world total) almost exactly balances the excess evaporation from the sea, so that the cycle repeats endlessly and a near-perfect land-ocean water balance is maintained during the usual climatic regime. If this were not the case, sea level would continually drop due to the excess of evaporation over precipitation.

During periods of climatic perturbations, such as the Pleistocene, the earth's water balance is drastically altered. In the cooler pluvial climates of the glacial maximum, the great ice caps that grew in North America and Eurasia caused the sea level to drop about 330 feet at the height of the last large-scale advance some 17,000 years ago. Should the mean world temperature increase by as little as 3° C, the present world ice caps would melt and release the 7 million cubic miles of stored water, inundating most of the major coastal cities of the world in the process.

In polar areas, the land precipitation that does not quickly return to sea largely accumulates as seasonal snow fields which may consolidate into permanent ice packs or glaciers. What soil is present is underlain with permafrost which prevents water entry. In temperate and tropical regions, however, most of two-thirds of the precipitation that does not run off enters the soil to recharge soil moisture and groundwater supplies. Such soils are porous mediums which accept precipitation, allowing some to pass through to the permanently saturated zone below, while retaining some like a giant sponge. The amount of water accepted and held varies greatly, depending upon the vegetative cover, depth and kind of organic layers on top of the soil, and the soil texture and slope of the land. When the soil becomes saturated from precipitation, additional water can enter only at the rate at which water percolates downward due to gravitational force. Precipitation on saturated soils in excess of the percolation rate runs off the land to streams and rivers to be returned to the sea.

The water within the soil profile is either picked up by plant roots, moves downward as gravitational water, or is held against gravity in the soil interstices as capillary water by the forces of adhesion and cohesion. These films of capillary water provide necessary moisture for vegetation between rains. Permanent wilting of plants occurs when the capillary water is exhausted and only water in the vapor state or hygroscopic water remains in the soil. Even most hygroscopic water is lost from the soil during periods of extreme drought or in very arid regions. Although coarse textured soils such as sands or sandy loams accept more water by infiltration of precipitation, relatively little is held within the root zone. Fine textured clay and clay loam soils experience a greater percentage of runoff, but hold more water against gravity. Water availability decreases in very fine textured soils as moisture is tied up within the clay lattices.

The gravitational water continues downward through the soil until it reaches the water table, which is the top of the permanently saturated zone or groundwater layer. In moist climates, the water table lies just a few feet below the soil surface and may be reached easily with a dug well. In some areas, perched or hanging water tables are locally common when an impervious stratum holds a shallow layer of water at some distance above the true groundwater. Shallow wells in such areas seasonally run dry as the suspended groundwater is quickly exhausted. Very dry regions may have no water table within the unconsolidated material, but water may be held in deep strata under impervious rocks. The enormous pressure exerted by the overlying rocks causes wells drilled into these subterranean aquifers (water-bearing substrates) to spout water several feet into the air. Such artesian wells continue to flow until the pressure is released. Artesian wells were common in the Dakotas, Florida, and other areas in years past. Recently glaciated areas, such as in North Central United States, frequently have large deposits of permanently saturated sands and gravels at rather shallow depths. These aquifers contain some of the best groundwater supplies in the world. Limestone substrates under hilly topography are the best sources of fresh water springs. Water moves laterally through rock fissures and caves to flow freely at the surface where the rock outcrops. The Ozark Mountains have some of the finest springs in the world. The Current River originates at Big Spring in Missouri where 60 million gallons of cold spring water gush daily from a limestone cliff. Lateral seepage or flow of spring water into stream courses helps maintain stream flow during dry periods or in arid regions.

The complete evaporation-precipitation cycle spans a very brief period in terms of average replacement time of the water vapor in the atmosphere. Since the total amount of water present in the atmosphere at any given time is estimated at about 3100 cubic miles (Table 8.1), and the daily precipitation is about 265 cubic miles, on the average all the water vapor in the air falls and is replaced about once in 12 days. Different water circuits within the hydrosphere move at varying paces, however. Water evaporated from the Pacific may fall back to sea within minutes as an oceanic rainshower. In another situation, water from a violent summer thunderstorm in Indiana may reach the Gulf of Mexico in a few days. Michigan rain may soak into the soil, percolate to the water table and be held for decades before it is pumped from the ground by a suburban Detroiter, then take additional decades to move from his automatic washer through the Great Lakes to the Atlantic. Yet another drop of water may fall as a snowflake on the Greenland ice cap and be imprisoned in a glacier for millennia before the iceberg calves into the North Atlantic and melts in the warm Gulf Stream current. The important thing is that water does move throughout the biosphere, and that man's water supply is continually replenished.

The annual water budget for both the world and the United States is itemized in Table 8.1. The vast majority (97.2 percent) of the earth's water is stored in the oceans in a form that is nonusable for most human needs without desalinization. Of the remaining 2.8 percent that comprises the fresh water supply of the world, 2.15 percent is immobilized in ice caps and glaciers. With present technology, water contained as glacial ice is essentially unavailable for wide scale use. Only 0.65 percent of the 326 million cubic miles of water on earth remains for potential use to man, with over 95 percent of this remaining fraction being held as ground water. Subtracting that in saline lakes and inland seas (0.008 percent), which is generally less readily usable even than ocean water, and the soil moisture which is extractable in quantity only by plants, only the 0.0091 percent or 30,300 cubic miles held in fresh water lakes and streams is potentially available for man's use at present. Upon this slender, silver, liquid thread dangles all terrestrial life of the biosphere, including man. It is the endless cycle of sea to cloud to land to man, endlessly repetitive and fortunately cleansing, rather than the total quantity, that makes life possible on the planet earth.

Water and Life

> We are tied to the ocean.
> And when we go back to the sea. . .
> we are going back from whence we came.
>
> JOHN F. KENNEDY

With the photographs from the relatively recent space flights, man is able for the first time to view the planet earth in proper perspective. The most striking features of these photographs are the abundant water and the evidence of water in motion. The deep blue ocean expanses and red-brown continents softened by atmospheric haze are in vivid contrast to the wreaths of white clouds of the enormous weather systems. This is a striking visual portrayal of the hydrologic cycle in operation. Perhaps only from his observation platform in space can man fully grasp the true meaning and concept of the biosphere as a functioning unit. Viewed in such manner, man now understands that water is truly the life blood of the earth.

The contrast is obvious when the moon is viewed from the same vantage point. The solar glare in the absence of an atmosphere outlines the lifeless crags and craters that have never known the erosional touch of water. Except for water, earthscape becomes moonscape.

The utter human dependence on water is exemplified by the fact that man can survive for five weeks without food, but he dies after only five days without water. Water is part of the basic structure of all cells and tissues. Our bodies average 70 percent water and vary from 2 percent for tooth enamel, to 22 percent for bone, to 75 percent for muscle and 83 percent for kidney tissue. Our minimal daily need is a quart and a half just to maintain our water balance. If the water supply of a human being is cut off for just a few days, he will die of dehydration long before the water is completely evaporated from his body. Water vapor is lost with every breath, and we cannot avoid breathing.

Plants move enormous quantities of water through the organism relative to the amount actually incorporated into tissues by the various growth processes. Usually no more than 0.1 to 0.3 percent of the water taken in by the roots is actually stored in the plant within oven-dried tissue. For example, an acre of corn may transpire 4000 gallons of water in a single day, or 500,000 gallons (4,300,000 pounds) in a growth season, yet yield only 8000 pounds dry weight of grain. The dried fodder would weigh even less. Hardwood forest trees will easily pump over one acre-inch of water from the soil per month during normal summer weather.

The transpiration stream cools the plant by dissipating the solar energy impinging on the leaves in vaporizing the large quantities of water moving through the plant body. However, transpiration in excess of the amount useful in cooling the leaves may become detrimental to the plant. When this occurs, an unfavorable water balance is created, that is, losses are greater than the intake. Not only does growth stop under such conditions, but the protoplasm may be desiccated to the point that leaves are killed. The exceptionally tall coastal redwoods of California are able to overcome a potentially unfavorable water balance because fogs move in from the Pacific and prevent transpiration losses from becoming excessive. In fact, the trees would not be able to lift water 350 feet to the needle tips in defiance of gravity if it were not for the exceptional ability of water molecules to cohere and to adhere to the cell walls of the small diameter tracheids in the xylem.

Water is also a raw material used in photosynthesis—the elemental process of food manufacture in the biosphere. Plants have the capacity to convert six molecules each of carbon dioxide and water to one molecule of sugar plus six molecules of oxygen. Chlorophyll in the leaf catalyzes this reaction in the presence of sunlight which serves as the energy source. By using isotopes of oxygen as a label, it has been demonstrated that the oxygen stored in the plant carbohydrates by the photosynthetic process originated from the water rather than from the carbon dioxide. Thus, water contributes in an additional way to the energy available at the first trophic level of most food chains in the ecosystem.

In addition to its roles in growth and energy storage in plants and animals, water carries the required metabolic nutrients into and through organisms. Because of its excellent solvent qualities, phosphorus, nitrates, potassium, and almost all other required macrometabolic and micrometabolic nutrients enter organisms in solution through the roots, cell walls, or from ingested food in the case of animals. The excellent solvent property of water contributes to its efficiency in removing wastes and other metabolic by-products from living organisms. In the absence of sufficient water moving through the excretory organs, the internal conditions of an organism would soon become too contaminated to sustain life.

Adaptations to water stress are best expressed in desert forms. Most species of both plants and animals are either drought-escaping or drought-enduring. Nocturnalism in animals and dry-season dormancy in plants are prevalent drought-escaping mechanisms. The kangaroo rat never drinks water, relying instead on metabolic water obtained from dry seeds. Since most excretory wastes are resorbed, only concentrated urine and very dry fecal pellets are discharged. The drought-enduring mechanisms of cacti are well known. Several desert plant species have growth inhibitors on their seeds that are dissolved only by sufficient rainfall to insure germination and growth of seedlings. This reproduction strategy avoids spending the seed "capital" following a light shower.

Life began in water and, even after 2 billion years of evolution, few, if any, species have become entirely divorced from their dependence upon water, at least during some stage of their life cycle. Land plants seemingly require water only to supply the transpiration stream, but on close examination it is evident that many are only amphibious, rather than truly terrestrial. Most mosses and ferns, for example, require water drops or films to effect fertilization, returning figuratively to the water to "breed" as do amphibious animals.

When life first evolved, the primordial organisms were continuously bathed by the nutrient-rich sea water that not only provided food materials, but also supplied the physical conditions crucial for maintaining living protoplasm. This situation continues today for most marine dwellers. Each cell is supplied with a slightly saline medium which maintains a favorable water balance in the protoplast, plus shielding the cell from environmental changes. Terrestrial organisms solved this problem through millions of years of evolution by building cell membranes to contain the protoplasm, an integument to shield the organism from drying, and a circulatory system whose pulse beat is reminiscent of the rhythm of the sea as it brings the slightly saline, food-supplying medium called blood to each cell. It is of interest that the saline content of the blood is about 0.85 percent of that of sea water, reflecting, perhaps, the reduced salinity of the sea at the time life evolved. Despite 2 billion years of evolution, we have not really escaped our aquatic origin in the primeval sea. We merely carry the remnant of our once-close association with the ocean with us within our water-tight skin.

The Demand for Water

> The problem is not whether water supplies are running out, but whether people are outrunning the supplies. Water supplies have finite limits, but the demand of people on the supplies have no known limit.
>
> RAYMOND L. NACE,
> Research Hydrologist of the U.S. Geological Survey

Geologists and hydrologists tell us that there has been the same quantity of water available on earth throughout all of recent geologic time. Pristine water is brought up by volcanic eruptions which emit more water vapor than any other gas. Water also becomes tied up in rocks as the water of hydration or bound water. Apparently, the water exchanged by the two processes approximately balances, keeping the available world supply at the same level.

As indicated in Table 8.1, only the smallest fraction of the total global supply is readily exploitable for man's use at present levels of technology. The atmosphere, where water occurs as clouds, rain, and water vapor; in lakes, rivers and streams, where rainfall or snow-melt has accumulated; or in the soil and groundwater, beneath where rain or melted snow have percolated downward are the only easily obtained water sources. These combined sources total considerably less than 1 percent of the earth's water supply.

Two basic water supply problems confront man. The first is as old as the human species, that of lack of uniformity in precipitation and fresh water supply. Man's history has been inextricably linked to his search for a dependable water supply. From times immemorial, nomadic man either followed wild game or domestic herds in response to drought cycles or settled in river valleys where water was in plentiful supply. The human population is now too large and many of the preferred habitats too dry to meet water demands from local sources. Since almost all of the habitable parts of the world are populated, man, for the most part, can no longer follow precipitation cycles. We are now bringing water to the people on a scale that would have staggered the imagination only 50 years ago. The insatiable human thirst is the subject of this section.

The second water supply problem is more recent in origin, at least on a global scale, and stems from deteriorating water quality. The reduction of a usable water supply due to pollution is a product of the human population size and man's runaway technological capacity which collectively exceed the carrying capacity of the local ecosystem. Water pollution is considered in detail in the following section.

In the first century A.D. the Romans obtained 300 million gallons of water per day by 14 aqueducts which collectively totaled over 1300 miles. The remnants of these engineering marvels are still standing in Italy today. Lavish use of water for everything from the sumptuous public baths to flushing the copious wastes through the *Cloaca Maxima* was a hallmark of the Roman civilization. The contribution of the deterioration of the water distribution system and the attendant water-related diseases to the eventual fall of the Roman Empire is a moot point, but they most probably were substantial.

The "new Romans" in America not only use water at incredible rates, but they demand that adequate supplies for every conceivable purpose be instantly at their disposal no matter where they choose to live. Present-day water shuffling and water management by American engineers dwarf Roman efforts by several orders of magnitude. Entire river systems have been "tamed." Even the magnificent Columbia has been controlled by dams and other structures to the point that only one 50-mile section remains as a free-flowing stream. The Chicago River was forced by the U.S. Army Corps of Engineers to flow backward to carry Chicago sewage to the Mississippi River rather than allow it to flow into Lake Michigan. And the California Water Project, with a series of 21 dams storing nearly 7 million acre-feet of water, is inadequate to slake the thirst of even part of the Golden State. At this writing, most of the readily accessible water supplies south of Canada are intensively used, and serious preparations have just begun to meet the soaring water demands that are inevitable before 2000 A.D. Engineers and hydrologists have even advanced proposals for transferring water from the Great Lakes, Canada, and Alaska to meet projected needs in western United States.

Americans use water for every conceivable purpose: water for drinking and for growing crops; to process beef and to can peaches; for showering and swimming; for transporting ships and sewage; to generate electricity; to cool houses and nuclear reactors; to clean cars and dishes; to water lawns and dogs; to refine steel and make newsprint; to dilute scotch and to manufacture "Uncola"; to fish in and to ski on; to fight fire and on occasions to quell riots; and even to christen children. Fortunately, water is a renewable resource that can be reused almost without limit. And reuse we must to meet the ever-growing demand.

When we evaluate water demands we commonly think only of the amount of water used directly to meet domestic needs—for drinking, bathing, cleaning, and so on. Far greater quantities are needed to produce the food, clothing, soaps, automobiles, metals, paper, and countless other items used daily in almost every business and household.

Water for transportation, cooling, or recreation is almost totally reusable.

Even water for cleansing and carrying domestic and industrial wastes is reusable with proper treatment. However, water for food production is consumptive in that, once used, ground or soil water can be replenished only by new precipitation and ground water movements via the hydrologic cycle. For example, wheat has a transpiration ratio of 500-to-1, which means that about 500 pounds of water are required for each pound of dry weight of wheat plants. Since the wheat grain is only about half the total weight of the wheat plant, it takes about 1000 pounds of water to produce 1 pound of flour. To provide a 2 1/2 pound loaf of bread, therefore, at least 2500 pounds or about 300 gallons of water are needed.

Meat production requires even larger quantities. Assuming that a steer consumes 25 pounds of alfalfa and 12 gallons of water per day, when transpiration and animal water losses are accounted for, about 2300 gallons are required to supply 1 pound of beef for the daily diet of one person. Adding some 200 gallons to produce the vegetable matter required per person each day, the total daily water requirement to produce food for one person in America becomes about 2500 gallons.

The production of other commodities requires even greater quantities of water. Cotton production for clothing requires about 800,000 gallons per acre. Manufacture of synthetic cloth takes 200 or more gallons per pound. Petroleum refining requires about 775 gallons per barrel, paper manufacture 20-25 gallons per pound, and a ton of finished steel about 30,000 gallons of water.

The United States leads the world in metered water consumption with nearly 400 billion gallons being used daily. This represents a per capita daily use of nearly 2000 gallons. In addition, it has been calculated that agricultural and forest lands require about 14,000 gallons of water daily per person to produce our food and forest products. The annual rainfall for the United States averages about 30 inches, for an average total precipitation yield of around 5000 billion gallons daily. This collective daily per capita water use of about 16,000 gallons represents a total daily water use of nearly 3300 billion gallons for 208 million Americans, or nearly two-thirds of the total daily precipitation received in the United States (exclusive of Hawaii and Alaska). Based on precipitation sources alone, a constant per-capita daily use of 16,000 gallons would allow for a population increase of only about 50 percent to 310 million. Or conversely, if our present U.S. population were stabilized at 208 million, the per capita daily use could be increased to about 24,000 gallons.

When the total water demand per capita is considered, it becomes evident that water supply problems of the future are among the most serious of those facing man. Projected metered water needs for 1980 have been estimated to be 600 billion gallons per day, and a forecast of a daily water demand of 2000 billion gallons by 2000 A.D. is frequently made. As more subhumid and arid lands are irrigated, water use for food and fiber production will also increase substantially.

Obviously, we are not tied to a single use of precipitation as our only source of water. For one thing, water can be and is reused, some of it many times, before it is returned to a river or to the sea. For example, cooling water used in steel refinishing may be recycled several times before being discharged from the plant. Sewage effluent from one city becomes the water supply (after dilution with river water) for a downstream city. It has been estimated that water in the Ohio River is reused an average of four times between Pittsburgh and Cairo, Illinois. As water demands increase, water withdrawal, followed by treatment and reuse, will, of necessity, greatly increase.

Water used in food and forest production usually cannot be reclaimed and reused. Irrigation waters are largely not reusable due to silt and salt uptake from alkaline soils. Some recharge of ground water may result from continued irrigation, but the water table is too deep and evaporation and transpiration losses too great in most arid regions to permit much percolation of irrigation waters.

Another way that the water demand is presently being met is by rapid extraction of ground water. Man has been overdrawing his ground water "bank account" for decades and the rate at which ground water is "mined" to meet current demands is accelerating. Except for certain limestone aquifers and a few situations in which huge beds of glacial gravels contain ground water that moves relatively rapidly, most ground water flows very slowly—usually not more than a few meters per day. As wells are located more closely and pumping techniques become more efficient, groundwater is withdrawn much faster than it can be recharged. For example, the water table in parts of Texas has been dropping at an average rate of nearly 10 feet per year with withdrawal rates in some localities exceeding the recharge rates by 140 times. Nebraska has lowered its fossil water table by 15 feet since World War II, and, even in humid Ohio, some sectors have witnessed a ground water decline of 1 foot per year. In some cases, water has been successfully pumped back into aquifers to maintain ground water levels, but these efforts by no means replace withdrawals.

Non-withdrawal demands for water are probably increasing more rapidly than withdrawal and consumptive uses. Such uses include fresh-water-based transportation; swimming, boating, fishing and other water-based recreation; and streams and lakes for waste water discharge by industry, municipalities, and agriculture. The transportation and recreational uses, though phenomenal in scope, do relatively little to impair water quality. We transport over 100 billion ton-miles of commercial freight on our lakes and rivers each year with few pollution problems unless oil spills occur. Recreational uses of water are even greater. The average American, from baby to oldster, goes swimming about five times a year. Swimming pools, which increase in number by over 70,000 per year, are consumptive users due to high evaporation rates, particularly in the arid areas where they are more common. Most water-based activities, however, permit perpetual use of water with little reduction in quantity or quality. Since there are over 8 million boat owners, 70 million swimmers, and countless fishermen and other water users, water-based recreation is doubtless the largest leisure-time activity in America. As work weeks shorten, these water demands will rapidly accelerate. Fortunately, most recreational uses are largely nonpolluting; however, oil contamination from outboard motors is becoming a serious problem on many lakes.

The non-withdrawal use of water for waste dilution and sewage removal is another story. Such water use has caused rapid aging of Lake Erie, turned the Mississippi River into the "cloaca of mid-America," and has seriously impaired the water quality of the majority of our lakes and streams.

Water Pollution Criteria

> Mother, may I go out to swim?
> Yes, my darling daughter
> Hang your clothes on a hickory limb
> But don't go near the water.
>
> ANONYMOUS

Pure water in the strictest sense never occurs in nature and is produced only under exacting laboratory conditions. Almost all water has small amounts of dissolved minerals and gases which impart its distinctive taste. Since pure water is seldom encountered, it is difficult to decide at what point water is considered to be polluted.

Pollution is usually defined as the presence of undesirable foreign matter in such quantities that the quality of water or its usefulness for beneficial purposes is diminished. Changed physical conditions of water, such as elevated temperatures, are also considered to be pollutants. Manifestations of water pollution include oil slicks, floating solid waste, unsightliness, reduced oxygen levels, altered pH, turbidity, algal mats, weed growth, no swimming signs, decline of game fish, and hepatitis outbreaks.

Perhaps the most appropriate concept of a nonpolluted environment is that it approaches a steady state ecosystem. Pollution introduces perturbations which alter nutrient cycling, energy pathways, and species diversity. An unpolluted environment is not static, but rather approaches equilibrium conditions through complex feedback mechanisms. Nutrient cycling is more nearly complete, preventing fertility buildups that induce algal blooms or increases in other troublesome species. Species diversity is usually higher in unpolluted systems.

Natural pollution of water is common even in wilderness areas. Mountain streams frequently carry ground-up rock fragments called glacial flour which greatly increase their turbidity and reduce light penetration. Volcanic discharges or hot spring effluents may contain large quantities of minerals which substantially alter the waters of streams flowing from such areas. Leaves and other plant materials which blow or fall into lakes and streams add, among other things, tannins, resins and cellulose. High mountain lakes contain excretory wastes from fish and insects, in addition to countless microorganisms. Such pollutants usually are of little concern since they normally do not impair water quality to the point that its aesthetic value or utility is reduced.

The water pollutants that are of increasing concern today largely result from man's use of the landscape. Water pollution increases in direct proportion to the combined proliferation of human population density, energy consumption, and the manufacture of products by our technology. Wastes are discharged into aquatic systems at rates that greatly exceed the capacity of water to cleanse and regenerate itself naturally. This action results in serious water pollution levels. In addition, many synthetic substances, such as resistant pesticides, certain detergents, heavy metals, plastics, and radioactive substances are almost immune to organism attack and breakdown. Since such synthetics are recent additions to aquatic ecosystems, few, if any, organisms have evolved the necessary enzymes for their degradation. Due to their frequently high mobility in ecological systems and their long half-life they tend to circulate and accumulate in the biosphere, thereby intensifying the extent and seriousness of water pollution.

Natural waters have remarkable capacities to purify themselves when organic matter or other degradable pollutants are added. This self-purification process involves biochemical breakdown by microorganisms, flocculation and sedimentation of inorganic solids, oxygenation due to wind action or turbulent flow, and dilution of concentrated chemicals such as acids. The process of recovery from pollution is complex and specific for each aquatic ecosystem and for each combination of pollutants, but the tendency of all systems is toward stabilization at a dynamic equilibrium.

A diagrammatic representation of the physical and biological changes associated with stream pollution is shown in Figure 8.2. Organic additives, such as sewage inflows, represent an ecosystemic perturbation that creates temporary instability. Increased nutrient levels trigger rapid population expansions in a series of microorganisms, including floating algae. Biochemical oxidation of the organic input by the numerous microbes, plus decay of the organisms as they die, rapidly deplete the water of dissolved oxygen. Invertebrate populations also gain ascendency in sequence according to their tolerance to low oxygen levels and other physical conditions resulting from pollution. Tubificid worms (sludge worms) occupy the zone of most intense pollution, followed by *Chironomus* worms (blood worms) and aquatic sow bugs as water conditions improve downstream. With continued stream improvement, clean water invertebrates become dominant and algae and bacterial populations decrease. The purification process serves as corrective feedback which readjusts nutrients, suspended solids, oxygen, species composition, and population sizes to pre-pollution levels.

At one time in our nation's history, families or villages simply moved upstream whenever the water supply became unusable due to contamination or

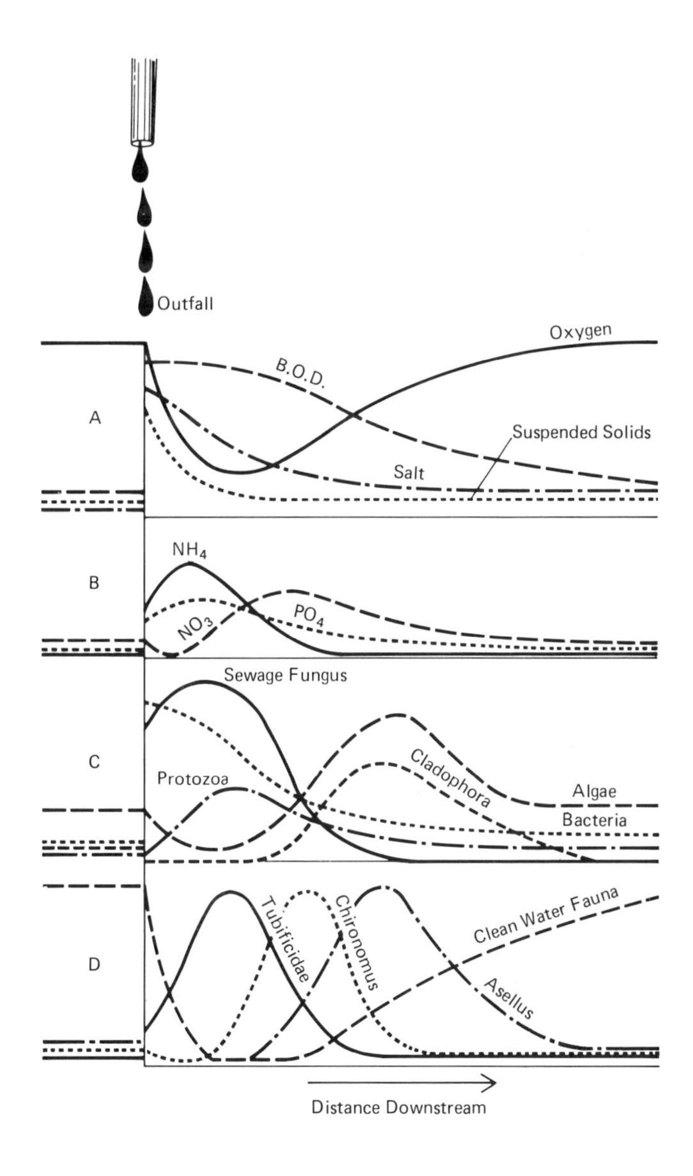

Figure 8.2 Diagrammatic Representation of the Physical and Biological Changes Associated with Stream Pollution

SOURCE: Modified from H.B.N. Hynes, *The Biology of Polluted Waters*, (Liverpool, Eng.: Liverpool U. Press, 1963), p. 94.

disease. It has now become impossible for even a fraction of our population to live upstream from pollution sources as urbanization and technology have increased to the point that reuse of water before the stream can purify itself naturally is inevitable. In heavily populated regions sewage outlets are dangerously close to water supply intakes, and downstream communities are continually treating and using waste water discharged by upstream communities. Waters from rivers, such as the Hudson, that carry excessive pollution loads can no longer be treated economically and flow unused to the sea while the municipalities along their lower shores must look elsewhere for their water supplies. As aquatic system after system becomes polluted beyond use, not only has mankind sacrificed a portion of his readily available fresh water supply, but he has also increased the likelihood that the corrective capabilities of the biosphere may be exceeded and the operation of the hydrologic cycle altered.

At some time, almost every conceivable type of material finds its way into bodies of water and causes water pollution. Such diversity of materials, from automobile bodies to anions, almost defies categorization, but several methods of classification are both convenient and instructive. Pollutants are most commonly listed according to source of contamination, for example, agricultural, industrial, and municipal or urban. Agricultural pollutants include mineral and organic soil particles, animal wastes, fertilizers, pesticides, insecticides, and crop residues. Industrial wastes are extremely varied and commonly include acids, bases, oils, tars, resins, food wastes, steel washings, paper mill wastes, heavy metals, mine tailings, residues from manufacturing plastics and other synthetics, tannery wastes, slaughter-house offal, radioactive materials, and high-temperature coolant waters. Urban areas mainly contribute domestic sewage, household wastes, salt and oil from street washings, detergents, insecticides, and solid wastes.

Less commonly used, though appropriate for many purposes, is a classification based on the size of foreign particles introduced into water. Although a continuum of particle sizes exists, the spectrum is frequently divided into three classes: suspended, colloidal, and dissolved matter. Suspended particles usually are larger than 100 microns and settle out rather quickly due to gravity. Included are soil particles, bacteria, algae, sand, and dusts. An abundance of such particles results in a muddied condition that readily extinguishes light and narrows the photosynthetic zone. Intermediate-sized colloidal particles settle very slowly and are filtered with difficulty. Colloidal clay particles from soil erosion are the most common materials in this size range, although colloidal-sized pollens, viruses, and dusts also contaminate water. Dissolved materials are less than 1 millimicron in diameter. They are either molecules or ions which neither settle nor become lodged by filtering. Special chemical treatment is required for their removal. Examples of dissolved materials include nitrates, carbon or sulfur dioxide, sugar, table salt, antifreeze, phosphates, oxygen, or fluoride. Very large objects also contaminate water, but do not fit conveniently into any of the above three size classes. Such solid wastes as plastic containers, glass bottles, drink cans, tires, wood debris, automobiles, appliances, and wire foul far too many of our streams, lakes and beaches.

Grouping pollutants according to degradability is important in terms of pollution control. Most nondegradable or very slowly degradable pollutants were mentioned earlier and need not be detailed here. In general, products of biological origin are readily degradable; those of manufactured origin, less so. To date, some 500,000 substances have been produced, most of which eventually find their way into water. This number increases by several thousand yearly.

The important point concerning water pollution is not categorizing pollutants, but rather the identification of their end effects through representative, easily administered tests and embarking on workable corrective programs. For far too long the attempted solution to pollution has been dilution. We now need a control program that is aimed at preventing the causes rather than treating the symptoms.

The most useful single indicator of the overall condition of a body of water is the dissolved oxygen content or D.O. Dissolved oxygen is required in the purification process to oxidize organic matter and other pollutants. A high D.O. value indicates that a body of water is capable of rapid self-purification toward a balanced aquatic system. The maximum dissolved oxygen (saturation concentration) at standard temperature and pressure (20° C and 1 atm) is 9 mg of O_2 per liter of water. Relatively unpolluted streams and lakes usually have from 5 to 7 mg of dissolved oxygen/liter. When D.O. values drop to 3 mg/liter or below, serious organic pollution levels are indicated.

The most commonly employed measure of oxidizable organic matter in a quantity of water is the biochemical oxygen demand test or B.O.D. The test itself is a quantitative chemical determination of the amount of oxygen used up by a water sample during a 5-day incubation period at a standard temperature of 20° C. Oxygenated water is added to the samples which are kept in dark conditions for the 5-day period to avoid photosynthetic production of oxygen. The difference between initial and final oxygen levels is a measure of the amount of dissolved oxygen used (B.O.D.) by the biological processes that degrade the organic matter that has entered the water. The utility of this test is to obtain relative amounts of oxygen required to oxidize different pollutants which may enter a given body of water. Residential sewage effluents normally vary between 150 to 250 mg/liter. Milk processing and food canning wastes commonly reach 5000 mg/liter, whereas highly organic wastes from paper mills may have B.O.D. values as high as 15,000 mg/liter. The majority of the B.O.D. load in American waters is of industrial origin. Presently more than 300,000 water-using factories in the United States discharge more than four times as much oxygen-demanding wastes as the entire sewered population.

Waters suspected of being contaminated by fecal material or sewage effluent are examined for the presence of coliform bacteria. These bacteria are not disease-inducing themselves but live in saprophytic association with other intestinal organisms in the colon of man. Their presence in water, however, indicates rather recent fecal contamination and the possibility that pathogenic organisms, which cause infectious hepatitis, typhoid fever, cholera, amoebic dysentery, or other water-transmitted diseases, may also be present. The most common indicator coliform bacteria are *Escherichia coli* and *Aerobacter aerogenes,* although the latter is a non-fecal coliform type. Drinking water is considered safe only if it contains fewer than 10 intestinal bacteria per liter. Water for swimming or non-domestic uses has less rigid coliform standards. It should be stressed that low coliform counts do not automatically imply water safety. Viral diseases in particular have been transmitted by water which meets drinking water standards according to bacterial tests.

Regardless of the measure employed in determining pollution levels, the really important consideration is the effect of pollution on human health and stream environment. As pollution levels increase, corresponding structural changes occur in the stream community as reflected in the kinds, distribution, and diversity of organisms present. Dr. Ruth Patrick of the Philadelphia Academy of Sciences has worked out a pollution index based on the number of species present in each of several groups of organisms. The species are grouped according to their tolerance to polluted conditions such as low oxygen levels. When pollution levels increase, the number of species and individuals in groups which are not tolerant of low oxygen levels is greatly reduced. Conversely, the number of species, as well as the number of individuals of tolerant groups may increase, particularly in the absence of competition from the eliminated species. These relationships are illustrated in Figure 8.2. Reduction in number of species in a stream community or the presence of *Tubifex* worms or carp tells the experienced limnologist more about water conditions than perhaps is determined by any series of laboratory tests.

Effects of Pollution

Man is but a reed,
the most feeble thing in nature . . .
The entire universe need not arm
itself to crush him, . . .
a drop of water suffices to kill him.

PASCAL,
Pensées

Aquatic ecosystems have developmental stages and limited life spans. Newly formed lakes, ponds, and streams characteristically begin as cold, clear, essentially sterile bodies of water. Nutrients eroded naturally from the watershed area gradually increase the fertility of the water. As phosphates, nitrates, and other metabolic nutrients accumulate, a series of organisms colonize the water, gain biological supremacy, then wane in importance as subtle changes occur in the aquatic system. As the system ages, organic and mineral deposits accumulate on the bottom, the water shallows, temperatures increase, oxygen levels decrease, and the lake or pond becomes a marsh. This process of natural aging and biological enrichment is termed *eutrophication*. For a large, deep lake, such as Lake Superior, this process may take millions of years; a small, shallow farm pond may disappear in less than two decades.

As these physical changes occur in the aquatic environment, the biota changes in corresponding fashion. For example, fish populations in a large northern lake may shift from northern pike to bass to perch to carp to mudminnows as the lake matures and shallows.

A major effect of pollution is an acceleration of the eutrophication process. An often-cited example is the changes that have occurred in Lake Erie during the past two decades. The Great Lakes are comparatively very young as they are products of the recent glacial advances. Lake Erie was only in a formative stage in the developmental sequence before the onset of heavy industrialization and urbanization along its shores. Only a few species of fish had successfully colonized its waters. Lake Erie is the fourth largest (9940 square miles) and the shallowest (its average depth is less than 60 feet) of the five Great Lakes. As such, Lake Erie's volume (less than 100 cubic miles) is much too small to dilute the enormous volume of pollutants that it receives from a surrounding population of over 5 million. For example, about 2 billion gallons of seriously polluted water are discharged into the lake daily by the Detroit River alone. Pollution levels have risen very sharply in the post-World War II period when industrial and population growth increased most rapidly. By some estimates, Lake Erie has aged more during the last 20 years due to man-induced changes than would be expected to occur naturally in 5,000 years.

Pollution increases plus invasion of the sea lamprey and alewife (a small relative of the herring) via the St. Lawrence Seaway and the Welland Canal have drastically changed the ecology of the lake. The first species to be seriously affected were the commercially valuable game fish that feed near the top of aquatic food chains. The take of blue pike plummeted from nearly 19 million pounds in 1956 to less than 500 pounds in 1965; walleye catches declined from about 15.5 million pounds to about 800,000 pounds during the same period. There were corresponding rises in catches of carp, smelt, yellow perch, freshwater drum, and other species that feed lower on the food chain.

Reduction of D.O. levels and silting of the lake bottom have caused the virtual disappearance of mayfly larvae, one of the major food sources of desirable fish species. Nutrient enrichment has favored phenomenal growth of huge beds of coarse floating algae which further depletes the oxygen as it dies and decays. According to Dr. Barry Commoner of Washington University at St. Louis, if it were not for a peculiar power of ferric iron to form insoluble complexes in the water with the organic residues from the algae and other pollution sources, in effect, sealing these oxygen-demanding materials in the bottom mud, the oxygen would long since have been depleted and most aquatic species would have been asphyxiated.

Lake Erie is not biologically dead as many have proclaimed. It is in such serious condition, however, that the Federal Water Pollution Control Adminstration has estimated that it will cost $1.1 billion in pollution-control projects in the next 20 years to arrest the accelerated eutrophication. Meanwhile, the Cuyahoga River, which flows into Lake Erie at Cleveland, contains so much filth and oil residues that it caught fire in 1969. A few hundred yards from Cleveland's sewage outfall, a sign proclaims that a "clean beach is a fun beach."

Far and away the greatest source of water pollutants in terms of total quantity is soil erosion from the farms, grazing lands, and construction sites of America. The volume of suspended solids reaching America's waters is more than 700 times as great as the total sewage discharge load. Silt concentrations of 250,000 parts per million have been recorded in corn-belt streams following heavy rains. In other words, the runoff water under such conditions is one-fourth soil. Erosion rates at highway construction sites during a rainstorm average about 10 times greater than for cultivated land, 200 times more than for grassed areas, and 2000 times greater than from forested areas, which occupy similar topographic situations. No instrument or practice devised by man will hold soil as effectively as will plant roots. Water from almost all of the more than 3000 water supply reservoirs in the United States requires special water treatment to remove suspended soil particles.

Silt affects aquatic environments in several ways. It fills channels and reservoirs, reducing their capacity and covering fish spawning beds and bottom-dwelling organisms. The excessive turbidity from suspended clay particles reduces light penetration to a degree that photosynthesis is restricted to the upper water layers. As photosynthesis is reduced, oxygen levels drop and fish kills become probable. Soil pollutants also supply nutrients which enrich waters, thereby causing premature eutrophication. Nitrates, phosphates, pesticides, and other agricultural additives are carried into the water as soil erodes. Nitrates washed from farm fields have reached such levels in the waters of Lake Decatur in central Illinois that a human health hazard has been documented (particularly by causing methemoglobinemia in children). Sediment-laden waters may cause mechanical damage as they pass over the gills of fish and filter-feeding invertebrates such as oysters and scallops. Continual deposition of sediments prevents successful nesting of several fish species by interfering with nest construction or by preventing successful aeration of eggs. Several water pollutants endanger not only aquatic organisms but terrestrial species as well because of their propensity for concentration at higher trophic levels. This was one of the points underscored by Rachel Carson in her book *Silent Spring*, which many scientists feel began the current thrust toward environmental awareness. Certain insecticides, such as DDT, are magnified in concentration at each step in the food chain to an extent that top aquatic carnivores, such as ospreys and eagles, are unable to reproduce. The mechanism of DDT action is an interference with calcium metabolism to such a degree that proper egg shell formation is prevented. The long-lived adults may survive for years since mature individuals are apparently not affected by even high concentrations of DDT, but failure to produce offspring dooms a species as surely as if excessive adult mortality occurred. Fish-eating species, such as mergansers, grebes, bald eagles, and ospreys, are particularly vulnerable because of the relatively greater length of many aquatic food chains. Water concentrations of only 0.02 parts per million DDT in the water of Clear Lake, California, were increased at succeeding trophic levels until grebes, which died from eating contaminated fish, contained 1600 ppm in certain tissues. This represents a magnification of 80,000 times.

Radioactive materials produced by nuclear weapons testing or as waste products from nuclear power generating plants reach water supplies via precipitation or by percolation into groundwater reservoirs. Radioisotopes may also be magnified to dangerous concentrations at higher trophic levels. Strontium 90, for example, may substitute for calcium in bone or other tissue and remain throughout the lifetime of the organism since its half life is 28 years. The ability of organisms to concentrate such materials should be given more careful consideration when setting maximum safe levels for discharging pesticides or radioactive materials into the environment.

Heavy metals, such as lead, arsenic, cadmium, and mercury, are highly toxic and are extremely serious water pollutants. The first two are cumulative poisons which may lead to serious illness or death. Lead may enter aquatic systems from battery manufacturing, lead-bearing paint or, as strange as it seems, from the millions of lead shot which are rained into marshes by duck hunters. Waterfowl contact lead poisoning when they ingest quantities of shot as they strain the bottom mud for invertebrate food. The iron shot now being advocated, though more abrasive to shotgun barrels, is easier on the ecosystem. Lead arsenate sprays are commonly used insecticides which may contaminate water supplies with both lead and arsenic.

Mercury occurs throughout the biosphere in minute concentrations and may even be essential to many organisms, including man, in miniscule quantities. Metallic mercury is used in dental fillings and is not particularly hazardous when ingested. The recent alarm over mercury contamination arose when several cases of mercury poisoning were discovered in certain parts of the world. Symptoms include loss of muscular control and mental retardation. Among other industrial uses mercury has been used in the manufacture of hats (hence, the "mad hatter" of *Alice in Wonderland*) and in the manufacture of chlorine. Mercury discharged into waters in industrial wastes may be converted to methyl mercury

$H_3C{\cdot}Hg{\cdot}CH_3$ by certain bottom-dwelling anaerobic microorganisms. Methylated mercury is highly toxic and is passed through aquatic food chains which often terminate at man's table. Tuna fish were found to contain dangerous concentrations of mercury recently and large shipments were banned from sale. Industrial wastes obviously should be treated for mercury recovery before discharge.

Some of the most serious effects on aquatic organisms are caused by drastic changes in pH levels of streams or lakes. Most bodies of water in midwestern United States are near neutral or slightly alkaline due to the ease of removal of basic cations, such as calcium, sodium, potassium, and magnesium, by erosion of watershed soils. Natural waters have a relatively high buffering capacity which normally resists changes in pH through gaseous exchange with the atmosphere, chemical binding of acidic or basic ions, and so on. However, strip-mine drainage waters are frequently strongly acid from sulfur-bearing rocks which produce sulfuric acid in the presence of water. In the Appalachian region alone some 10,500 miles of streams are seriously affected by runoff from strip-mined lands. Such streams are frequently almost devoid of organisms, other than acidophilic microbes, if the pH drops to 3.5 or lower. Fish venturing into such streams to spawn may suffer severe deterioration and death from contact with strongly acidic waters.

Pollution of waterways by petroleum products is extremely damaging to aquatic life. Since many oil spills are marine, the effects of pollution from this source are discussed in a later section on the sea.

Detergents are one of the most hotly debated pollution topics. A few years ago the major issue concerned degradability by bacterial activity. Soaps, being natural substances, are readily decomposed by microorganisms and cause few problems in aquatic ecosystems. Synthetic detergents, which contain alkyl-benzene-sulfonate (ABS), are very stable and highly resistant to microbiological breakdown. When nondegradable detergents were widely used, water treatment plants became saturated with detergent scum, harbors were churned into mountains of foam by tugboat action, and water from detergent-contaminated wells foamed at the tap. Not only was this aesthetically unappealing, fish growth was sometimes retarded and the surface tension of the water was reduced, creating serious problems for surface organisms. Sale of nondegradable detergents was banned by law and linear alkyl sulfonate (LAS), which is more susceptible to bacterial decomposition, is now used in detergent manufacture.

The present detergent argument concerns the ecological effects of phosphates which were added to detergents to increase cleaning efficiency, particularly in hard water. Many laundry pre-soaks, for example, are quite high in phosphates. A question that needs to be resolved is the actual contribution of high levels of phosphate additions to the accelerated eutrophication of lakes and streams. Phosphorus is a major metabolic nutrient that encourages enormous algal build-ups under certain conditions. Phosphate-rich detergent waters apparently aggravate this situation. Some research findings indicate that unless phosphorus is the limiting factor in the aquatic system in question, the addition of waste waters containing phosphated detergents does not add to the eutrophication problems. Instead, they consider nitrogen to be the culprit in causing algal build-ups. Nonetheless, alternate cleaners which lack phosphate are available, but manufacturers contend, among other things, that these substances are more toxic than phosphated detergents when ingested by children. One might observe that all cleansers should be kept from children unless a thorough mouthwash is in order, then soap should be used!

Thermal additions to water systems also come in for their share of controversy. Approximately two-thirds of all water used in industry is for cooling purposes. Both coal-fired and nuclear-powered electrical generating plants use enormous quantities of water for cooling. Electrical energy is generated by a steam-powered turbine which drives the generator. The difference is in the energy source used to heat the water to produce high pressure steam. Once the steam is spent, it must be recondensed into water—a physical change that gives up 539 calories of heat energy per gram of water. Cool water for condensing the steam is taken from a natural water supply, then re-discharged at a temperature that ranges from 10 to 20° F higher than at intake.

Engineers frequently refer to warm-water discharges into lakes or streams as "thermal enrichment" and view the effects as beneficial rather than detrimental. From a standpoint of conservation of energy, warm effluents should be put to practical use. In some cases, harbors or shipping lanes could be kept ice free or the growth season of agricultural crops could be extended by warm-water irrigation. Some warm-water fish species may be benefited by the warmer discharges.

The ecological changes wrought by such temperature changes are devastating to many aquatic forms, however. Because of their high specific heat, bodies of water change temperature slowly under natural conditions. This gives cold-blooded aquatic forms time to adjust, rather than being exposed to accelerated metabolism or even death as temperatures rapidly increase. Warm water has lower oxygen-holding capacity, forcing fish and other aquatic forms into respiratory stress. Cold-water species, such as lake trout, cannot tolerate temperatures above 77° F, and experience reproductive failure at temperatures above 48° F. Thermal additions that keep a lake at 60° F would exterminate the species even though the adults did not immediately succumb. Hot water discharges may continually impose a barrier across a stream that prevents migration and spawning of temperature-sensitive fish species, such as grayling and salmon, thereby reducing their fresh-water distribution and their breeding success. Furthermore, increased water temperatures often create ecological imbalance due to algal buildups, particularly if nutrient enrichment simultaneously occurs.

Electrical power needs are increasing at about 10 percent per year, creating enormous thermal threats to most water systems in the decades ahead. By 1980, the electrical power industry is expected to require the equivalent of 45 percent of the total fresh water runoff of the United States to cool its turbines and reactors. Safe methods of disposing of this heat energy must be designed. One promising possibility for coastal cities is to use the excess heat energy from power plants in evaporating sea water to alleviate growing fresh water shortages. Cooling towers are in current use, but may create fog or icing problems under certain weather conditions. As power plants proliferate to meet electricity needs, solutions to thermal pollution must follow close behind.

All of the above-mentioned forms of pollution lower water quality and, potentially at least, affect human well-being. The most serious effect of pollution is the direct threat to human health by water-borne diseases, although epidemics of dysentery, typhoid fever, or cholera are no longer prevalent in the United States. It is our generally effective water treatment that reduces the danger of infection, rather than the total eradication of the diseases. Even in this country we are at times walking a very thin tightrope between control and serious outbreaks. Viral diseases, such as infectious hepatitis, are increasingly difficult to contain. On the world scene, diseases and and parasites transmitted by water cause horrible suffering. An estimated 180 million residents of the tropics and subtropics (5 percent of the world population) suffer from schistosomiasis, which is caused by an aquatic blood fluke. Even more are infected with the nematode which causes filariasis. Cholera still takes thousands of lives in India and Bengladesh during the monsoons. A drop of water can still suffice to kill man.

Restoring Water Quality

> The 1970s absolutely must be the years when America pays its debt to the past by reclaiming the purity of its air, its waters, and our living environment.
>
> RICHARD M. NIXON,
> January 1970

A tidy definition of water quality that satisfies all contingencies is as elusive as an all-encompassing definition of water pollution. One reason for this is that the image of water quality, like beauty, changes in terms of the beholder. Dirty effluents may be viewed as increased profits by an industrialist, while the downstream fisherman bristles at the prospect of massive fish kills. Another reason is that water quality standards have relevance only in terms of intended or potential water use. General agreement *has* been reached concerning quality standards for drinking water supplies. Not only do we expect and demand water that is safe from infectious diseases, but it must also flow gin-clear from our taps without objectionable tastes or odors. Less clearly specified, and even less stringently enforced are standards for permissible levels of animal wastes in streams in rural Ohio, or coliform counts that preclude water skiing on a Kansas reservoir, or maximum allowable radioactive accumulations in snow-melt waters for irrigation of peach orchards in Utah. Some people even contend that clean-up of the nation's waterways is both unwise and economically infeasible. The opinion has even been expressed that since we cannot use our lakes and streams and retain pristine waters, we are justified in turning them into open sewers.

Water pollution is increasing in severity as indicated by increased rates of discharge of nutrients, accelerated sediment runoff largely due to massive, poorly planned construction practices, and the rapid increases in the kinds and toxicities of new chemical contaminants during the past decade. In addition, reported fish kills increased from 6 million in 1960 to 23 million in 1970. The public reaction to mercury in our waters peaked and subsided but the level of pollution from this deadly metal in our waters did not. The only positive note among these dismal pollution reports was that improvement in waste treatment controls during the 1960s had nearly kept pace with increases in organic waste levels in our waters.

Restoration and maintenance of water quality to a level that is acceptable to most water users and for most water uses is both technologically possible and economically justifiable. To accomplish this end, several facets of the problem need simultaneous attention. Among these are: research to fully determine the extent, nature, and effects of water pollution; water quality standards that will guarantee that all water supplies are safely usable for their intended purpose; legislation that will provide penalties proportional to the severity of the pollution offense; enforcement of laws and prosecution of violators; a surveillance system that is not influenced by political or economic clout; incentive awards such as tax breaks to reward those persons or agencies that voluntarily clean up waste waters; new technological advances to solve both water quality and water quantity problems of the future; and, finally, a financial commitment that dwarfs present pollution control expenditures to fund the clean up of our lakes, streams, bays, and estuaries.

Many people who seriously concern themselves with thinking about our water pollution problems feel that funding and construction of new and improvement of old sewage treatment facilities are all that is necessary to solve our water quality problems. Although statistics indicating the backlog of needed improvements to sewage treatment facilities are indeed awesome (only 60 percent of the sewered population has adequate treatment facilities), this approach to alleviating the nation's water pollution load is inadequate on several counts.

First, the notion that domestic sewage is the major source of the organic load dumped into the nation's waters is patently unfounded. For example, there are over half as many cattle in the United States as people and the daily B.O.D. production of waste per cow is nearly five times as great as for a human. On this basis, a feedlot with 10,000 cattle generates a sewage load equivalent to a city of about 45,000 people. Wastes from the majority of the nation's 100 million plus cattle, as well as the 100 million hogs, sheep, and horses and the 500 million poultry, flow untreated or with only token treatment into our surface waters. This creates a total animal waste load equivalent to that produced by over 2 billion humans.

Many of these animal concentrations are placed (seemingly deliberately) on steep topographic situations that receive sufficient precipitation runoff that the wastes are carried away with each rain. Not only is the stench unbearable near many of these areas, but thousands of miles of streams are rendered unswimmable and practically unusable throughout much of the Midwest.

Field spreading of manure is becoming less practicable in many areas as urban fringes extend into farm lands. Anaerobic degradation of animal wastes in sewage lagoons has been the most widely used approach whenever treatment is employed. Unless properly designed, sewage lagoons intensify odor problems, create mosquito breeding havens, and on porous substrates may contaminate groundwater. Any method of disposing of animal wastes should include recycling of the nutrients involved rather than overloading aquatic ecosystems. One possible solution is to use the liquified wastes as a hydroponic solution for the production of algae, which could in turn be used as a protein source.

In addition, pollution from sediments has always occurred under certain conditions, but man's intensive use of the landscape has drastically increased the sediment load of streams throughout the nation. Erosion from construction sites often can be reduced at the source by building temporary catchment lagoons for entrapping silt before it reaches streams and lakes. On some occasions, these small ponds could be retained after construction was completed to provide water-based recreation. Proper soil and water conservation measures have long been available for reducing soil losses on agricultural and grazing lands. A major problem is with enforcement of conservation practices either by law or by incentive award programs. It is not a matter of "know-how," it is a matter of implementation. The dilemma was aptly summarized by the farmer who commented that "I am not now farming nearly as well as I know how to."

Acid mine waters oxidize iron sulfide or iron pyrite to form sulfuric acids and other troublesome compounds. Since moist, well-aerated conditions increase the reaction rates, one method of control is to flood or seal the mines to prevent air from entering. Lime may also be added to streams containing mine seepage to neutralize the acid. Ion exchange or flash distillation of the mine flowage waters to produce potable municipal supplies is also possible, though presently costly.

Phosphate detergents presently contribute about 2 billion pounds or about 50 percent of the total phosphate load to our waters each year. Phosphates, like mine pollution, should be stopped at the source. Alternate cleansers, including soap, which is effective if the water is softened, and detergents in which sodium nitrilotriacetate (NTA) is substituted for phosphate, could be marketed to reduce phosphorus pollution. The multibillion dollar detergent industry will likely not make the changes, however, unless forced to do so.

Improved municipal waste-water treatment is both technologically and economically feasible at the present time. Major problems stem from the fact that most municipalities have combined storm and sanitary sewers which permit waste waters to flow untreated into the river during periods of heavy rain. In addition, about one-fifth of the communities have no treatment and nearly one-third of the waste water receives only primary treatment.

Sewage treatment is designed to simulate, but speed up, the action of natural purification of water in flowing streams. Several of the processes, including aeration to increase oxygen levels, flocculation to precipitate suspended solids, sand or gravel filtering, and biological oxidation are similar in nature and in the treatment plant. Sewage treatment accomplishes many of the same ends as the stream, but does so in a shorter time and distance.

Primary treatment, which consists of grit removal, screening, aeration, grinding, flocculation, sedimentation, and post-chlorination, removes perhaps 60 percent of the suspended solids, only about 35 percent of the B.O.D., and less than 20 percent of either nitrogen or phosphorus.

Secondary treatment involves biological oxidation of organic materials by either trickling filter or activated sludge methods. Secondary treatment removes about 90 percent of both suspended solids and B.O.D., plus up to 50 percent of the nitrogen and phosphorus. Secondary treatment costs, however, are about double those involving only primary treatment. Existing water quality standards, if enforced, would require secondary treatment for an estimated 90 percent of the urban population by 1973. Projected expenditures to meet water pollution control needs between 1970 and 1976 were established by the U.S. Conference of Mayors to be $33 to $37 billion. Actual federal funds spent for waste water treatment construction during 1970 and 1971 totalled only $740 million. At present, America is a long way from paying its debt to the past by reclaiming the purity of its waters.

Then, too, secondary treatment does not solve the eutrophication problem, as about one-half of the nutrients in municipal waste water would still reach surface or ground waters. These nutrients then foster accelerated algal growth which robs the waters of oxygen when it decays. B.O.D. and organic levels then rise again and the water is in a condition similar to that before treatment. This problem, coupled with the vast amounts of nutrient input from runoff enriched by agricultural fertilizers, makes the water quality problem seem hopeless.

Advanced (tertiary) treatment is necessary to remove excess nutrients plus tastes and odors from water so that eutrophication is halted and the treated water is reusable for industrial and sometimes even domestic supplies. Several techniques are possible. Suspended solids may be removed by microscreening or filtration through diatomaceous earth; dissolved organics by oxidation; dissolved inorganics by electrodialysis, ion exhange, or reverse osmosis; phosphorus by chemical precipitation; nitrogen by denitrifying bacteria; viruses and bacteria by chlorine, iodine, ozone, or other rigorous disinfectants; and tastes and odors by filtering through activated carbon. Such complex treatment is expensive. At experimental plants now in service it costs at least 1 1/2 times as much as combined primary and secondary treatment. If we ascribe to the notion that pollutants should be considered as resources out of place, rather than nuisances to cope with, a portion of advanced waste-water treatment costs could be offset by recovery of nutrients, metals, gases, chemicals, or other substances. Since treatment is essential anyway for a quality environment, any return from salvageable materials would serve to reduce the level of public and private expenditures. In any event, advanced waste treatment must become a reality before we can entertain any serious hope of restoring the nation's water quality.

The clean-up costs may be high but they must be borne by both the water polluter and the water user. They ultimately become one and the same (the water user) since pollution control expenditures by industries are passed on to the consumer in the form of increased product costs. Plentiful supplies of fresh water at very low costs are the greatest bargain in America today. If rates were tripled, water would still be a bargain in terms of price increases on almost all other commodities in recent years. Everyone will benefit from safer, cleaner water, and clean-up will cost far less now than in the future.

Funding of pollution abatement may be more readily obtainable than suitable legislation and a surveillance agency that enforces the laws without showing partiality. One of the few statutes that has been used effectively against irresponsible polluters is the Refuse Act of 1899. This law was "rediscovered" in 1969 after a 70-year period of lying largely ignored and unenforced. This law makes it a federal crime to put "any refuse matter of any kind whatsoever into the navigable waters of the United States without a permit from the U.S. Army Corps of Engineers." Violators may receive fines of up to $2500 or jail sentences of up to one year per violation. Despite the fact that American industry generates more than four times the B.O.D. of our combined population, and despite the fact that an estimated 300,000 factories discharge wastes into navigable waterways, by January 1972 injunctions had been filed against less than 100 by the Environmental Protection Agency (EPA), according to the Nader Task Force on Water Pollution. The old 1899 Act could provide the initial impetus toward water quality control, but it remains largely unenforced. Similarly, the Federal Water Pollution Control Act of 1948 and its subsequent amendments and the 1965 Clean Water Act all have loop-holes that render them largely impotent and unenforceable.

The Senate Subcommittee on Air and Water drafted a new water pollution law in June 1971 that would have provided for the attainment of truly clean water. This bill called for the elimination of pollution discharges into the navigable waterways by 1985. This no-discharge provision was intended to stimulate efforts toward recycling and reclamation of wastes rather than allowing them to further degrade other water supplies. Although this bill passed the Senate by an 86-0 vote in November 1971, it was badly emasculated in the House of Representatives, and finally passed as another impotent pollution control law that is filled with loop-holes that favor continued pollution.

Americans can have water quality restored to almost all of the nation's surface waters if they want it badly enough. It is both technologically feasible and within reach economically. But the people must demand it and must be willing to pay for it.

Water, Land, and Man

> Near by is the graceful loop of an old dry creek bed. The new creek bed is ditched straight as a ruler; it has been "uncurled" by the county engineer to hurry the run-off. On the hill in the background are contoured strip-crops; they have been "curled" by the erosion engineer to retard the run-off. The water must be confused by so much advice.
>
> ALDO LEOPOLD,
> *A Sand County Almanac*

At one time the hydrologic cycle operated without restraint. Water fell on the land and flowed unchecked to the sea. Obviously, not all regions were equally well watered. Some island and coastal areas received an average of 1/2 inch per day or more; some inland areas received an average of 1/2 inch per year or less. Life was lush or sparse in direct proportion to the amount and effectiveness of precipitation. Originally man also had to adapt to the available water supply and prospered or perished accordingly.

With the beginnings of civilization, the world's most successful biped began rearranging things, at first slowly, then on an ever-increasing scale. The earliest civilizations were based on cereal grain culture in the fertile valleys, wheat in the Nile and Tigris-Euphrates, rice in the Indus, maize in Mexico. The rivers brought both water and nutrients to man. As populations grew and civilizations spread, less well-regenerated local water and nutrient sources were used until they would no longer meet the demand, then importation was necessary or the civilization declined.

Man's manipulations of the land and water provided a more predictable level of productivity, thereby permitting greater population densities. Urbanization quickly resulted in populations that outgrew local water and food sources, particularly as contamination reduced the usable supplies. Human preferences for living in poorly watered subtropical areas with Mediterranean climates have intensified the water supply problems. The mass migration to Southern California, Arizona, and other warm areas in recent decades has brought unparalleled demands for water, most of which is imported over long distances. The rate of moving needed water to people and keeping excess water from other people will accelerate until the man-land-water equation becomes balanced and local supplies are made fit to use.

Floodplain occupation was the obvious choice as the United States was settled. Rivers simplified frontier travel, provided abundant fresh water, powered mills, and hauled freight. Infrequently flooded river bottom soils are fertile and easily provided food for growing settlements. However, continued development of floodplain lands is no longer necessary; it now runs counter to man's best interest.

There have always been floods, and floods will doubtlessly always occur. Flooding is an act of nature, but flood damages are caused by man. Floodplains are the safety valves of watercourses; they were carved out and deposited by the streams and, on occasion, streams reclaim their valleys. Man has aggravated both flooding heights and frequencies and flood losses or damages. The former has resulted from increased runoff due to widespread drainage of wetlands, forest clearing, paving surfaces, fallowing farm lands, and straightening stream channels. Flood losses have mounted as industrial, residential, and agricultural developments have crept onto the floodplains in the wake of flood control structures which "insured" flood prevention. In no area of environmental mismanagement has the public been so duped as it has with respect to flood "control." Economist Kenneth Boulding expressed the essence of the dilemma when he stated:

> The truth is that what we call "flood control" means the eradication of little disasters every ten years or so at the cost or of a really big disaster every fifty or 100 years . . . no flood-control program is able to protect a floodplain against the 100-year flood. After all, that is why the floodplain is there. . . .

Each time a major flooding disaster does occur, the clamor begins anew for the construction of more large dams. Not only is this approach to solving flooding problems inordinately expensive, it usually is only marginally effective and is frequently ecologically disastrous. The U.S. Army Corps of Engineers and the Bureau of Reclamation in the Department of Interior have the responsibility for construction of the large dams which flood vast areas. They also drain wetlands for agricultural production and irrigate arid regions of the West. Many times these efforts serve cross purposes, both of which are financed by tax dollars. One might observe that their motto should be: If an area is wet, drain it; if it is dry, flood it!

There are basically two ways of controlling flood *damage;* both should be employed simultaneously. The structural approach involves keeping the water away from man by holding a portion of potential flood waters behind levees, floodwalls, dikes, and dams which vary in size from small farm ponds to immense reservoirs. The floodplain zoning approach involves keeping man away from the floodwaters by preventing the development of floodplain lands for industrial, residential, or intensive agricultural uses. In addition, structures such as highways, electrical transmission lines, or pipe lines that must cross floodplain lands should be elevated above the highest recorded flood levels, or incorporate other flood insurance measures.

Flood control structures have a place in river basin management, but expenditures, for large dams in particular, have gone far beyond what was essential. The best dam sites have already been appropriated and future construction should be halted. Usually the dams are justified as multiple-use water management structures. Such benefits as those accruing from flood control, hydroelectric power generation, pollution dilution, irrigation, and recreation are all figured into a benefit-cost ratio. These ratios, though widely used to justify Corps of Engineers projects, have been shown to be suspect by several impartial studies. The amount of electricity generated from "harnessed" rivers is an insignificant percentage of the total electrical energy output and will become less important in the future. Pollution should be controlled at the source instead of being diluted and washed downstream. Continued irrigation of marginal lands is questionable when thousands of acres of the best farm land are lost from production annually to suburban sprawl, highway construction, and other "crystallized" land uses. Recreation potential of large reservoirs is frequently overestimated by including every person within the reservoir area as a possible user for every recreational day, even though there may be several other recreational possibilities open to the same potential visitor. In fact, when Robert Haveman, an economist at Duke University, studied every Corps of Engineers project constructed in 10 states during a 16-year period, he found that 63 of the 147 projects could not pass any of 5 standard economic tests for showing whether costs were greater than benefits.

Not only are most dams not justifiable on economic grounds, they frequently are environmental disasters. One of the most-cited drawbacks is their vulnerability to loss of capacity due to silting. About 2000 dams in the United States have already silted in and have been converted to alluvial plains. All but a few of the large ones will follow suit within the next 200 years, a relatively brief time when compared to the historical span of man. Classic examples of premature silting of reservoirs occur on the Santa Ynez River in California. Not one, but two water supply reservoirs for Santa Barbara are silting so rapidly that a third dam is already under construction. Reservoirs in arid areas increase water losses because the evaporative surface becomes greater. Downstream users are often deprived of water or have their supply reduced. Dams also change the entire stream ecology by converting it from a free-flowing body to a lake. Recreational use then changes from canoeing, rafting, and trout fishing to the flat-water sports of water-skiing and pan fishing. The few wild rivers that remain badly need to be preserved.

Despite these and other valid objections, the pace of dam building shows no evidence of abating. The California Water Plan alone provides for the construction of 600 more dams in northern California. When completed, every major water course in that region would be obstructed except for a few creeks. The Colorado River has been controlled to the point that the demand for water is greater than the river can supply, even with reuse from one dam to another. The

Bridge Canyon and Marble Canyon Dams scheduled to flood parts of Grand Canyon National Park were killed by massive opposition by conservationists, but, on occasion, these proposals still rear their ugly heads.

The ecological changes following the building of the Aswan High Dam on the Nile are instructive. The United Arab Republic anticipated nothing but improvements in its economic situation after Lake Nasser filled behind the largest dam in the world. Doubled electrical power output and increased crop production due to irrigation were forecast. Instead, a harvest of problems has resulted. The lake is rapidly filling with silt that once enriched the Nile Valley soils at flood time and provided nutrients for marine food chains in the Mediterranean Sea. Sardine harvest has declined in the Nile delta region and shrimp fishing may be similarly affected. The continued use of mineral fertilizers and irrigation waters may cause salinization of the Valley soils, decreasing rather than increasing agricultural production. Perhaps the most serious problem is the continuing spread of a dread snail-borne disease called schistosomiasis via the irrigation waters. A significant percentage of Egypt's population now has this debilitating disease.

Engineers contemplating such monstrosities for the Amazon, Mekong, or other tropical rivers should reflect on these water-related ills. Enormous dams that have been proposed for the Yukon or MacKenzie Rivers of the north should be reconsidered in terms of their earthquake vulnerability. Some scientists are concerned that reservoirs may even precipitate earthquakes due to stresses placed on underlying rocks by the enormous additional weight.

Restriction of floodplain development involves returning part or all of the floodplain to the service of the river, or to uses not unduly damaged by flooding. Agricultural crops are suitable on soils that are infrequently flooded, providing that the farmer rather than the taxpayers assumes the risk. The floodplain farmer enjoys a competitive economic advantage over his upland counterpart by virtue of the lower fertility needs of alluvial soils. Forestry is also an excellent floodplain use. Fast-growing species, such as sycamore and cottonwood, can be harvested every three or four years as "silage" for pulp production. Green belts along river courses bring nature into the heart of urban areas. Such parks bring diversity of living experience to the townsman and suffer very little damage should flooding occur.

A practical solution to flooding problems and soil losses is to hold raindrops where they fall. Proper land conservation practices—such as contour strip-cropping, terracing, sod waterways and farm ponds, slow runoff—increase infiltration into the soil, and reduce erosion. Several million farm ponds have been built, largely through the help of the Soil and Water Conservation Service. Many small impoundments make more sense than a few large ones in terms of restricting runoff. Ponds also provide more wildlife habitat and recreational opportunity per tax dollar invested.

The image of the Soil and Water Conservation Service as a public service agency has been unparalleled during most of its tenure. The SCS brought about a new wave of conservation techniques to replace the disastrous farming practices of the 1930s. This assignment of preventing erosion, preserving soil fertility, and promoting land conservation was carried out by working with individual farmers through the local Soil Conservation District. Then in 1954 Public Law 566 gave the SCS the job of water and land management in small watershed areas. They continued the constructive land treatment on individual farms, but added the sometimes questionable practice of constructing small reservoirs on the tributaries of most creeks to hold water on the land, and the frequently destructive technique of channelizing streams into straight ditches to rush water from the land. Their image has changed into a miniature Corps of Engineers in this respect. Over 4000 miles of streams have been channelized and the SCS has plans approved for some 12,000 additional miles. Such action changes streams into ruler-straight drainage ditches that are largely devoid of wildlife. Environmentalists are understandably alarmed over channelization practices.

Stream color is an excellent indication of the quality of land use in the watershed. For example, near my university, the Wabash River (named for the Indian word Oubache, which means clear or white) is usually "too thick to drink, but too thin to plow!" Its muddy color broadcasts to all who will see (and smell) that all is not well on the highly productive agricultural lands of the Wabash Basin. The river color derives from fall plowing and continuous row cropping. Much of the better farm land never has vegetative cover, other than corn or soybeans during a four- to five-month growing season.

I frequently ask my students how long it would take for the Wabash River to run clear or white again if man and his influences were completely removed from the watershed. Estimates vary from a few months to 100 years. I then ask what environmental changes would occur in the basin and how would they aid stream recovery. There follows a discussion of biotic succession in the watershed, elimination of industrial and domestic wastes, dispersion of fertilizers and pesticides, river-bank stabilization, limnological improvements, and river bottom cleansing. It quickly emerges that the time span for remarkable stream improvement would be on the order of a few months as successional advances covered the farm fields. A complete return to pristine conditions would take decades. There is little likelihood that man will voluntarily leave the Wabash or any other basin, but the discussion helps students comprehend the remarkable resiliency of nature and the vast recuperative powers of streams if they are given half a chance. The river will never again be so clear that "the white sand glittered through the clear water at a depth of several feet" as the history of Terre Haute described the early Wabash. Our intensive use of the landscape precludes that level of recovery. The point is that the Wabash and thousands of other American streams can be dramatically improved if we only care enough and show enough wisdom to apply the land and water management know-how that we already possess.

Fate of the Sea

> Thus the coasts of Africa and America were almost equally distant when we suddenly entered another area which was so polluted that we had to be attentive of washing ourselves . . .
>
> THOR HEYERDAHL of his "Ra" Expedition,
> *Biological Conservation*, **3**, pp. 164–167

For thousands of years man has viewed the sea as illimitable and eternal. The sea which supplies the water, moderates the climate, shapes the surface, and husbands much of the life of the Earth is incredibly large. But it *is* finite. This concept of the sea is harder for man to grasp than is his concept of the land. Land has delimiting landmarks. What are the seamarks? Land can be staked off and owned, but most of the sea, like the air, belongs to no one, hence to everyone. As such, the sea has become the ultimate sump—the sink into which almost all pollutants eventually flow.

It was difficult for man to accept the fact that the vast land areas could be ravished by human occupation. To most, the sea remains impenetrable, unassailable. But many land ecosystems are already "on the ropes"; it is now apparent to those who look closely that the sea is also in jeopardy.

DDT and PCB (polychlorinated biphenyls) are now found everywhere that man looks, even the most remote recesses of the sea. Despite the possible environmental hazards from these chemicals, their omnipresence in the sea lends credence to the fact that the oceans are, in fact, a single ecosystem through which materials and energy flow. Both chemicals cause serious ecological concern because of their effects on the reproduction of fish-eating birds and on the productivity of marine food fish. It has even been suggested that DDT could reduce photosynthesis, thereby lowering marine productivity. Due to their stability, persistence, and the large quantities already present as land and fresh water contaminants, the effects of these chemicals in marine environments would likely worsen for several years even if an immediate ban was imposed on a world-wide scale.

The effects of pesticide and herbicide pollution on the marine environment are mostly biological. Oil pollution, however, is even more devastating as it affects both the physical and biological components of the ecosystem. Although minor spills and marine dumping of oil had been commonplace for over a half century, when the wrecked *Torrey Canyon* dumped 36 million gallons of crude oil into the sea off the south coast of Great Britain, world attention was focused on the potential danger to the sea from massive oil pollution. An estimated 10,000 spills of oil and other dangerous substances pour into the navigable waters of the United States each year. The sea also eventually receives much of the 350 million gallons of used oil that is disposed of annually by gasoline service stations.

Sea birds, marine mammals, oysters, clams, zooplankton and other marine life all suffer ill effects and many (especially birds) die as the oil coats their bodies, fouls their respiratory apparatus, or seals off their oxygen supply. Since oil and water are immiscible, slicks form on the ocean surface which prevent light penetration and gas exchange, thereby halting photosynthesis. Beaches, rocks, and submerged plants are coated in estuary and tidal zones. In short, oil wreaks havoc on most ecosystem functions.

Fortunately, some components of the crude oil serve as an energy source for microorganisms which break down the straight-chain hydrocarbons. Branched-chain hydrocarbons with higher molecular weight tend to form tarry lumps which are more persistent and are now found in all ocean areas. In the mid-Atlantic, Heyerdahl reported "seemingly endless quantities of oil-clots of sizes ranging from that of a wheatgrain or pea to that of a sandwich." Detergents applied to aid in "clean-up" of oil spills apparently increase the toxicity of the crude oil.

Oil spills are occurring more frequently as ocean shipping of petroleum increases. Presently about 1 billion tons—about half of the yearly production—is shipped at sea creating routine losses of about 0.1 percent or annual spills of about 1 million tons. If the oil from Alaska's North Slope is conveyed by tankers from the pipeline terminus at Valdez to California refineries as proposed, it is inevitable that spills will contaminate parts of the west coast of North America. Leaks and small spills at the pipeline terminus could easily change the ecology of the beautiful Valdez harbor. Even with the North Slope oil, if petroleum use increases as projected, the United States may be importing nearly 40 percent of its petroleum products by 1980. Chances of numerous spills at sea are very likely. Since the sea is not "owned," who is responsible? Who makes reparations to whom? It is obvious that the biosphere sustains the cost.

Ownership or user rights of offshore reaches are the most disputed of any part of the sea. It is here that the best fishing areas occur. Also, the most productive new oil fields are in offshore locations. As continental reserves are depleted, offshore production will increase. Leaks from offshore wells like the Santa Barbara incident of 1969, which pumped over a million gallons of crude oil into the sea, could well become more common.

Both solid waste and waste water have been discharged into the sea for centuries. For example, garbage from New York City was dumped into the Atlantic for decades. The North Atlantic receives a disproportionate amount of the world's oceanic pollution because it is encircled by the highly industralized nations of North America and western Europe. Although the dilution capacity of the ocean has not as yet been approached, the environmental impact of massive nutrient and organic additions is becoming more significant each year. Highly productive estuaries may be especially sensitive to additional nutrient inputs. Solid wastes windrow many of the world's beaches and are observed floating throughout the seas.

Estuaries are the most productive of the major ecosystem types on earth.

Mixing of fresh and saline waters, due to water density differences, wind action, and tidal wash in these coastal habitats, creates a nutrient trap that maintains high levels of water fertility. As a result, the amount of organic matter fixed per unit area of estuary is about 15 to 30 times as great as the productivity of the open ocean. Estuarine productivity is two to three times as great as in moist agricultural regions. Estuaries also serve as nurseries for great numbers of species which come to the beaches or rocky inlets to lay eggs, spawn, or bear young. To destroy the estuaries is to destroy the regenerative capacity of many marine species, thereby dooming even the pelagic adult populations.

No habitat type, however, has suffered such massive recent alteration or outright destruction. These changes take many forms. When the flow of rivers entering the estuary is reduced, salinity gradients where fresh water and sea water meet are steepened to the disadvantage of species adapted to low salinity zones. Upstream dams can also reduce sediment discharge which is essential for the success of the larval forms of several deep-water species. Excess organic inputs increase the B.O.D. of coastal waters, reducing the survival of fish and shellfish. Heavy metals, radioactive wastes, or toxic chemicals, which are frequently discharged with industrial wastes, often poison seafood species directly or are concentrated in them to problem quantities, such as in the recent cases of mercury poisoning at Minimata, Japan. Most serious of all is the direct destruction of the estuaries themselves by filling, dredging, or diking along the seaward edge. In a 30-year period beginning in the early 1920s about one-fourth of the salt marshes of the United States were destroyed by these processes. This activity continues throughout the world. If the estuaries disappear, so does the majority of oceanic productivity.

The sea has been highly touted as the source of an almost unlimited food source to supply the world's rapidly growing population. Underdeveloped nations, in particular, have looked to the sea as a protein source for their underfed millions. At present, about 86 percent of the world's fish catch comes from the sea. This 65 million-ton harvest supplies about 18 percent of the world's protein. Some 90 percent of this is fin fish, the rest is comprised largely of whales, oysters, clams, lobsters, crabs, and shrimp. A few marine areas are presently underharvested, but most waters are being fished at near capacity. Most cautious estimates place the maximum sustained annual yield at about 100 million tons. Additional exploitation would result in mining the resource and would trigger skidding fish populations, such as have already occurred for most species of whales. Serious pollution, particularly of the fish-rich coastal reaches of the sea, would further jeopardize harvests by reducing fish populations or by rendering huge quantities unfit for human consumption due to contamination.

The complex interactions of the immense marine ecosystem are only beginning to be understood. For far too long we have largely neglected this 7/10 of the world's surface. In fact, we probably know more about space than we do about the sea. Will we continue to jeopardize the sea which serves as the "governor of the biosphere"? Without question, more oceanographic research is needed to answer all kinds of questions such as, what would be the long-term ecological effects of a sea-level canal across the Isthmus of Panama? Interest in marine science is at an all-time high, and the formation of the National Oceanic and Atmospheric Agency (NOAA) is a promising step. What is needed now is the designation of an International Hydrologic Decade to enable scientists of all nations to arrive at a long-term plan for global water management. Unfortunately, our past record on environmental concerns has been to act first, then examine the consequences. To continue our present course of contaminating the sea is fatal folly. Not only is marine pollution potentially disastrous, we cannot afford to lose the nutrients and other minerals that are thoughtlessly discharged.

Beyond Tomorrow

> Must we always try to bring water to people, no matter where and in what inhospitable regions they may choose to wander? Is a man entitled to buy up, settle, or promote a chunk of desert and then demand that his government bring water to him from the general direction of the North Pole?
>
> MAJOR GEN. JACKSON GRAHAM (retired),
> U.S. Army Corps of Engineers

California is the prototype of the American future. Whatever happens in the Golden State has a way of repeating itself nationally five to ten years hence. American water planners should turn their eyes to the West and take a good look at what is happening in our most populous and thirstiest state. Already the Colorado River and the Owens Valley have been essentially milked dry to slake the thirst of the City of Angels; the incomparably beautiful Hetch-Hetchy of Yosemite and a dozen other canyons were drowned to supply San Francisco and the northern valley; Californians are now enviously watching the waters of the Columbia, the Peace, the Fraser, the Yukon, and even the icebergs that float north from Antarctica.

Not that they are not planning for the future growth in water demand. Far from it. The California Water Plan is the most ambitious aquatic undertaking ever devised by man. It has been called (among other things) a "master plan for the control, conservation, protection, and distribution of the waters of California, to meet present and future needs for all beneficial uses and purposes in all areas of the state to the maximum feasible extent . . ." (Boyle, Graves, and Watkins, p. 161). This colossal intricate web of waterworks already includes or has plans for 21 dams and reservoirs with a total capacity of nearly 7 million acre-feet of storage and over 500 miles of shoreline, 6 major aqueducts with a combined length of nearly 700 miles, 22 pumping stations, and 6 power plants with a generating capacity of over 5 billion kwh. This is only the beginning and the Water Plan is already in deep financial trouble. The engineers are now eyeing the 30 million acre-feet of water that flow untrammeled to the sea from northern California streams. An additonal 600 dams could pockmark that water-rich region.

California's water shuffle hinges on two major problems. The first is supplying the enormous agricultural sponge. About 90 percent of the total water used goes for irrigation. Water used by agricultural crops is consumed and must be resupplied since it returns to the atmosphere after use rather than to the waterways.

The second problem is that the state's exploding population wants to live where the water isn't. Abundant precipitation does not enhance outdoor barbecues or sunbathing. This already-serious problem of human distribution will worsen as the 1970 population of about 19 million nears the expected 50 million mark by 2020. If present trends remain unchanged, most new additions will be in the already water-short areas. Per capita water use is also increasing.

Is the California Water Plan of ever manipulating surface and flowing waters to meet ever-growing needs the best long-run solution for California or anywhere else? It certainly is not the most economical, even at present. Projected delivery costs range from $40 to $60 per acre-foot; whereas, waste water is presently being reclaimed in some southern California locations at costs ranging from $20 to $30 an acre-foot. Desalinization of sea water is presently too costly to be competitive, but will become so in the future.

The same water problems that beset California arc in store for much of the rest of the nation, as some localities have already run the same gamut of water troubles. Even the water-rich New York City area has experienced water shortages periodically, largely due to lack of metering and undercharging which create exhorbitant waste. New York's plumbing system has been described as the "leakiest faucet in the world."

Throughout the nation, the surface waters have grown dirtier yearly and their clean-up more costly. "Fossil" groundwater that accumulated during the last ice age of some 15,000 years ago is dropping toward the limits of the pumps in every dry region where irrigation is common. For the entire United States, some 95 percent of the consumed (not used and replaced) water goes into irrigating crops. Efforts of the Green Revolution to feed the projected future billions will most likely be stayed, not by suitable hybrid crop varieties nor by fertilizer supplies, but by the limited availability of water supplies to irrigate the millions of acres of crop land required for such food production.

Metered water for industrial and domestic use in America that totaled only 40 billion gallons per day in 1900 now approaches 400 billion gallons daily, and is expected to reach 800 to 1000 billion gallons by the year 2000. Yet Congress determined in 1963 that even with optimum advances in purification and engineering, surface and ground water sources could be expected to supply not more than 650 billion gallons per day. The problem of meeting water demands in most U.S. localities is likely to reach crisis proportions within a decade.

Several possible solutions to this growing water dilemma are in store. Improved technology has lowered the price of desalinized water from about $1600 to about $320 per acre-foot during the past 20 years. Widespread use of nuclear energy could lower costs dramatically. Dr. Glenn T. Seaborg, former Chairman of the Atomic Energy Commission, feels that such applications of nuclear energy have the greatest potential for meeting water needs of any water management tool in human history. Electricity generated by breeder reactors would greatly increase in quantity and decrease in cost so that sea water and waste water could be economically purified for reuse by cities and agriculture.

Lest desalinization be adopted as the panacea for all water needs, some problems need serious thought. A formidable one is what to do with the vast quantities of salt removed from the saline waters. (It appears that we have come a full cycle in 200 years. Daniel Boone's tiny Kentucky colony boiled hundreds of gallons of mineral-laden water at Blue Licks to obtain a few pounds of salt!) Most elements from sea water could be used as nutrient or chemical sources if the salt supply produced was not so overwhelming. Dumping the salt back into the sea is not the answer. For example, if a plant desalted water at the rate of 100 million gallons per day, about 15,000 tons of salt would be recovered. Dumping this amount near shore would likely kill all fish and plant life in the vicinity. Besides, since increasing quantities of salt would have to be removed from each successive batch of water, a point of diminishing returns would soon be reached. In addition, the movement of vast quantities of desalinized water from coastal areas uphill to the arid interiors of continents presents an engineering project that rivals the current, massive largely downhill fresh water movements.

Tapping the fresh water supplies contained in icebergs or glaciers is not as preposterous as it sounds. Since icebergs float about anyway, they could theoretically be towed northward from Antarctica via the cold Humboldt Current along the west coast of South America to West Coast cities. Since fresh water floats temporarily on sea water, extraction of water from melting icebergs could possibly be accomplished by surface pumping.

The "lower 48" have looked thirstily to Alaska and Canada's abundant water supply for several years. Massive amounts could be moved southward by colossal conduits extending through mountain valleys or river basins. Reversing the flow of major rivers which flow into the Arctic Ocean has been proposed by both Soviet and American engineers. Interbasin transfers of such magnitude should be held in abeyance until other sources of supply have been thoroughly investigated. Ecological damage from such schemes is likely to be enormous and irreparable.

The most promising source of meeting future water needs is frequently mentioned last, if at all. This obvious choice is the purification and reuse of our waste waters. Not only can we ill afford to lose the nutrients and other resources that they contain, but waste water treatment represents the most locally available and likely the most economically justifiable water source for most communities.

The notion that treated waste water should be esthetically unappealing or a threat to health is contrary to ecological understanding. After all, most water that we drink or use has previously passed through an organism, or has washed a product or person at one time in its recent history. Water is cleansed by nature as it revolves through the hydrologic cycle. Man can apply the same processes, thereby shortening the cycle. Municipalities that take their water from surface waters apply treatment that is nearly identical to advanced waste water treatment. (In fact, it may not be as successful in removing viruses or nutrients as some tertiary waste water treatment facilities.) The next step is to retain the water at the city and reuse the effluent rather than allow it to move downstream to the next city which reuses it after the river partially dilutes the wastes. The Willamette River in Oregon has been almost completely restored by such intensive water treatment measures. We now need to do likewise, nationally and throughout the world.

One point that was made abundantly clear at the recent United Nations Conference on the Human Environment at Stockholm, Sweden is that most nations have awakened to the fact that we do indeed belong to one world. The problems of this world community must be solved for continued human existence. Since it is the common use of the air and water that binds the biosphere and its peoples together, perhaps we should begin with international efforts to monitor, then alleviate, the threats to our water supply—the life blood of the land.

References

American Chemical Society, *Cleaning Our Environment: The Chemical Basis for Action.* Report by Subcommittee on Environmental Improvement, Committee on Chemistry and Public Affairs, American Chemical Society, Washington, D.C., 1969, p. 95–162.

BERRY, R. STEPHEN, "Only One World: An Awakening." *Bulletin of the Atomic Scientists*, 1972, **28** (7), p. 17–20.

BOYLE, ROBERT H., JOHN GRAVES, and T. H. WATKINS, *The Water Hustlers.* San Francisco: Sierra Club, 1971.

BRADLEY, CHARLES C., "Human Water Needs and Water Use in America." *Science*, 1962, **138** (3539), p. 489–491.

BROWN, ALISON LEADLEY, *Ecology of Fresh Water.* Cambridge, Mass.: Harvard University Press, 1971.

BUSWELL, ARTHUR M., and WORTH H. RODEBUSH. "Water." *Scientific American*, April 1956.

Council on Environmental Quality. *Environmental Quality.* First annual report of Council on Environmental Quality U.S.G.P.O., Washington, D.C., 1970, p. 29–59.

COX, JAMES L., "DDT Increasing in the Marine Environment." *Biological Conservation*, 1971, **3,** pp. 272–273.

FAIRBRIDGE, RHODES W., "The Changing Level of the Sea." *Scientific American*, 1960, **202** (5), pp. 70–79,

HERFINDAHL, ORRIS C., and ALLEN V. KNEESE, *Quality of the Environment: An Economic Approach to Some Problems in Using Land, Water, and Air.* Washington, D.C.: Resources for the Future, Inc., 1965, p. 10–23.

HEYERDAHL, THOR, "Atlantic Ocean Pollution and Biota Observed by the 'Ra' Expeditions." *Biological Conservation,* 1971, **3** (3); pp. 164–167.

HYNES, H. B. N. *The Biology of Polluted Waters.* Liverpool, England: Liverpool University Press, 1960.

Institute of Ecology, *Man in the Living Environment.* Madison: University of Wisconsin Press, 1972, p. 219–267.

MAXWELL, JOHN C., "Will There Be Enough Water?" *American Scientist,* 1965, **53** (1), pp. 97–103.

MURRAY, C.R., "Estimated Use of Water in the United States: 1965." *Geol. Survey Circ. 556,* 1968.

NACE, R. L., *Water and Man: A World View.* UNESCO, 1969, p. 46.

OGBURN, CHARLTON, JR., "Why the Global Income Gap Grows Wider." *Population Bulletin,* 1970, **26** (2), pp. 3–33.

PIPER, A. M., "Has the United States Enough Water?" *Geological Survey Water-Supply Paper 1797,* U.S.G.P.O., Washington, D.C., 1965, p. 1–27.

POWERS, CHARLES F., and ANDREW ROBERTSON, "The Aging Great Lakes." *Scientific American,* 1966, **215** (5), pp. 95–104.

SAYRE, A.N., "Ground Water." *Scientific American,* November 1950.

SCHNEIDER, W.J., and A.M. SPIEKER, "Water for the Cities—The Outlook." *Geol. Survey Circ. 601–A,* 1969, p. 6.

SCHRAG, PETER, "Life on a Dying Lake." *Saturday Review,* September 20, 1969, pp. 19–21, 55–56.

SMITH, F.G. WALTON, "What the Ocean Means to Man." *American Scientist,* 1972, **60,** pp. 16–19.

TURK, TURK & WITTES, "Water Pollution," in: *Ecology-Pollution-Environment.* Philadelphia, Pa.: W.B. Saunders Company, 1971, pp. 109–134.

VAN HYLCKAMA, T.E.A., "The Water Balance of the Earth." *Publ. Climatology,* 1956, **9,** pp. 58–177.

WARREN, CHARLES E., *Biology and Water Pollution Control.* Philadelphia, Pa.: W.B. Saunders Company, 1971, p. 3–63 and 322–372.

WOODWELL, G.M., "Toxic Substances and Ecological Cycles." *Scientific American,* 1967, **216,** pp. 24–31.

9

The Land Resource Base

ALTON A. LINDSEY
Purdue University

Alton A. Lindsey was born in Pittsburgh, Pennsylvania, in 1907. He received his B.S. from Allegheny College. In 1937, Cornell University conferred his Ph.D. in the fields of botany, ornithology, and animal ecology. He spent 1933–1935 as a biologist with the Byrd Antarctic Expedition II in Little America. He has taught at universities in the District of Columbia, California, and New Mexico. Since 1947 he has taught biology at Purdue University. He is managing editor of *Ecology* and *Ecological Monographs*.

Dr. Lindsey's special interest is plant ecology, particularly the vegetation of Indiana, ecological methodology, and natural area preservation. His books include *Natural Areas in Indiana and Their Preservation, Natural Features of Indiana,* and *Vegetation of the Life Zones in Costa Rica.* He has published 60 technical papers.

Science and *The Book of Genesis* agree that man is formed ". . . of the dust of the ground." Soil and its contents are indispensable parts of our life support system; activities based on land resources include farming, grazing, forestry, mining, water use, transportation and real estate development. Intelligent land management is essential to economic life. But man does not live by economics alone; land and its natural productions are also subjects of scientific understanding, recreational and esthetic enjoyment, and, in some cultures, religious veneration. Land is discussed here chiefly from the biological viewpoint, with emphasis on its role as environment for our species.

Table 9.1 Two of the Three Major Types of Biotic Systems and Appropriate Types of Conservation Activity*

Dependent Biotic Systems	Independent Biotic Systems
(Cities, Towns, Suburbs; Cropland, Farm Pastures and Woodlots, Monoculture Tree Plantations, Orchards, Reservoirs)	(Natural Areas, Nature Preserves, National Parks, Monuments and Landmarks; Wildlife Sanctuaries, Wilderness Areas; Seas in Part)
Few species, many individuals	Many species having few individuals
Less stable, growth phase	Steady-state, mature
Intensive use and development	Exploitation limited, often by law
High labor requirement	Little or no human effort
Require fossil fuel energy	Self-maintaining with solar energy
Steep energy gradients, few steps	Gentle gradients, many steps
Highly productive	Low net productivity
Vulnerable, poorly buffered	Protective and healing
Tend to unbalance world ecosystem	Help to balance other systems
Export CO_2 worldwide	Export O_2 and H_2O
Source of various pollutants	Pollutant sink and neutralizer
High in values extrinsic to land	High in intrinsic values (natural beauty, and so on)
Improved by introducing Nature	Not improved by artificiality
Economics (now) before Ecology	Ecology before economics
Area being expanded	Area "progress"ively reduced
Intensive *conservation* techniques must be actively applied for protection, maintenance, pollution control, materials recycling and beautification, while restraining abusive technologies and population increase.	*Conservation*–Preservation for use in *perceptive-recreation* (hiking, riding, canoeing, fishing, adventure, creative loafing) and in *interpretation* (esthetic, scientific, educational, spiritual, artistic, literary)

**Compromise Systems* are intermediate, in the above respects, to the other two. They include public and private forests in multiple-use, rangeland, most wetlands and intensively used recreation areas as state parks and public hunting grounds, and the seas, in part.

Conservation here depends on both use and management being light enough to take advantage of Nature's protective buffering and restorative powers.

From this ecological standpoint, land may be classified on the basis of the traits shown in Table 9.1. The right-hand column describes biotic communities at the end of the plant and animal succession, which are essentially natural because they are either completely undisturbed by man or are only lightly managed. This "old nature" is in general balance with the climate and other environmental determinants. When pioneers cleared the forests or broke prairie sod, they pushed the system back to resemble the "young nature" of early succession. One indication of this is the population of the same annual weeds occurring both in cultivated fields and on fresh soil after a landslide or emergence of a river bar. The farmer must manage his field intensively to hold back natural succession; the crop monoculture corresponds with an early or growth stage, ecologically.

Dependent biotic systems are produced by man's influence or kept immature by his management. Cities, suburbs and farms belong here; the city, insofar as it is a biotic system, is the most dependent one. The contrasts with *independent* wildlands are shown by the paired traits in Table 9.1. A third kind, the *compromise system,* is intermediate between the other two. In the continuous gradation from city to wilderness, the extremes show diametrically opposite properties, making it clear that the same philosophy and approach to conservation and management would not fit the entire continuum.

Reservoirs with widely fluctuating levels should be considered dependent, but those maintained nearly constant develop into compromise systems, in time. In the less developed parts of the world, some forms of subsistence agriculture, taken in context, are essentially compromise, as primitive agriculture has been historically.

The protective, stable systems of "old nature" are not only strongly buffered against their own deterioration, but provide water and gas exchanges, waste metabolism, and nutrients for the areas far beyond their borders. They thus provide overall balance in the planetary ecosystem and give recreational, esthetic, and spiritual benefits essential in an increasingly crowded world.

The subterranean resources are largely minerals; these geological requirements are exploited by mining and are nonrenewable. Subsoil resources will be considered here only insofar as their development substantially alters or deteriorates the land surface.

On urban-industrial lands the nature and fertility of the soil itself is relatively unimportant compared with the location, local topography, drainage, water supply, underlying geology, and climate. The same is only slightly less true of suburban areas, which for the present purpose may therefore be considered a subdivision of the urban-industrial land type.

Independent Biotic Systems

National Parks

The park and wilderness lands that have high protective but low productive importance are often considered "undeveloped." This is true enough of wilderness areas accessible at most only by trails for hikers and horsemen, but it holds only partially and relatively for the great national parks, which are suitably developed for a number of intangible and nonconsumptive public uses.

This is not to say that land serving for recreation and preservation constitutes a "noneconomic set-aside" or a "nonproductive lock-up" as some claim. In 1967 our national park visitors contributed $6.4 billion to the national economy, generating $4.8 billion in personal income on which an estimated $950 million in federal taxes were paid. To equal the $4.8 billion income, an ordinary industry would have to invest $119 billion at a 4 percent annual return, but the amount was realized from a Congressional annual appropriation of only $102 million for the parks, a remarkable benefit/cost ratio. The calculations are extremely conservative, omitting 25 percent of the 140 million visitors because they were classed as nearby residents, day visitors, or otherwise minor spenders. About 3 million foreign visitors spent an estimated $320 million visiting our national park system in 1968. Obviously, there are roads, campgrounds, lodges, museums, and trails to accommodate the public, but these developments even now are so diluted within the 24,676,000 acres of the national parks and national monuments as to permit classing such lands in general as independent systems, largely withdrawn from the more exploitative aspects of "multiple use." Appropriately, the National Park Service is a branch of the Department of the Interior. Each park was selected as the outstanding example of the type of natural features it represents. Lodging, buses, and so on, are provided by private concessioners. The visitors in these parks should "take nothing but photographs and leave nothing but footprints."

The national park idea was conceived by Cornelius Hedges and was agreed to by others of the first expedition in what is now Yellowstone National Park. This first national park in the world was established in 1872. Our system now comprises 38 national parks, 82 national monuments, and 13 national lakeshores and seashores. The "monuments" are natural rather than artificial ones; the average area of a national monument is 120,198 acres.

The United Nations lists 93 countries as having 1205 national parks and equivalent reserves, occupying about 360,000 square miles. Seven countries have 5 percent or more of their area in such parks; scenic New Zealand has 8 percent. The United States has about 3 percent, including state parks and wildlands under all federal agencies.

National parks serve two major purposes, which their popularity in a growing United States has recently rendered somewhat incompatible. The 1872 Congress provided for ". . . the preservation from injury or spoilation of all timber, mineral deposits, natural curiosities or wonders . . . and their retention in their natural condition." The act also declared Yellowstone "dedicated and set apart as a public park or pleasuring ground for the benefit and enjoyment of the people." The dilemma of the Park Service is to maintain a reasonable balance between its two functions. With increasing recreational demands without correspondingly higher funding for needed protection, the use (and abuse) aspect has come to dominate. As early as 1950, Bernard deVoto proposed temporarily closing our national parks to permit recovery. At Yosemite and a few other parks, campgrounds became outdoor slums. While extreme concentration tends to protect the finite park lands from undue sprawl of facilities, the congestion greatly detracts from enjoyable outdoor experiences.

Some parks have separated trailers and tents into different campgrounds. National park campgrounds are larger and the camp sites much closer together than in national forests; the latter have many scattered, small, attractive campgrounds with more "primitive" arrangements and tend to appeal more to those outdoorsmen who do not regard travel-trailer life as camping. The influx of uncontrollable numbers of would-be outdoorsmen has spread discarded trash over America's finest scenic wonders, trampled vegetation, eroded trails and overwhelmed facilities. In 1972 there were 211,621,000 visits to the national parks, a 5.5 percent increase over 1971. The numbers of people make it impossible to administer the parks "in such manner and by such means as will leave them unimpaired for the enjoyment of future generations." Drastic direct limitations on public use of public property seem out of place in a democracy but may be preferable to the accelerating wear and tear on these national assets. Use zoning is commonplace in Europe with full public approval; people there have long had large populations on too little land.

For millennia England's pastoral agriculture has been so intensive that no wilderness and very little compromise land existed there. Can a nation, then, do without these systems? The seafaring British had wide access to less dependent systems by importing raw materials from and exporting their young people to the wilderness and compromise regions of the colonies. British scientists and adventurers sought intangible satisfactions far from their tamed country-side—Wallace collecting new species in the Malay Archipelago, young Darwin aboard the Beagle, Shackleton and Scott in Antarctica, and Hunt on Mount Everest.

Most wilderness in America is publicly owned and administered by the Forest Service, Park Service, or Bureau of Land Management. A real wilderness offers the experience of long-continued isolation in nature which is regarded very highly by persons who find that, due to modern communication and transportation, the greatest luxury today is just the chance to be alone. Wilderness offers a degree of remoteness from the excesses and tensions of civilization, a very extensive natural landscape and the presence of native big game animals and large predators. (Deer do not require wilderness.) The term is often used in a figurative sense, in literary writing, for natural areas which lack the above mentioned qualities but within which subjective feelings approaching a wilderness experience may be possible, or where the young may get a start toward later appreciation of true wilderness values.

Much biological and geological research on these outstanding areas has been published, but relatively little of it was by park staff members. A great city museum considers its research at least as important as its educational function, but Congress has never encouraged science in our national "outdoor museums." On the other hand public educational work has been very effectively done by the park naturalist staffs in evening lectures, all-day hikes, auto caravans, interpretive nature trails, and splendid visitor centers (museums). People on vacations are eager to learn about the superlative environments in America's show places from enthusiastic and well-qualified young men and women employed as seasonal ranger-naturalists. This promotes perceptive-recreation that is constructive, whereas many forms of mass recreation cause attrition and deterioration of outdoor resources, consuming their base. Types of recreation perfectly legitimate elsewhere are distinctly inappropriate to national parks; hence, meeting the needs for increased facilities for those nearby but outside the parks would alleviate some unnecessary pressure on park quality. State parks serve, on a more modest scale, the same recreational, esthetic, and scientific purposes as national parks. While several states have fine park systems, many state parks now suffer from overuse even more than national parks. The idea that merely getting more people into existing parklands promotes conservation is a fallacy propagated by more than a few public agencies. Land suitable for parks is generally rough and unsuited for cropland; it is important to set aside soon as much of this land near population centers as possible. Having such facilities nearer home would make outdoor recreation available to more people more often, while helping to relieve the remote wilder parklands from overcrowding.

The problem of abuse of national and other parks is like that of the human habitat in general; both are soluble only by greatly improved environmental education by the school and home from very early life, and abandonment of the American illusion that exponential growth can continue indefinitely.

WILDERNESS IN NATIONAL FORESTS

Although the great bulk of land under the Forest Service is under multiple-use management and timber is harvested, 88 units totalling 14.6 million acres in 14 states are protected as wilderness. In addition, a very minor fraction of Forest Service land is preserved as small natural areas, scattered through agricultural-industrial regions.

In the Wilderness Act of 1964, Congress recognized the remaining wilderness resource as a valued part of the nation's heritage, and established the National Wilderness Preservation System under the Departments of Agriculture and Interior. In effect, the Act confirmed the wilderness area concept which the Forest Service pioneered in 1924 by administrative action, and prescribed a 10-year review period for adding present primitive areas in the national forests and potential areas under the Park Service and Fish and Wildlife Service. When a tract comes under the Wilderness Act, it continues under the same administrative agency as before, but limitations on resource use and physical developments become better assured. Commercial timber harvesting, roads, motor travel, or any permanent developments are prohibited except where necessary for administrative needs. All the national forest wildernesses are open to hunting and fishing in season under state laws. Clear lakes and pure streams provide excellent fishing in most.

Among wilderness wildlife (Table 9.2) we find the largest and wariest of large mammals and birds, as well as the black bear and deer that often occur in compromise systems also. And, for the student of nature, the geology and ecology provide fascinating subjects in undisturbed settings. Backpackers pit

Table 9.2. Some American Wildlife Species and the Major (H) and Minor (h) Biotic System in Which Each Has Its Characteristic Habitat, Especially in Its Breeding Season. A blank does not necessarily signify the animal is completely and always absent, since some animals spread widely during migration.

	Dependent Systems	Compromise Systems	Independent Systems
Ungulates			
Antelope, pronghorn		H	
Bighorn sheep		h	H
Bison			H
Caribou			H
Deer, mule		H	h
Deer, whitetail	h	H	h
Elk		h	H
Musk ox			H
Peccary		H	h
Large carnivores			
Bear, black		H	h
Bear, grizzly			H
Bear, polar			H
Cougar		h	H
Coyote	h	H	h
Wolf			H
Wolverine			H
Fur bearers			
Beaver		H	h
Ermine		h	H
Fox	H	h	h
Marten		h	H
Mink	h	H	h
Muskrat	h	H	h
Opossum	H	h	
Otter		h	H
Raccoon	h	H	h
Skunk	H	h	
Farm game			
Bob-white quail	H		
Cottontail	H	h	
Dove, mourning	H	h	
Partridge, Hungarian	H		
Pheasant, ringneck	H		
Squirrel, fox	H	h	
Squirrel, gray	h	H	
Raptorial birds			
Condor, California			H
Eagle species		h	H
Falcon, peregrine		H	h
Hawk, red-tailed	H	h	
Hawk, rough-legged		H	h
Osprey		H	h
Owl, barred	h	H	h

	Dependent Systems	Compromise Systems	Independent Systems
Owl, great-horned	h	H	h
Owl, snowy		h	H
Other birds			
Crane, whooping			H
Duck, many species	H	H	H
Geese		h	H
Loon		h	H
Grouse, ruffed	h	H	
Grouse, sharptail		H	h
Prairie chicken	h	H	
Ptarmigan			H
Songbirds	H	h	h
Swan, trumpeter			H
Woodcock	h	H	

their stamina and skill against mountain trails; others obtain saddle horses, pack animals, and professional guides. Detailed maps are supplied free at ranger stations, and, for those who still go astray or meet with accidents, search and rescue operations utilize aircraft if necessary to meet these emergencies. During 1968, about 5 million visitor-days of use occurred in U.S. wilderness and primitive areas; as the number of people enjoying the wilderness scenery and solitude increases yearly, the solitude itself tends to decrease.

Natural Areas and Nature Preserves

Man easily and lightly destroys natural areas but cannot, with all his technology, create one. He could, however, prevent the devastation of the few small remnants of undisturbed nature which have chanced to escape development within heavily populated regions. This effort was initiated by a group of ecologists, and is being pushed vigorously by several private citizens groups, especially the Nature Conservancy. Eight of the states have governmental programs.

A natural area is any outdoor site that contains an unusual biological, geological, or scenic feature, or else illustrates common principles of ecology uncommonly well. A nature preserve is a natural area which has been formally dedicated or legally committed as such, in what is intended to be permanent status, by a private, governmental, or corporate owner. These areas range up to a few hundred acres, but most of them are much smaller. Many are owned by colleges and universities. State departments of conservation are supported by hunting and fishing license fees, and they tend to serve sportsmen's and mass recreation interests almost exclusively. Nature preserves are too small and vulnerable for ordinary mass recreational use such as hunting, camping, or picnicking. Suitable uses are for outdoor laboratories in science and education, preservation of rare and endangered organisms (of possibly great future practical value), and perceptive-recreation involving nature hobbies and esthetic appreciation. Nature seen from the vantage point of the civilized life has for many people spiritual values at least equivalent to great art, and the vision of the wild as the Creator's handiwork is quite sufficient motivation for those who wish to protect it from destruction by power lines, highways, or subdivisions. Examples of original ecosystems are needed also as base-line, untreated controls for comparing with and evaluating the results of land management operations in practical technologies such as forestry, range management, and soil conservation. By retaining environmental diversity, natural areas reduce monotony and make human experience more interesting. They are cultural assets equal to planetaria, aquaria, zoos and botanical gardens, but differ in that, once destroyed, natural areas cannot be replaced. Hence, preserving them holds some land-use options open for decision by future, perhaps wiser, generations. It is ironic, and may prove tragic, that we have just come to recognize the crucial importance of living systems at that point in history when so many of them are in imminent danger of being squeezed to death by the encroachments of civilization.

Compromise Biotic Systems

Timberlands

Multiple use and sustained yield Forests that supply our needs for wood are compromise systems in that they are, or should be, managed by foresters not for one purpose solely but rather on the principle of multiple use. (Multiple use of natural resources is not applicable to nearly the same degree in either dependent biotic systems, which yield chiefly economic values, or independent systems which yield mainly noneconomic values.) Compromise or multiple-use systems serve well for both tangible and intangible values; they stand between the other two systems with respect to the traits listed in Table 9.1. The land manager in a compromise system aims for the best *combination* of uses. Since not all possible uses are compatible in the same place at the same time, the total forest area consists of a mosaic of differently used subareas in some of which the uses shift with time, and in others, as nature preserves, use should remain constant perpetually. Forestry is defined as "scientific management of forests for the continuous production of goods and services." A key word here is "continuous" which implies the objective of sustained yield. Timber is a renewable resource and forestry stresses growing wood, not mining it. Under sustained-yield management, no more lumber or fiber is harvested annually than is added by tree growth. Other forest values—water, wildlife, forage, and fun—are also on a sustained-yield basis in the national forests and lands of the more progressive, usually larger, timber-producing companies. The services and products are to be used or taken only at a level that can be continued indefinitely without harming the land's ability to produce. Unfortunately, many private owners of forest lands still manage them very poorly for the long-term productivity of their land and for the forest values other than wood. Modern forestry is a complicated field, not yet fully appreciated by owners who may fail to get, or accept, competent professional advice. This is available free through state extension foresters.

The wood-using industries in this country depend on government-owned land for more than one-fourth of their raw material. Standing timber grown in national forests is sold to private lumbermen who do the cutting. The Forest Service, U.S. Department of Agriculture, administers 154 national forests and 19 national grasslands in 41 states and Puerto Rico, comprising 187 million acres. It is noteworthy that this is about eight times the combined area of the national parks and national monuments.

Forest cover protects watersheds, promoting infiltration of water into the soil and preventing erosion on steep slopes. It tends to reduce floods and stabilize flows of clean surface and groundwater supplies for domestic and industrial needs in hundreds of cities. Snow survey teams measure the valuable snowpack in mountain forests, where conifer trees prolong its melting, and predict the water that will become available for irrigation in valley croplands.

National forests and grasslands and areas under the Bureau of Land Management provide habitat for big and small game, game birds, and fish. Much of the nation's finest hunting and fishing is available there, subject to state conservation laws. Visiting sportsmen employ guides and packers, and inject funds into local economies in exchange for lodging, food, equipment, and gasoline.

The forage grown in national forest areas is handled by range management experts, who lease use of lands for grazing to private stockmen for rather nominal fees. About 1,300,000 head of cattle and twice as many sheep are run on these public lands. Some compromise between uses enters in, since trampling and grazing in excess of carrying capacity (which may be interpreted differently by stockmen and range management officials) can inhibit forest tree reproduction and bring about loss of topsoil and increase runoff.

These lands are open the year round for public enjoyment, and are accessible to most citizens since few states lack a national forest. Accommodations are numerous in privately operated resorts and towns in or adjacent to the forests. The Forest Service provides 7665 campgrounds and picnic areas. Campfire permits are required in some forests. Forest recreation includes scenic drives, hiking, swimming, skiing, boating, nature lore, camping, hunting, and fishing. Contributing to the visitor's understanding and pleasure is the interpretive service, which provides self-guiding trails and auto tours, campfire programs, roadside overlooks, visitor centers, signs, and exhibits. This program promotes the public appreciation that reduces forest fires, vandalism, littering, and other abuses stemming from uninformed boorishness.

Other special areas besides wilderness are the 120 scenic areas in 49 states that total 810,000 acres. There are also historical, archaeological, geological, botanical, memorial, and virgin areas. Some of these special areas are really nature preserves and therefore also represent independent systems forming parts of the varied land-use mosaic within an overall compromise biotic system, contributing to the multiple use of the larger area. This is quite proper, since multiple use obviously cannot be rigidly applied to every acre. These special high quality portions are restricted to their best use rather than put to all possible commonplace uses. In our homes, multiple use applies to the house as a whole, but we do not use each room for every purpose. The term "multiple use," like "conservation," is often misused; those who call for "multiple use" of wilderness areas, national parks, and preserves wish to destroy them as such, by changing these independent systems into compromise systems through opening them to commercial lumbering and over-exploitation.

Enlightened federal and state land use policy is needed for maintaining nearly all aspects of environmental quality, as well as the nation's physical, social and economic development. Even those land use questions of obviously wider significance still are being treated as a responsibility of local government. The resulting type of land development is often just another form of pollution. Zoning changes and variances are in local politics, but land use planning has been nearly isolated from the political process, hence ineffective. A modest start in meeting public concern with controlling and channeling future growth is now before Congress as the Land Use Policy and Planning Assistance Act of 1973.

Political scientist L. K. Caldwell proposed, in a legal journal, that criteria for public land policy be derived by the ecosystems approach. He wrote, "If human demands upon the natural environment continue to mount, it will become necessary as a matter of welfare and survival to abandon present land policy assumptions for a policy of public management of human environment on ecologically valid principles. An ecosystems approach is essentially a total systems approach."[1] This would have a holistic emphasis, use scientific knowledge to determine goals as well as techniques, and apply administrative means in preference to adjudication. Its substance is to identify, to protect, and, in the interest of human welfare, to manage the natural ecosystems upon whose continuing viability human welfare depends. "So far as feasible, an ecosystems approach allows natural processes to carry on the work of self-renewal unassisted by human effort."

Techniques of forestry Like agriculture, forestry is an applied science or "practical art." Its practice is based on understanding of forest ecology,

[1] "The ecosystem as a criterion for public land policy." *Natural Resources Journal* 1970, **10,** University of New Mexico School of Law, pp. 203–221.

silviculture, entomology, wildlife, meteorology, geology, soils, genetics, economics, sociology, machinery, forest utilization, and wood technology. These subjects, and more, are studied in the forestry curriculum. Qualified foresters are employed mostly by wood products firms, federal agencies, state departments of conservation, and forestry schools, or set themselves up as consulting foresters. The profession attracts young men, and a few women, who like outdoor life.

The dynamics of forest development and the natural distribution of forest types are significant in forestry, as in wildlife and soil conservation. Henry David Thoreau was the first American observer to write about succession after land disturbance. A sequence of slow changes progresses on the same site from a pioneer stage, tolerant of extreme and difficult site conditions. The cause of succession lies in the changes induced in the substrate and microclimate as a given stage develops. Its plant and animal community produces physical and chemical effects that make the habitat less suitable for that stage and more so for a different community that invades and eventually replaces it. The final or "climax" stage is static compared with earlier ones, since the requirements and tolerances of its component species fit the matured environment, which the changes throughout the series had been influential in producing.

Understanding succession is a key to successful forest management. Most coniferous evergreen species which are of high commercial value belong chiefly in stages before the climax. To keep such forests productive, ecological principles dictate management that holds succession in the productive earlier stage. In the southeastern pine forests, the ultimate state would be nonproductive scrubby cover of oak and other hardwoods. Early successional species such as longleaf and other pines are fabulously productive and form the basis of forestry in the deep South. For centuries, frequent wildfires and occasional hurricanes kept the landscape unstable and encouraged the pineries. Proper silviculture follows suit, using controlled burning every few years as a management technique.

Recovery of the landscape to productive and protective conditions following an atomic holocaust would depend on succession. Unfortunately the pine genus, which is so important in early woody plant stages, is among the most susceptible of plants to radiation damage, being comparable in vulnerability to man and other mammals.

While fire can be a useful tool of the experienced forester under certain circumstances, random wildfires caused by lightning or careless tourists (or Smokey Bear sneaking behind a tree for a smoke) destructively burned 4,232,000 acres of timber in 1968. The decrease from the 26.4 million burned in 1941 was due to enlightened public opinion, improved protection methods, and a five-fold increase in state and private funds originally stimulated by a federal subsidy program. An interesting spin-off of fire research by Forest Service physicists is knowledge of the incredible fire-storms which would be the most destructive result of high altitude detonations during nuclear warfare, in urban and suburban areas as well as in rural ones.

Man causes about 90 percent of forest fires, which suggests the importance of further preventive indoctrination publicity. In federal forests under any administering agency only 40 percent of the fires are man caused, accounting for only 15 percent of the burned acreage. Smoke-jumpers operating from aircraft have caught numerous small fires before they could spread. In the north-central states, the leading cause of fires was loss of control of debris fires on private property, burning 400,000 acres; the next cause was deliberate incendiarism.

Of the 5.6 billion cubic feet of mortality in growing stock each year, 42 percent is caused by insects and disease as compared to 6 percent lost to fire. Intensive research is needed to develop more effective methods of detection and ecologically acceptable methods of control. Multi-spectral remote sensing from aircraft and orbiting satellites, with results later printed out by machines, is used for detection of tree disease epidemics, insect attacks and resource inventory in forestry and agriculture. Biological and better chemical methods are being developed for controlling pests—specific approaches that will work more like a rifle than a shotgun in the fragile ecosystems.

Progress in fire prevention and suppression has resulted in improved natural restocking, but millions of acres are still in need of reforestation. Tree planting on all ownerships has tripled since 1950. Nurseries produce seedlings for transplanting by machine or hand. A more recent and cheaper method is direct seeding by helicopter or planting drill, after coating the seeds with a repellent to prevent their loss to rodents and birds.

Foresters apply three main methods of harvesting, depending on the forest type: (1) *Clear-cutting* is used where the major tree species require much sunlight for proper growth, reproduce by wind-blown seeds, and are so shallowly rooted that wind is a serious hazard to trees left exposed. Clear-cutting by blocks, customary in the Douglas-fir region, clears out distinct patches of less than 100 acres. (2) *The seed-tree method* is suitable where trees require abundant sunlight and have wind-disseminated seeds, but produce strong root systems so that the isolated seed trees left uncut are not susceptible to windthrow. This is the management technique for the pineries of the southeastern states. (3) *Selection cutting* is adapted to trees that grow well in partial shade, that is, the climax or near-climax species. It periodically takes out only individual mature trees or small clumps of them, leaving very small openings. Suitable forests for selection cutting are the hardwoods of the central and northern states, western yellow pine, red spruce of the northeastern states, and Pacific coast redwood.

The Forest Service has recently promoted clear-cutting in types where selection cutting has been preferred, causing a controversy with environmentalists.

Continuous forest inventory is being aided by computerized methods to supplement the old-style growth measurements by sample plot, diameter tape, and increment corer. Long-term land policies and classifications should be based on surveys that give essential data on land resources and potentials. Continual research in all fields of forestry is being done by nine federal forest experiment stations and many experimental forest tracts, and by the larger commercial firms.

There is currently enough of a backlog of maintenance and development work in the recreational lands of the national forests, national parks, and state parks to justify an effort as great as the Civilian Conservation Corps of the 1930s. Expenditures could serve environmental welfare at the same time as they serve direct human needs. Congress provided for fiscal 1971 $2.5 million for a new youth Conservation Corps for youths of both sexes, age 15-18.

Rangeland

The western range differs from pasture in having precipitation too low for cropland so that all its terrain types are grazed, whereas, in humid regions, pasture is the proper land use for moderate slopes, crops for fairly level ground, and forests on steep slopes or problem sites of other kinds. The regions most appropriately used for range are the high plains or short grass country east of the Rockies, the high Great Basin plateau between the Rockies and the Sierra-Cascades, and much of the Southwest. Cattle do not reach the desired condition while on the range; many are therefore shipped to the Midwest for feeding on grain before slaughtering. As grain farming encroaches on the western range, the South is coming forward in beef production.

Depending on precipitation, site, and grazing pressure, three life forms of dominant grasses are short grass, mid-grass, and tall grass. Short-grass species are potentially knee high, mid-grasses waist high, and tall grasses up to 10 feet tall. The tall-grass or prairie lands lie east of the plains in a region of higher precipitation; in fact, they could have supported forest had it not been for recurrent grass fires and, often, soil either too wet or too well drained for tree growth. The original tall-grass land is not rangeland today, but is cultivated for corn, soybeans, small grains, and so on. Just as the climate in the prairie could have supported forest, that of much of the Great Plains could have sustained a mid-grass climax but for the intensive grazing by bison, antelope, and domestic cattle. Much of this great landscape type is seriously overgrazed due to increasing demands for beef, and much former grazing land has been taken over for wheat production by special "dry farming" methods, or by mining the groundwater for irrigating crops.

Since range management is applied ecology, the successful manager must not only know the herbaceous plants but also needs a thorough knowledge of plant succession dynamics. When the vegetation is degraded by grazing pressure it reverts from near climax to a pioneer condition with dominance of unpalatable annual weeds. The range manager must judge the safe carrying capacity of his range for grazing animals. Weeds do not cause poor range, but are merely a symptom of overuse. Jackrabbits and ground squirrels are animal "weeds" having higher populations in early successional stages. Using herbicides to kill annual weedy plants on rangelands is foolish, since weeds represent an advance over bare erodable ground and start the successional process. However, mesquite is an aggressive woody plant in the Southwest that invades rangelands and shades out grasses. Removing it by herbicide treatment may be necessary for restoring productive grass cover.

Grassland ecosystems have a fascinating diversity of living things, but there are far too few protected natural areas. The kit foxes, coyotes, hawks, and owls are predators useful for holding the rodent hordes in check. Man's "predator control" and "rodent control" are unecological, inhumane, and usually uneconomic activities in range country, yet only starting in 1972 were funds drastically cut for this type of activity, and probably for the wrong reason.

In much of the Great Plains, where one year's precipitation is too low to yield a crop, "dry farming" is practiced. To exploit this area for wheat, farmers alternate strips of planted ground with fallow strips in large fields. The fallow strips store most of that year's precipitation within the soil, and are planted the following year, so that each strip yields every other year one crop from two years' moisture supply.

In the early 1930s high grain prices combined with the preceding wet years to encourage plowing up the protective sod to make wheatfields. The drought years 1935 and 1936 brought widespread wind erosion and the "grapes of wrath" in the Dust Bowl country of Oklahoma, Kansas, Nebraska, and Texas in the eastern belt of the plains. (See *Deserts on the March* by Sears.) This exemplified the disastrous consequences of pushing a dependent system into lands suitable for a compromise system. Such ecological lessons that the earth is finite are short-lived because human greed and need appear to be infinite.

A decade of wet, cool (high effective moisture) years starting in 1959 brought about still greater expansion of cropland in the high plains. If the alternating fallow and planted strips run at right angles to the prevailing wind direction, and crop residues are left to mulch the ground, wind erosion is less damaging. Some speculators in Great Plains farming expect to make killings in moist periods and relinquish the land in the unfavorable part of the weather cycle, matching the "cut out and get out" viewpoint of the early timber barons in the Lake States. Newer knowledge of wind microclimatology and protective techniques may give an adequate basis in theory for preventing future dust-storm catastrophes. However, because of the gap between agronomic recommendations and grass-roots practice, and the artificial economic incentives that are traditionally provided agriculture by government, other Dust Bowls are certain to recur, given certain conjunctions of circumstances.

Wetlands

A simple test of a person's outdoor understanding is his reaction to the word "swamp" or "bog." To the ecologically naive, it appears that wetlands are

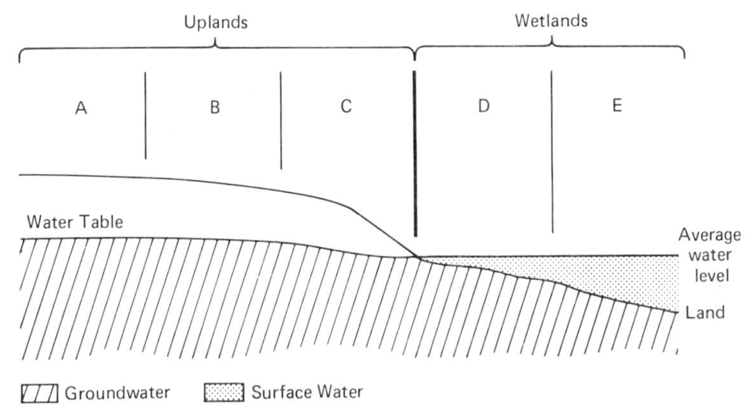

Figure 9.1 Soil Drainage-Aeration Profiles. On Uplands, A Marks Level or Nearly Level Upland Soils of the *Imperfectly Drained* Profile, B Is the *Well-Drained* Profile from 2 to 15 percent Slope, and C Is *Excessively Well-Drained* Soil of 15 to 55 percent. The Level of Surface Water Fluctuates Seasonally, and the Level (Water Table) of the Groundwater Reservoir to a Less Degree. Lowland Soils Marked D Are *Seasonally Ponded,* Being Under Water Six Months or More during a Year of Normal Precipitation, unless Drained Artificially. Those Marked E Are Called *Permanently Ponded,* Being Submerged Eleven Months or More during a Normal Year.

landscape nuisances and obviously should be eliminated so that the ground may be used for some purpose other than producing mosquitos. To the scientist, a wetland may be the most fascinating segment of the landscape, revealing and illustrating earth history, full of extremely unusual species and communities, and important in the water cycle.

Figure 9.1 shows the five drainage-aeration profiles determined chiefly by topography and partly by marked differences in soil texture and permeability. Water from precipitation generally runs off the high ground and tends to collect on low ground. This obvious difference in drainage, added to effects of climate, parent material and vegetation, produces a particular identifiable soil series when sufficient time elapses. The figure legend explains the water relations of the surface soils.

The classes of wetlands are named according to the dominant life-form of their vegetation. *Wet prairie* is a relatively shallow wet area covered chiefly by tall grasses and sedges. It is ponded a shorter time each year than other wetlands, and occurs geographically in prairie country. A *marsh* is also herbaceous, but develops where the water level is more dependable. Typical marsh cover is of cattails, reeds, and rushes. A *swamp* has woody plants as dominants, so that two kinds are tree swamps (for example, cypress) and shrub swamps. Deposits of organic, largely plant, remains may be built up in marshes and swamps, but peat is much better preserved in *bogs,* which are undrained and hence quite permanently ponded, with low oxygen supply, usually highly acid reaction, and a quaking mat of floating, intertwined plants. These conditions inhibit decay-microbes and the direct oxidation of the organic matter, so that plant parts are recognizable. Pollen grains (of the past plant life) that blew into bogs from surrounding lands can be analyzed to show the history of forest change since the glaciers melted. The unusual flora and fauna include species found on uplands only hundreds of miles farther north, even some typical of arctic tundra. This is caused by the cool summer temperature and the peculiar chemical and physical conditions in bogs.

The fertile black lands of the formerly timbered, or eastern, part of the Corn Belt were opened to agriculture by artificial drainage. These are the low ground soils, especially Class D of Figure 9.1, that are rich in organic matter and nutrients. In Class E, when brown peat is exposed to air by lowering the water table, it is oxidized to black muck. Muck soils are used in the recently glaciated areas to produce mint, potatoes, and truck crops.

Wetlands act as blotters to absorb water, releasing it gradually, thus helping stabilize stream flows, lake and pond levels, and the underground water table. Wetlands are essential as breeding grounds for a variety of fish, waterfowl, and fur-bearers, and serve as resting and feeding areas for migrating ducks and geese. Salt marshes and estuaries along our coasts are richly productive for the fisheries and oyster industry, but have suffered severely from pollution and encroachment for real estate use. Too many wetland areas have been drained for agriculture or otherwise exploited, and the programs of different government agencies have worked at cross purposes. On balance, many governmental units should be included in the list of the ecologically naive who promote wetland over-exploitation and destruction.

Although all three biotic systems contain wetlands, they are discussed only under compromise systems because of their combination of great natural diversity with high economic and ecological value.

The Wildlife Resource

The ecological fundamentals that underlie wildlife management are now well understood by conservation professionals, but not sufficiently by their clientele—the sportsmen with guns, rods, and votes who sometimes dictate continuation of fruitless or harmful traditional policies by state game departments.

The nature of the specific ecosystem best able to support the mammal, bird, or fish species; the animal's needs and role therein; and the principles of its population dynamics and behavior are the essentials of this conservation field. The game manager's task is not only to help nature provide an adequate annual harvest during the hunting season but also to insure sustained yield for the future by the carry-over of sufficient breeding stock.

Each game species has its special niche in the ecosystem, in that its functional role or ecological business involves the transfer of materials and energy through the trophic levels in a food chain that links green plants to the ultimate predator and reducer organisms.

The most crucial factor in the welfare of wilderness, compromise area, or farm wildlife (Table 9.2) is having sufficient high quality habitat. Animals need suitable protective cover, food, water, and nesting materials. Many of them have psychological requirements for a particular landscape type or or even for specific dominant plants. Often instinctive territorial needs will enforce limits on the density of breeding pairs possible in an available space. Vegetation is extremely important in determining the habitat and supplying the animals' needs therein, directly and indirectly. Because of a wild animal's high reproductive potential, the carrying capacity of the habitat is the major limiting factor for population numbers; carrying capacity can be improved by management techniques applied to the landscape rather than to the game species directly. Given abundant good habitat, neither the pressure from natural predators nor human hunters can reduce the game population below a safe level of breeding stock. The higher the density, the higher the loss rate to enemies, diseases, and hunters, so that as the surplus population of the year is taken, the principle of diminishing returns for hunting success becomes more and more effective in protecting any breeding stock that enjoys a good habitat. Game crops do not store up from year to year; they may be taken when ripe like other crops, otherwise the overly dense populations will be reduced by natural factors such as predators, diseases, competition for food, and winter storms. In an extensive and suitable habitat, it is not normally possible for hunters to reduce game populations to unsafe levels. Nevertheless, hunting is unacceptable in national and other parks and nature preserves for other sufficient reasons. Conscientious objections to killing animals should also be respected.

Predators (Table 9.2) play an important role in modifying excessive population increases; without them the herbivores would cause earlier and greater damage, during the build-up phase of population cycles, to the vegetational support base. Habitat destruction has often resulted from man's persecuting wolves, coyotes, bobcats, or cougars. Predators most readily capture weak, injured, diseased, or stupid prey, or the young of such females, and thus selectively improve the breeding stock in vigor, health, and genetic endowment. Bounties to encourage predator kills over any general area (as a state) are invariably unwise. Local situations where artificial control is clearly justified can be handled individually though even there it is better to have the foxes, coyotes, or even cougars hunted from an interest in them as quarry rather than for money. Bounty hunting is more a government subsidy program to a few individuals than an effective form of game management. Shooting vulnerable animals from aircraft is easy, unsporting, and abhorrent to public conscience, and should be stopped. Also, general persecution of hawks, eagles, and owls, which are among the most attractive and interesting of birds, is reprehensible. The species shot are usually the large, slow, conspicuous hawks that are decidedly useful predators.

Dependent Systems

CROPLAND

Soil has been termed the "universal catalyst" because almost everything that remains in contact with it for a long time is changed into more soil. But when we bring a sample of soil into the laboratory, it changes into "dirt." We have left outside the organized soil system, a part of the natural ecosystem. While the soil scientist should, of course, analyze certain properties and components in detail by laboratory methods, the most important features—its organization, interrelationships and evolution—must be studied outdoors where the processes are going on.

From the practical standpoint, soil is the normal medium for plant roots, but a better definition focuses on the soil itself. Soil is a natural body on the earth's surface characterized by layers (horizons) parallel to the surface, resulting from modification of parent materials by physical, chemical, and biological forces under the influence of certain conditions during various periods of time. These conditions are mainly determined by parent material, topography, climate, vegetation, and animal life—including man. Soil consists of mineral matter, air, water, living plant and animal microbes, and dead organic matter.

In terrestrial biotic systems, the indispensable elements are solar energy, soil, and atmosphere, to which modern mechanized agriculture adds the requirement of energy from fossil fuels. Applied research in government experiment stations has brought out techniques that have made agriculture highly productive per man-hour. Farms that sell at least $10,000 worth of crops a year are considered commercial. About one-third of our farms are commercial ones, but they produce most of the crops and get 87 percent of all farm income.

The embarrassment of productivity suffered by our farm industry until recently has been relieved by new legislation and by economic and political factors, both domestic and international. A peculiarity of food economics is that a 5 percent increase in production causes a 30 percent decline in food prices. When the supply decreases, prices zoom disproportionately. The 1973 farm legislation deemphasized subsidies, but put a floor under prices.

In 1830 there were 1 billion people on earth, by 1930 there were 2 billion, and, by 1960, 3 billion. Now there are more than 3.8 billion. Worldwide, agricultural production is not keeping pace with increased demands, as many countries of Latin America, Asia and Africa now grow ten times as fast as a century ago. Genetic applications such as development of high-lysine corn help counter the widespread protein malnutrition. Nobel-prize winner Norman Borlaug, a leader in agronomic improvements for underdeveloped nations, stressed the limitations of the "Green Revolution" and the importance of population control. According to President Nixon's July 18, 1969, Message to Congress, "The promise for increased production and better distribution of food is great, but not great enough to counter these bleak realities." The visionaries who proposed mass emigration to other planets have been discouraged by NASA findings of the inhospitality elsewhere in the solar system and by the fact that keeping a steady-state population on earth by this means would now call for shooting more than 200,000 people into space every day.

The part of the earth suitable for cropland is finite and the saturation level is not far into the future. In fact, much former (poor) cropland even in the humid eastern United States has been abandoned and is reverting to forest cover. Wet and moist climates in the tropics cannot support intensive agriculture because of rapid leaching out of nutrients by rainwater, decomposition of humus at the high temperature, and proliferation of destructive insects.

Agricultural scientists have given little thought to a strategy that differs strikingly from traditional row-crop farming. This "detritus agriculture" as discussed by Dr. Eugene Odum in *Science* may someday increase food productivity of the forested tropics and shift temperate cropland toward compromise status, avoiding a number of current problems:

> There is no reason why man cannot make greater use of detritus and thus obtain food or other products from the more protective type of ecosystem. . . . The short-term yield could not be as great as the yield obtained by direct exploitation of the grazing food chain. A detritus agriculture, however, would have some compensating advantages. Present agricultural strategy is based on selection for rapid growth and edibility in food plants, which, of course, make them vulnerable to attack by insects and disease. Consequently, the more we select for succulence and growth the more effort we must invest in the chemical control of pests; this effort, in turn, increases the likelihood of our poisoning useful organisms, not to mention ourselves. Why not also practice the reverse strategy—that is, select plants which are essentially unpalatable, or which produce their own systemic insecticides while they are growing, and then convert the net production into edible products by microbial and chemical enrichment in food factories? We could then devote our biochemical genius to the enrichment process instead of fouling up our living space with chemical poisons!
>
> By tapping the detritus food chain man can also obtain an appreciable harvest from many natural systems without greatly modifying them or destroying their protective and esthetic value.

Conservation of Soil, Water, and Wildlife

The essence of conservation in an independent system is preservation, or letting nature alone except for enjoyment, study, or inspiration. In a compromise system conservation promotes protective buffering of the environment when light management is guided by ecological principles. In the more artificial or dependent croplands (Table 9.1), man guides nature with a tight rein, keeping it immature in order to obtain high net production. But the boom and bust instability typical of early succession with its disturbance effects and few species makes cropland systems vulnerable to many hazards. These include disease epidemics of fungi, bacteria or viruses, insect and other animal pests, extremes of precipitation, damage to soil structure, loss of nutrients by leaching, and loss of the topsoil itself by wind and water erosion. The more natural the solution to such problems, the less the chance for unforeseen side-effects and the better the prospects for long-term welfare of our environment. The techniques of soil and water conservation apply to rangelands and especially to the dependent systems, since lands heavily exploited by man lack adequate protective feedback mechanisms for self-maintenance. Hence, ecological understanding is even more vital for managing cropland than for compromise systems, which nature's buffering makes less liable to damage from misuse and better able to recover unassisted by man.

When the first soil conservation districts were organized in the United States in 1937, the concept of soil conservation was not new. George Washington and Thomas Jefferson practiced it on their lands and wrote about it. The soil conservation movement has become so popular at the grass roots that many people think ''conservation'' means soil conservation alone.

One aspect of soil conservation entails abandoning practices that have caused soil to lose its quality and its ability to produce crops in quantity. Proper soil management considers the populations within the living soil itself, and the organisms dependent on the production from that soil from the lowest elements of the food chain to the highest—man. Science-based inventory and description aid the manager, for lands possess characteristic energy systems and potential for production. But those who own and manage the land are responsible for its proper use. Not many are farming as well as they know how to, from the long view.

Research in humid regions has conclusively shown that multiple use does not work in farm woodlots. Grazing and continued wood production do not mix, because cattle eat up the young trees and trample the soil. If a farmer desires pasture, he should remove the trees; if he desires woods, domestic animals should be kept out.

Crops, and especially row crops, should not be raised on land that slopes steeply or is vulnerable to erosion for other reasons. Bare soil lacks the protection of foliage and the mulch of litter. By far the most important water pollutant is sediment from the land. Soil particles are dislodged and moved away by raindrops (splash erosion), or by runoff water through the headward advance of gullies, or uniformly over sloping ground by sheet erosion. Gully erosion is conspicuous, the other forms are insidious, but all remove valuable topsoil and ruin cropland or thinly-grassed pasture. Good management promotes infiltration to deeper levels for moistening of the root zone and recharge of groundwater supply. The water table is the (somewhat fluctuating) upper surface of the groundwater reservoir, which is a vital, though hidden, natural resource.

Where gullies exist, the first step in restoring the land is to trap sediments above small ''check dams'' built across them. To avoid the start of gullies or other erosion, low broad terraces are constructed on the contour to retard runoff and promote infiltration. Crop rotation of grasses and legumes with row crops has been used to minimize erosion. Legumes are less needed now that commercial nitrogen is cheap.

Conservation methods and the efficient use of modern equipment must be compatible. Size and shape of fields are being altered, and the topography is being changed by land grading to permit installing erosion control systems, to provide smoother fields for large machinery, and to insure more uniform application of water or more orderly removal of excess water. Reducing runoff and erosion simultaneously reduces movement of sediment and other pollutants into the streams. On some farms, grassed waterways for terrace outlets are being replaced by buried tile lines running downslope with vertical perforated risers in low spots in the terrace channels. This fits in with herbicide agriculture.

Many farmers think that "no-till" methods of planting with correct use of herbicides, or at least minimum tillage, is revolutionizing row-crop farming more than anything else since the introduction of hybrid corn. Good yields are obtained while reducing labor cost, cutting soil losses up to 90 percent, and maintaining good soil tilth. No-till planting eliminates plowing, disking, and harrowing, thereby halving tillage costs. Special planters place seeds in soil under the previous crop residue (organic litter mulch), or beneath a cover of grasses or legumes; both methods reduce erosion by water or wind. Herbicide application substitutes for cultivation for controlling weeds.

Farm ponds, used for watering stock, fishing, or supplying water for spray diluent or irrigation, retard runoff and help recharge the groundwater storage.

Flood plains were formed by flooding, and make up natural portions of stream valleys where flood waters should be able to encroach harmlessly, and so pass down the valley more gradually. Since great floods are meteorological phenomena which occurred even in presettlement times when the watersheds were completely forested, neither soil conservation practices nor anything else man can do will eradicate periodic floods. Good conservation starts at the points where raindrops fall, and small scale but universal practices on the land could reduce flood crests and, especially, reduce soil loss upstream which would otherwise cause sedimentation downstream.

"Flood control" by major engineering works can usefully affect only small floods; it cannot substantially reduce the less frequent but disastrous great floods. It is the latter which cause damage to property and loss of life. Effective protection against great floods nowhere exists and would be inordinately expensive and impractical over-insurance. We should be concerned with preventing, not floods, but flood damage. Flood control measures may actually increase flood damage, through the mores of advertising and promotion, by giving the public a false sense of security and promoting construction which will be swept away by the inevitable great flood. Thus, the more flood control we have, the more we "need" beyond that. Forestry and recreation are the best uses of river flood plains, and risk agriculture (without buildings there) is next, but industry and residential uses are insupportable. In towns and cities, flood plains should be reserved for greenbelts, parks, picknicking, bridle paths, hiking trails, and water sports. The flood plain should be given back to the river and used by man more as a compromise system and less as a dependent one. Failing this, improved watershed practices, zoning, warning systems, flood-proofing of structures, and so on, must be combined with the traditional but unsuccessful engineering "flood-control" approach.

Farm wildlife (Table 9.2), not that of the more remote wildlands, provides the most man-hours of hunting, hence has the most economic impact. Farm game includes cottontail, squirrels, pheasant, bob-white quail, some duck species and the gentle mourning dove. Wild game belongs not to the individual landowners but to the states, which license hunters and set seasons and bag limits in accord, ideally, with ecology and sound management principles. Surplus game produced over and above an essential breeding stock is considered a crop harvestable by the public, but subject to permission of landowners.

Many farm boys run trap lines in winter, and sell the pelts of muskrat, mink, raccoon, skunk, opossum, and weasel to the fur industry. Fishing is the most popular of all participant outdoor sports except walking for pleasure. Bird watching is a bloodless form of perceptive-recreation that is rapidly gaining popularity and can be pursued in habitats ranging from one's front yard to the remotest wildlands. National organizations of sportsmen, for example, the Izaak Walton League and the National Wildlife Federation, and outdoor groups like the Audubon Society and Sierra Club, are active in broad and well-conceived conservation and environmental quality programs, and in support of enlightened legislation.

Urban-Industrial and Suburban Areas

When artificial developments invade our few remaining wildernesses this only "adds water to the already thin soup," but the obverse, or bringing nature into suburbs, towns and cities, improves and humanizes these dependent systems. A strong trend in city planning and landscape architecture is to design ecologically, and to light up urban environments by open space, greenbelts, parks, ponds, and streamside recreation spots. This brings "man-made nature" and increased species diversity into a sparse biotic community dominated by people, English sparrows, pigeons, starlings, and rodents. Although naturalistic plantings only simulate natural communities, most Americans prefer the effect over that of the formal gardens favored on the European continent.

The kinds of trees that do best in urban environments are those native to river flood plains, where the periodic floods prevent accumulation of a decomposing organic layer on the soil surface. In a natural forest on upland, the feeder roots are close beneath the rich mat of leaf litter, but this is not true of flood plain species like elm, sycamore, bur oak, green ash, and silver maple. The latter, for example, seldom survives to maturity on upland sites in nature, but, when artificially planted on upland soils in town, it is better adapted to the current absence of leaf mat than the species native to those soils. Besides, mowing keeps out the trees and herbs that would compete under natural conditions there. The reason silver maple does not survive in natural upland situations is not that it cannot tolerate physical conditions there, but that it is smothered out by the upland species which are better suited to those conditions.

The floral beauty of English cities suprises and charms American travelers. Parks, street and sidewalk edges, parking lots, yards, odd spots, and the many windowboxes are alive with colorful cultivars; our drab American towns offer a painful contrast. The Englishman's love of the outdoors and commitment to a better environment are frequently expressed by pottering in the family flower garden after work. While the mild, moist English climate is helpful, the same results are feasible in American communities.

The unemployed in the United States usually number 4 to 5 percent of the labor force and are chiefly city dwellers. Can we not marshall the sociopolitical mechanisms to give many of these people meaningful jobs in cleaning up and beautifying our living space, in combination with recycling of now wasted materials for future needs? About $1 billion worth of minerals are lost annually in municipal waste alone. Men could be usefully employed in the public sector to cope with the 9 million old cars, trucks, and buses that are abandoned each year, by compressing those not now needed for scrap metal into bales and burying them on deposit for posterity's iron resources. Although too much governmental funding still goes into politically motivated pork-barrel projects that damage rather than improve the environment in the long view, great progress has been made recently toward adequate funding for sewage treatment facilities.

Soil erosion is not restricted to rural areas; in urban and suburban construction projects the loose soil may lie bare to erosion for long periods and contribute much sediment to streams. This form of pollution from commercial activity would be easy and inexpensive for municipalities to control, compared with pollution by sewage or smoke.

As one result of the current "ecology movement" many new public ecologists have sprung up like mushrooms, and some are no more substantial. Ecologists, in the original sense of those trained in ecology as a biological science, strongly oppose establishing new cities in the few remaining areas of the independent biotic type.

References

Advisory Committee on Predator Control, *Predator Control—1971*. Ann Arbor: Institute for Environmental Quality. University of Michigan, 1972, p. 207.

ALLEN, DURWARD L., *Our Wildlife Legacy* (revised edition). New York: Funk & Wagnalls, 1962.

ALLEN, DURWARD L., *The Life of Prairies and Plains*. New York: McGraw-Hill, 1967.

BAKELESS, JOHN, *The Eyes of Discovery: The Pageant of North America as Seen by the First Explorers*. New York: Dover Publications, 1961.

BROOKS, MAURICE, *The Life of the Mountains*. New York: McGraw-Hill, 1967.

MATTHEWS, WILLIAM H., III, *A Guide to the National Parks: Their Landscape and Geology* (2 vols). Garden City, New York: Natural History Press, 1968.

ODUM, E.P., "The Strategy of Ecosystem Development." *Science*, 1969, **164**, pp. 262–270.

PLATT, RUTHERFORD, *The Great American Forest*. Englewood Cliffs, N.J.: Prentice-Hall, 1971.

SEARS, PAUL B., *Deserts on the March*. Norman: University of Oklahoma Press, 1935.

SUTTON, ANN, and MYRON SUTTON, *The Life of the Desert*. New York: McGraw-Hill, 1966.

UDALL, STEWART L., *The Quiet Crisis*. New York: Holt, Rinehart and Winston, 1963.

WATTS, MAY T., *Reading the Landscape: An Adventure in Ecology*. New York: Macmillan, 1957.

WHITTAKER, ROBERT H., *Communities and Ecosystems*. New York: Macmillan, 1970.

10

Land Misuse and the Land Ethic

ALTON A. LINDSEY

Land Pollution

Fertilizer Pollution

The use of fertilizers in excess or at the wrong season is a common practice. The potash ingredient is not a problem, being held on soil particles. Whereas phosphates in sewage and detergents pollute water and age or eutrophy the streams and lakes, phosphate from fertilizer is nonmobile in soil. Most land pollution by agricultural fertilizers is caused by the nitrogen part.

Many farmers who have just harvested a good crop and sold it profitably use grossly excessive fertilization late in the same year to reduce income taxes, since the cost of the chemicals comes out of the good year's income. Even after a very rainy subsequent winter there may be enough nitrogen left for the next crop to make this tax dodge economical, but it is ecologic malpractice. (Another reason for fall fertilizing is to reduce the work load the next spring.) In heavy soils, 5 to 10 percent of the nitrogen may be leached down to the water table or carried away in surface runoff, but in porous sandy soils the loss is much greater. From a soil pollutant the chemical soon becomes a water pollutant, producing algal blooms and accelerating eutrophication in streams and lakes that belong to everyone.

Fallout and Thermal Pollution

Industrial smokestacks release a great variety of substances causing air and land pollution especially. Soot falls in cities in tons per square mile annually. Fly ash is usually siliceous or mixed material of large particle size, which drops relatively close to the source.

In dry climates, such as our Southwest, SO_2 largely remains in the air, damaging plant leaves. In humid climates the SO_2 unites with atmospheric humidity to form sulfuric acid. This damages crops more severely in the industrial Midwest than in other parts of America. The acid collects also on the soil surface, and often changes the soil reaction. About $10 million annual crop damage is attributed to acid fallout pollution in the United States. Smoke particles between 0.1 and 5 microns may serve as condensation nuclei, increasing rainfall and snowfall and bringing down acid from air to ground. Precipitation at LaPorte, Indiana, which is downwind from the Gary industrial area, has been correlated since 1925 with the rate of steel production in the latter area, and measured about 35 percent above the normal precipitation. Many metals in smoke accumulate on the ground and on vegetation.

Lichens are lower plants seen on rocks and tree bark in the country. Their absence in cities indicates pollution fallout and gases. *Ailanthus* trees (Chinese sumac) and ginkgo tolerate polluted cities better than other trees.

Weather modification, usually to increase rainfall by cloud seeding, is the first of the many modern techniques that could produce large scale environmental changes which is being approached with the necessary degree of caution. Many direct and indirect effects on ecosystems of changes in the amounts and regime of precipitation are foreseen, but only effects of the fallout of seeding agents need be mentioned here. Silver iodide is almost the only seeding agent being used for experimental cloud seeding; it can be divided readily into fine particulate smoke, and small quantities start ice crystal formation. Silver is much more toxic to fish than to land vertebrates; it interferes with gas exchange in the gills. It is only moderately harmful to mammals. Since most land plants do not take it up actively, land food chains are not likely to concentrate it at higher trophic levels. Aquatic organisms are reported to effectively concentrate silver and other heavy metals; oysters can build up a concentration of heavy metals 18,700 times that in the water around them. The most probable harm to ecosystems is through inhibition of growth of algae, fungi, and bacteria; this might affect sewage treatment processes.

Iodine is an essential element for animals and is fairly abundant in ecosystems; the amounts introduced by cloud seeding with AgI would be a negligible addition.

The lead from anti-knock gasolines is now a serious environmental pollutant. Its concentration is increasing in food, water, and air. Often its level in human blood is so high as to be associated with subacute toxic effects. Lead pollution, along with smog from exhaust hydrocarbons, underlies the current search for a practical substitute for the internal combustion engine and the encouraging of mass transport developments.

The Food and Drug Administration estimates that Americans are being exposed to 500,000 different alien substances, many of them over very long periods. Less than 10 percent of these have been analyzed so as to give a basis for determining their effects or assessing the potential hazards, including synergistic interactions between contaminants and drugs, food and drugs, and among different drugs. It seems obvious that the emphasis of technology must shift from exploitation to analysis and guidance of man's activities toward his long-term adaptation to a viable environment.

Radioactive isotopes from industrial, military, research, and power applications are cycling in our ecosystems in the same manner as the stable forms of the same elements, are taken up in food chains, and thus find their way into human bodies. They cannot be really "disposed of" since nothing we can do will change the rate of their nuclear disintegrations. Low-level wastes are diffused and dispersed throughout the natural environment, while intermediate and high level ones are buried at AEC reservations, or concentrated, confined and hidden in deep geologic strata in several ingenious and doubtfully safe ways.

The radioisotopes of most serious ecological consequences are ^{131}I (with a half-life of only eight days), ^{90}Sr (28 years), ^{137}Cs (27 years), and ^{14}C (with a 5568 year half-life). Since the United States and Soviet Union have banned their atmospheric tests of nuclear weapons, surface pollution from military sources has been greatly reduced, now coming mostly from unintended venting of underground explosions and accidental land pollution near plutonium plants.

Although the coolant water circulating through power reactors becomes somewhat radioactive from tritium (a genetic hazard), the chief reactor difficulties are accidents and thermal pollution. Calculations show that if the 7 percent per year increase in energy use were to continue, the 15° C increase to a world mean surface temperature of 30° C would take only about 130 years, after which period the earth would be uninhabitable for man and most other organisms.

PESTICIDES

Use and misuse Chemical methods of controlling insect pests have been widely used since World War II, when DDT proved effective in killing malarial mosquitos in the Pacific arena. Although control of pests is basically an ecological problem, economic and medical factors completely dominated public policy on pesticide use until about 1968.

Of the thousands of insect species that are more or less harmful to man's interests, several hundred are so destructive as to require continuous control measures. U.S. farmers spend $3 billion annually to reduce losses from animal and plant pest species, including weeds. Research for more efficient and less damaging methods, emphasizing biological and integrated chemical-biological methods, is somewhat belatedly in progress. Nearly one billion pounds of pesticides were either produced in or imported into the United States in 1964. The total U.S. consumption of pesticides in that year had a value of about $1 billion. In 1965, 100 million acres of agricultural land were treated with insecticides, while herbicides were used on 85 million acres. About one-third of our total cultivated crop acreage gets general applications of pesticides each year. Only 25 basic chemicals make up 90 percent of the pesticide use in this country.

It was through sheer good luck rather than intelligent foresight that wholesale dissemination of poisons throughout the world ecosystem did not disrupt the nitrogen cycle and unbalance the atmosphere by affecting essential soil microbes. As with most other contaminants with which modern technology has burdened the ecosystem, the pesticide policy was to shoot first and ask questions afterward, or, if economic profit was threatened, to hush up the questioners. Ecologically, there are no ''side effects'' from anything we do—there are only *effects*.

Insecticides and herbicides may be applied directly to the surface or drilled into the soil to control soil insects or as a pre-emergent treatment against weeds. Also, about half of sprayed chemicals end up in soil from missing the target or washing off the foliage. The chief insecticide residues in soils are chlorinated hydrocarbons, which degrade to nonpoisonous compounds only very slowly. This persistence is advantageous economically for that part remaining in the intended place, but disadvantageous ecologically when wind, surface water runoff, and leaching carry them into general circulation. One study showed these percentages still in the soil 3 years after application: DDT, 50 to 78 percent, dieldrin 40 percent, chlordane 15 percent, heptachlor 10 percent, and aldrin 5 percent. The maximum time these may persist in the environment is not known. These chlorinated hydrocarbons have non-polar molecules, hence they are fat-soluble rather than water-soluble, and are held by the organic fraction of the soil more than by the clays. The common organo-phosphorus insecticides are far less persistent, degrading in the soil in months or sooner. Arsenicals accumulate readily, especially in fine textured soils.

Pesticides attenuate or weaken in the soil environment much more readily than in bodies of water, thus they are more troublesome when washed over the surface, or when reaching the air by unsettled sprays or on dust or by codistillation, than when the chemical penetrates below the soil surface. Pesticides as a group attenuate through decay, sorption, and dilution. The great range in subsurface conditions brings about great variability in movement of contaminants. Changes in soil-water conditions cause the distances traveled to vary by several orders of magnitude, posing serious dilemmas for health and waste management officials enforcing often rigid laws and policies.

The persistent or hard pesticides are exemplified by DDT, whose degradation products DDE and DDD remain quite potent poisons. Chlorinated hydrocarbons apparently have a half-life in soil of 10 to 15 years. The least objectionable place for them is in the soil zone of aeration, certainly less so than in the zone of saturation or in streams. Hence the deeper the water table the less objectionable is use of hard pesticides. The latter travel vertically downward in the aeration zone, but after reaching the water table they move laterally in the ground water from which they may reach surface waters and the seas.

Most pesticides are unselective, killing useful insects like bees and the predatory insects that help keep down the pest populations. Many long-known destructive effects of DDT are on wildlife species. Although young fish and other aquatic animals are most highly susceptible, this discussion will consider land animals. Books and countless research papers have abundantly documented wildlife population losses and regional extinctions. Animals are affected either directly or by reduction of their food supply. The chlorinated hydrocarbon insecticide group to which DDT belongs is toxic to the central nervous system. In birds the substances also disrupt reproduction by reducing eggshell thickness.

Robin populations were hard hit by elm-spraying programs because earthworms accumulate chlorinated hydrocarbons. Where pesticide residues in Wisconsin made up 19 ppm of the dry soil, earthworms from the same soil contained eight times that concentration. Up to 250 ppm has been found in brain tissue of dead and dying robins. Due to spraying of a bottomland forest by 2 pounds of DDT annually for four years, American redstarts declined 44 percent, parula warblers 40 percent and red-eyed vireos 28 percent. The same rate of application of aldrin in Illinois against Japanese beetles caused nearly complete elimination of many songbird species, heavy mortality of gamebirds, and some mortality of mammals. Heptachlor and dieldrin have also caused heavy mortality among birds. The early federal poison campaign in our South against the fire ant (a mere nuisance) blanketed 2.5 million acres and practically eliminated bird life. Long-term and indirect population effects among vertebrates have also been demonstrated, including upset behavior patterns and learning ability, leading to speculation about the possibility of such effects on humans.

Species at the tops of food chains have proved the most vulnerable to DDT and its breakdown product DDE. This applies more strongly to carnivorous birds than to mammals. The latter have been killed locally, within the areas of application, but birds that eat fish or other birds suffer from the universal ecosystem contamination. The terrestrial peregrine falcon, sparrow hawk, prairie falcon and accipiters, and the fish-eaters, including osprey, bald eagle, brown pelican, double-crested cormorant, shearwaters, and Bermuda petrel, are among those known to be severely depopulated. The falcons and other hawks may be considered good biological indicators of environmental damage from pesticides. The fatty tissue of bald eagles contains up to 2800 ppm of DDE, which has one-fifth the toxicity of DDT on the central nervous system.

Starting in 1950, the majestic peregrine falcon, the most admired predatory bird, suffered a catastrophic population crash in North America and parts of Europe. The bird was completely extinct in the eastern United States by 1965. An international conference of specialists who met in 1965 to discuss this disaster produced an objective book giving ecosystem pollution by chlorinated hydrocarbon pesticides as the cause.

Results of experiments corroborate the field observations on the reduction of eggshell thickness and resulting egg breakage, with failure to lay any, or enough, eggs, disinclination to renest, and reduced viability of the young. Mallard ducks fed only 3 ppm of DDE laid eggs that were 13.5 percent thinner than those of control birds, were cracked or broken six times as often, and produced less than half as many healthy ducklings.

The relevance of the recent experiments is in the fact that the pesticides were administered at current environmental levels. The "breaking point" for actual population change in most wild birds is a 19 percent reduction of eggshell thickness. The thinner shells of eggs laid by brown pelicans accompanied the crash from 10,000 of these birds in Louisiana and Texas to practically none. The egg shells are affected by DDE inducing liver enzymes that cause hydroxylation of the birds' steroid sex hormones which regulate metabolism of the calcium.

Hard pesticides and human health Because DDT and DDE occur in the soil, waters (including rain and snow), and atmosphere even in the remotest regions, and are taken up by plants and passed up the food chain, all people carry a considerable body burden in fatty tissues, about 5 to 7 parts per million for persons in countries using considerable DDT. A massive inadvertent and uncontrolled experiment is being conducted worldwide, with all humans and all other animals serving as the experimental subjects.

The U.S. Food and Drug Administration set a limit of 0.05 ppm in milk shipped in interstate commerce. Human milk contains more DDT than cows milk. It seems remarkable that massivc information is available about effects of pesticides on wildlife, but very little solid research has been done on humans, although potential effects on infants, children, and women would seem especially important. It is easy to find alarming statements such as, "Of dieldrin, British and American babies consume about ten times the recommended maximum, and Australian ones up to thirty times." But direct effects on human health must be largely inferred from the undoubted harm done to laboratory mammals, since experiments on humans have been few and largely unconfirmed as yet. Concentrations now occurring in human fat are associated in rats with elevated levels of steroid hormone hydroxylases, so that such a change in liver enzymes may also occur in man. Even assuming that direct effects on human health do not take place, we are dependent on the health of the planetary ecosystem. The proven massive harm to wildlife shows that our planetary ecosystem is a diseased one, because of these and other excretions of civilization.

Restrictions and alternatives Public concern sparked by conservationist criticisms of (only) persistent pesticides brought about in 1967–1969 bans on DDT in Sweden, Denmark, Canada, Britain, Australia, and West Germany. In November 1969, seven years after Rachel Carson published *Silent Spring*, and again in August, 1970, the U.S. Department of Agriculture announced curbs that conveyed the public impression that the DDT problem was coming under control here. However, only the manufacturers, not the users, were ordered to adhere to federal standards; this loop-hole enabled users to spray it on crops other than those legally sanctioned. Also, legal appeals against the regulations enabled the manufacturers to postpone the curbs going into effect for at least two years. Even after regulations eventually took effect, U.S. manufacture and foreign sale continued, so that our chemicals return to us via the planetary circulations.

Some defenders of DDT have stated that this pesticide is unjustly blamed for damage caused by the polychlorinated biphenyls (PCBs). These relatively involatile industrial chemicals are found, strangely, in remote regions along with DDT. Laboratory findings reported in 1973, if confirmed for the atmosphere also, give evidence that the widespread PCBs in nature may be derived from DDT by degradation of the DDT derivative DDE vapor. This replacement of one environmental contaminant with another would help solve the mystery of what happened to some of the presently untraceable fraction of the 1.3 billion pounds of DDT dispersed into the biosphere since the early 1940s.

By June, 1972, the Environmental Protection Agency had canceled almost all uses of DDT within the country "except for emergencies" and had taken another step in the complex administrative procedures for outlawing the organochloride pesticides aldrin and dieldrin. Administrator Ruckelshaus declared before leaving this agency that dieldrin and DDT are "potential human carcinogens."

In matters involving pollution hazards, the USDA, AEC, and federal aviation authority have exemplified the fallacy of entrusting regulation to any agency whose chief mission is to promote the industry in question. Although one would expect the situation to improve by shifting pesticide regulation from the USDA, longtime friend of hard pesticides, to the EPA, the management there at the present writing illustrates the familiar administrative ploy of appointing the fox to look after the henhouse.

The attack on malaria by DDT spraying in certain tropical areas saved many human lives. The long-term results were not equally happy, since the resulting explosive population release brought congestion, starvation, and misery. Prominent ecologists called it immoral to drastically reduce a people's death rate without helping them limit the birth rate correspondingly.

Reliance on pesticides brings about genetic selection for resistance to the chemical; 200 insect species now show such resistance. Control agencies often respond by stepping up the rate of application, and thus of environmental pollution. Much of the worldwide DDT pollution results from spraying in the tropics and subtropical countries by foreign health agencies.

Thus, the worst problem with pesticides is not in misuse or accidents, but in their persistence and behavior in ecosystems. Even though DDT may be more economical and immediately effective than alternative methods, we can no longer afford its ecological costs. Consumers demand attractive, undamaged fruits and vegetables, regardless of invisible chemical residues. Much ingenuity is being applied in researching substitute approaches, especially where insecticides have failed. The monoculture method of modern food production makes crops especially vulnerable to pests, diseases and weeds.

Integrated control originally began by combining pesticide use with biological controls, but it now involves physical-mechanical controls, cultural control (sanitation, tillage, weed control), breeding for crop resistance against pests or microbial and virus diseases, or antimetabolites such as feeding deterrents and hormones. Integrated control is now primarily an ecological approach, using all available effective techniques to keep pest populations at their economic tolerance levels. It is based on the steady-state concept, and uses insecticides minimally or as a last resort. A chemical weapon in biological systems should be used as a scalpel rather than as a sword.

Biological control works best on imported pests. Parasites and predators are later brought from the native region, and tested one by one. A parasite usually attacks only one host insect, whereas a predator needs several victim species to complete its life cycle. Predators are more useful; their populations fluctuate less.

The screw worm of cattle was eradicated in Texas by x-ray sterilization of captive male populations and turning them loose to mate with females which thus became infertile.

Herbicides The management of vegetation on the roughly 70 million acres of land constituting roadsides and rights-of-way of power and telephone lines, and radio-transmitter sites is an ecological matter. Before the advent of herbicides in 1945, these lands were hand cut; today they are being badly mismanaged even though applicable scientific knowledge has been readily available since 1950. There are many types of roadside situations and suitable methods should be applied to each.

Herbicide chemicals such as 2, 4-D, 2, 4, 5-T, and ammonium sulfamate are somewhat selective in that they spare grasses and have been considered harmless to animals, although destruction of habitat can eliminate them as effectively as direct poisoning. These sprays when applied broadcast gave good kills of tall woody species to the ground surface, but the poor root kills permitted early resprouting. Such indiscriminate blanket spraying sears countless miles of roadsides with unsightly scorching of flowering herbs and shrubs. It is devastating, expensive, ineffective, and flouts ecological principles. Chemical manufacturers advertise their products on the basis of controlling "brush" which is undesirable, rather than dealing with dynamic plant communities that can and should be manipulated to the advantage of the public agency, utility company, the landowner, and the public. The work is usually directed by engineers innocent of botanical understanding.

Sound ecological principles dictate that the grass cover produced by broadcast spraying invites reinvasion by trees, whereas most needs can be met by a low shrub cover which remains much more stable. A shrub cover of sheepberry *Viburnum* in Connecticut, for example, is known to have maintained itself and resisted tree invasion for at least 25 years. Working in winter with small knapsack sprayers, operators spray selectively the basal root collar region of unwanted tree species. While slower, this results in much better root kill of trees, and produces not grassland but an attractive mixed cover of flowers and flowering and berry-bearing shrubs. The selective method might seem more expensive because of higher labor costs, but it is really much more economical in the long run and yields high conservation and natural beauty values besides. The roadside crisis must therefore be attributed to a negation of ecological science.

The same 2, 4, 5-T which has been used for years in the United States against plant life was tested on laboratory rats in 1966 for the National Institute of Health and found to produce a high rate of deformed embryos, prompting its removal from the domestic market. Its use as a crop-destroying chemical in South Vietnam was allegedly stopped in November 1969 following reports of deformed babies in villages sprayed with it, and its use as half (with 2-4D) of the defoliant spray called "agent Orange" was then reportedly restricted to "unpopulated" forested regions in the war zone.

The military use of herbicides for forest defoliation and crop destruction in Vietnam began in 1962; more than one-seventh of the country, or an area larger than Massachusetts, has been sprayed one or more times. A non-governmental investigation sponsored by the AAAS submitted a preliminary report. According to *Science* (January 8, 1971), "Their formal reports to the AAAS annual convention were guardedly conservative, but their findings added up to a charge that the military use of herbicides has been considerably more destructive than anyone had previously imagined." One-fifth to one-half of South Vietnam's mangrove forests, some 1400 square kilometers in all, have been "utterly destroyed," and even now, years after spraying, there is almost no sign of new life coming back. Roughly 35 percent of the country's 14 million acres of dense forest have been treated, and 6.2 billion board feet of merchantable timber killed, equivalent to South Vietnam's domestic timber needs for the next 31 years, worth $500 million to the local government in stumpage taxes. Recovery will long be inhibited because of the invading bamboo and grasses. The team claimed that nearly all the food crops destroyed (10 percent of the program) would actually have been consumed by civilian populations. They found no definite evidence of adverse health effects in the preliminary study, but further study is needed of a high rate of stillbirths and two kinds of birth defects coincident with large scale spraying. The scientists made no effort to assess the military usefulness of herbicides, but Army spokesmen stressed the saving of American lives through the defoliation. Coincident with the science meeting, the White House announced "a program for an orderly, yet rapid, phase-out of the herbicide operations."

Encroachment on Surface Space

Exponential growth of population and the flight from the central city produce undesigned urban sprawl. Urban development spreads radially from centers, interconnects and fills in, growing like a monstrous fungus colony. As the dependent system encroaches upon the surrounding compromise system, the latter in turn invades, diminishes, and deteriorates the more remote indepedent systems.

In 1800, over 90 percent (or 5.5 million) of the U.S. population lived in rural areas, whereas today only 25 percent do so. It is expected that by 2000 A.D. only 15 percent of the 300 million persons will be rural residents. Obviously this involves drastic shrinkage of rural land area. Although present cities may be no more densely populated than cities have been for centuries, they are infinitely larger and their rate of growth threatens the existence of open lands representing support systems for urban and other populations. Needed support goes beyond foods and raw material supplies to opportunities for recreation, refreshment, and isolation in nature. The correlations established between incidence of mental disorders and distance away from the center of the city raise questions about the ability of the human species, whose long evolution in green space instilled inherent territorial needs and psychological tolerance limits, to adapt genetically to congestion and regimentation in the urban-industrial world now being created.

The American dream was a log cabin on one's own homestead. It became a home in the city on a streetcar line, then a home and two automobiles in a suburb. Now it is two homes (one for vacations), or a home plus a recreational vehicle nearly as spacious and complex. Modern demands, combined with increasing population and higher affluence, gobble up farm and range lands at accelerating rates.

Land developers in this country are said to gross more than $6 billion in annual sales. There are now 9,000 vacation-land-development firms, *Business Week* reported that 10,000 entrepreneurs are selling land on an installment basis, often sight-unseen, to out-of-state residents, and that by the end of this century an estimated 18 million acres more of rural land will be urbanized. The assault on western rangelands by land developers, strip-miners of coal, and speculative wheat-growers makes one wonder where our future beefsteak dinners are to come from.

Some decades ago northerners were snapping up undrainable swampland in Florida. Now, tiny lots in huge subdivisions are being bought as "investments," largely by midwesterners, in the creosote-bush deserts of Arizona and New Mexico. With strictly limited or inaccessible groundwater supplies, such shadow subdivisions have gridded the lovely desert terrain with unused "streets" and cannot conceivably be profitable to anyone but the original "developers."

The oft-cited "balance of nature" can be fully expressed by independent biotic systems only. In these stable systems fluctuations are relatively minor. But within compromise systems, man cannot permit nature's type of independent equilibrium which offers only low productivity from his economic viewpoint. Instead, a quasi-balance *with* nature should be sustained. There is still greater human control within growth systems such as cultivated land. In urban-industrial areas artificiality reigns nearly supreme in that no *internal* natural balance is possible. Even there, an appropriate "balance with nature" is essential among the city and the several outlying systems for long-term survival of urban systems.

Roughly 70 percent of the O_2 in the world's atmosphere comes from photosynthesis by plant plankton in the seas, the rest from terrestrial vegetation. More generally recognized imports from the less dependent biotic systems are food energy, water and wood supplies without which urban productivity and viability would quickly wither. But cities exploit nature instead of reciprocating toward outlying systems, for example, the excess of CO_2 that they add by combustion tends to unbalance composition of the earth's atmosphere. The unbalancing effects of growth systems interacting with steady-state systems must be monitored because a reasonable degree of overall homeostasis is imperative. Cities and other dependent systems rely on a healthy planetary ecosystem, which in turn requires vast protective biotic systems of land and ocean. The ocean should be permanently managed as a largely protective rather than productive system. If sewage, land pollution, and industrial wastes kill Lake Erie and Lake Michigan, our species could conceivably survive that tragedy, but not the destruction of the protective oceanic ecosystem.

Present political jurisdictions affecting land use and development are numerous and uncoordinated. Zoning now resembles a 1900 model highway serving present model cars. The whole landscape will soon have to be zoned. As the concrete octopus devours good farm land even faster than poor, regional design should keep available, for cultivation, more of the richest soils on topography suitable for farm machinery.

Along the roads extending from a town, families build homes dispersed as "string developments" since land farther from town is less expensive. They soon demand improved roads. Utilities must reach out from town, and school bus routes are extended, so that utility rates and county taxes rise for everyone, subsidizing the buyers of the cheap land. Their parcels are of uneconomic size for commercial farming, and often the land is hilly and suited only for growing timber. But commercial forestry is now impossible because the acreage is cut up into small fields which farm subsidies made it profitable to keep out of timber production. Proper county design for forestry and recreational uses, and a better farm policy, could have kept the small towns from decay and protected the countryside for a more economic as well as more ecologic utilization, relieved the shortage of lumber, and reduced its price.

The pressure to maximize speed on all roads is rapidly changing the interesting scenic country roads to commonplace ones. Picturesque covered bridges, for example, are disappearing. The two-lane, winding Blue Ridge Parkway, planned for pleasure driving and closed to trucks, is renowned for its vegetational beauty and scenic vistas. These can be enjoyed because the speed limit is 45 mph. Modest country roads conform to this concept in part, and should be spared too much improvement. New roads built to bring the public into scenic areas not infrequently destroy their destinations. Despite the great value of superhighways, they cannot substitute for roads that lead to psychic satisfactions by providing amenities other than rapid movement.

When funds from federal taxpayers were being appropriated for construction of the vast interstate highway system, subversive elements attempted to prevent outdoor advertising close beside the rights-of-way, but the freedom of Americans to enjoy billboards in their countryside was successfully defended by lobbyists and Congress.

A study of the number of accidents per mile on the New York Thruway in areas with and without billboards showed that there were at least three times as many accidents in sections with billboards.

The industry has developed a new kind of sign for use in states having the most restrictive legislation. It is 24 times the size of a standard 24-sheet billboard, being 100 × 80 feet, and suspended between two aluminum towers. A development for night advertising can project messages hundreds of yards long against mountaintops and clouds.

Environmental impact statements Introduction of the recent requirement for studies of the adverse effects of proposed federal projects, grants or contracts on the human environment was a very important forward step. It focused needed attention on serious environmental problems, but it is not, in the present form, very effective in solving them. It considers each item in isolation, when we really need integrated studies of extensive areas because developments affect one another. Extremely few projects, even quite undesirable ones, have actually been brought to a dead stop because of unfavorable impact statements, even though this was the principal value foreseen for this governmental mechanism. Also, the impact statements are prepared by the proposing agency, not by an impartial outside group. Many of them have been incredibly poorly prepared. Evaluation is done ultimately in the Washington headquarters of the Council on Environmental Quality, where they are judged less on content than on meeting pro forma standards as legal documents.

Noise Pollution in Wildlands

Of the several types of off-road vehicles (ORVs), the most objectionable in their effects on the people who use wild lands legitimately, and on the lands themselves, are motorcycles and snowmobiles. Because outdoorsmen cherish quiet and solitude, machines that can go almost anywhere and can easily be heard over a mile away pose a serious conflict with proper recreational uses. Reasonable standards have proved unenforceable. Off-trail use disrupts or compacts forest soils, damages plants and tree reproduction, harmfully affects wildlife populations, promotes accidents, and emits pollutants from unburned hydrocarbons (especially from two-cycle engines), lead, and oxides of nitrogen.

Bad as it is on compromise-type lands, ORV use is completely incompatible with the wilderness concept.

The economic beneficiaries of the currently soft Forest Service policy are, overwhelmingly, the foreign motorcycle manufacturers. We have had the most severe trade imbalances in history, particularly with Japan. In efforts to alleviate these, we shipped out billions of board feet of lumber to Japan, creating a shortage at home; much of the petroleum to be drawn from Alaska's north slope is also to go there. The Administration asks for a 50–100 percent increase in the national forest timber cut. Opening up the forests to imported ORVs will tend to aggravate our balance of payments situation and force still more timber sales to Japan at a time when domestic homebuilding is being priced out of the market. These wider impacts of the proposed national ORV policy must not be overlooked. Trying to do too many things on too little land brings about sharp reductions in the total productivity and benefits of forests.

The national forests comprise only 7 percent of our national land area and should not be the dumping ground of every new open space demand that arises. Golfers and football fans do not expect the government to meet their needs for land. There is no good reason why ORV tourists cannot meet their own needs through private arrangements, so that the national forests, which have enough serious problems already, can continue to serve uses appropriate to them.

WASTE

The outer layers of this planet may soon be reclassified to include lithosphere, hydrosphere, littersphere, and atmosphere. A Navy admiral, aboard a submarine 50 miles off San Diego and 2450 feet down, peered through a porthole to gaze at the wonders of the undersea world. "The first thing he spotted, two feet away," says the Navy, "was a beer can."

This country produces 800 million pounds of trash a day, much of which ends up in our fields, parks, and forests. Waste is the ugly side of plenty. With only 5.7 percent of the world's population, we consume 40 percent of the world's production of natural resources. Each year we spread 48 billion rustproof cans and 26 billion nondegradable bottles over our landscape. Our foreign aid aims to help other countries get rich like us, but rich people occupy much more space, consume more of each natural resource, disturb the environment more, and create more land, air, water, chemical, thermal and radioactive pollution than poor people. The disposal of solid wastes is now our largest item of municipal expense after schools, welfare, and roads, because each city dweller produces 6 pounds of garbage daily, reproducing himself each month in garbage—a "garbage man." Solid waste disposal in the United States costs $4.5 billion a year, and is highly ineffective. Education, research, demonstration, and local and regional planning are essential for progress in this neglected area.

The first national survey of roadside litter by the National Academy of Sciences and 29 state highway departments reported that each year American motorists dump 16,000 pieces of litter per mile on primary highways. Glass bottles and aluminum cans (recognizable by the seamless bottom and sides) are particular nuisances. The report recommends educational and publicity efforts rather than additional punitive measures.

Solid waste comprises household garbage and trash and industrial, commercial, demolition, and construction left-overs. Of the solid municipal waste handled by collection agencies, 90 percent goes to 12,000 land disposal sites, of which only about 6 percent are sanitary landfills. Eight percent goes to 300 municipal incinerators. Municipal waste is scarcely recycled at all. The economic and tax systems at present encourage the exploitation of scarce natural resources.

A sanitary landfill calls for no open burning, no water pollution problems, and daily covering. Thin layers of refuse are compacted by bulldozers. When about 10 feet of compacted refuse has been built up, a thin layer of compacted earth is placed over it, then another layer of waste. The completed landfill is sealed over with 2 or 3 feet of compacted clean earth.

The technology we already have, if generally practiced, could greatly improve handling and disposal of municipal refuse, but better techniques are being sought. Use of garbage-reducing bacteria in composting has proved no panacea. Giant incinerators can possibly be much improved; so far they are expensive to operate, especially with needed anti-pollution devices. One approach is cutting down on the amount of trash by soluble "self-destruct" containers; food containers made of collagen are edible when dissolved.

Almost half the country's copper products are made from scrap, and over half of the lead, whereas the steel industry does less recycling now than ever before. Wastes from mining and processing are accumulating steadily, often representing a loss of resources as well as eyesores. In the future, changing economics will justify recovery of various minerals from them; meanwhile, suitable techniques for doing so should be under study.

Industry should go further in distinguishing between solid materials that can be recycled and those that are true solid wastes. The effort to upgrade solid wastes management, utilization, and disposal should be justified, according to the American Chemical Society, on the basis of esthetic values and control of pollution.

Animal manure was formerly in demand for fertilizer, but now largely constitutes a true waste material because of the low cost of chemical fertilizer. Land and water pollution problems arise from the nitrogenous wastes in feedlots and pigyards. Odors and visual pollution cause trouble with neighbors. Anaerobic decomposition worsens odor, whereas aerobic oxidation after dilution in the water of shallow lagoons is beneficial. Although considerable federal research is focusing on this increasingly serious problem, no ideal solution seems possible. Current recommendations are care in animal management, use of the land to the extent feasible, deodorizing by oxidation, and care in public relations.

Subsoil Resources

Fossil Fuels and Aggregates

Energy and agriculture To release the stored energy of sunlight fixed by photosynthesis in ages past, we mine coal and drill for petroleum and gas. These operations concern us here because they may modify the land surface, either directly, as in strip mining, or indirectly, by supplying energy for widespread agriculture and other activities on the land.

Mechanized agriculture makes available the needed calories of food energy not merely through photosynthesis but also at the expense of the calories in the fossil fuels which power this industry. This energy exchange negates much of the claim for the efficiency of modern agriculture (that so few people raise so much food on so little land). By subtracting from the caloric value of the final foodstuffs all the calories consumed in getting, refining, and shipping the fossil fuel to be used, mining ores, and manufacturing and distributing the farm machinery, running the machinery, producing and applying fertilizers and pesticides, irrigating or draining, and moving the foodstuffs to consumers, we would realize how largely the food industry is extractive, depending on nonrenewable energy sources which are inexorably being exhausted.

Effects of extraction In surface or strip mining, minerals are not removed through tunnels or shafts, but by moving the overburden and extracting the mineral from above. Open pit mining and quarrying are forms of surface mining that involve less areal extent but greater depth than strip mining. Coal, sand, gravel, and stone mines cover large areas while iron ore, copper, and phosphate production are more concentrated.

Most strip mining is for coal; a power shovel that may hold 200 cubic yards piles the overburden in long sharp ridges behind it. At the last site worked, there remain a flat area and a long deep pit where a pond usually forms.

On mountainsides, contour mining is practiced. In the Appalachians, mineral rights purchased for a pittance years earlier often legalize destruction that drives landowners away permanently. One type of machine has an auger six feet in diameter that drills horizontally into coal veins. Steep slopes are denuded of vegetation and soil, severe erosion from slopes is followed by clogging of valleys by deposited spoils, and landslides make the area dangerous. Most of the coal thus produced is used in steam power plants by the Tennessee Valley Authority, the nation's largest user of coal. The steep slopes reduce the effectiveness of any attempts at revegetation and reclamation. A federal field appraisal team surveyed ten states and reported in 1968, "Reclamation as currently practiced in the Appalachian Region does not adequately restore mined lands to minimal standards. . . ."

Strip-mining companies are also digging up the dead and burying the living in a new coal rush of gargantuan scope and implications, in a vast stretch of country from Montana and Wyoming to Arizona and New Mexico. A power panic triggered the chaotic, unguided, and ruthless development of very extensive low-sulfur coal deposits. This is sacrificing the rights of ranchers and their land, water, and vegetational base. It recalls the statement of one of our earliest conservationists, Gifford Pinchot: "Out in the great open spaces where Men were Men the domination of concentrated wealth over mere human beings was something to make you shudder. I saw it and fought it, and I know."

The deleterious results of surface mining include esthetic nuisance and damage, water pollution by sulfuric acid and sediment, air pollution, noise, vibration, impairment of health and safety, reduction of agricultural acreage and losses of the tax base. Although public opposition based largely on esthetics has been strong, more than four-fifths of the ores and solid fuels now mined in the United States come from surface mines.

Reclamation and regulation Since the topsoil is practically never saved, the spoil surface is very rocky, often acid, and subject to extremes of temperature and unfavorable moisture fluctuations. After a few years the loose material settles and weathering improves the surface a bit. In a national survey by the Department of the Interior 48 percent of the areas were too acid to grow good vegetation, and only 5 percent had been leveled enough for farm machinery.

Minimally, bulldozers should follow the huge shovel and knock the ridge down enough to leave a 10 to 15 foot flat top. The least leveling is needed for tree planting, more for pasturing stock on planted spoil, and the most for operating farm machinery in cropland use.

Local zoning ordinances and state regulation of reclamation apply to surface mining. Few states have been effective in reclamation, partly because state conservation departments are satisfied with tree establishment and recreational use even when more intensive use would be feasible.

Gravel, sand, and crushed stone needed for concrete and highways are costly to transport; a mining company operating close to its market can better afford to reduce its nuisance and reclaim lands for reuse in accordance with good regional planning. One aim of local zoning is to prevent excellent sources of aggregates from being made unavailable by other developments. Parks and lakes on the edges of cities can follow after surface mines. Surface mining offers a chance for multiple use in which the several uses are sequential in time rather than concurrent. The sand and gravel trade association announced in 1971 that remaining gravel resources would last for another 18 years.

Shallow Disposal

It seems likely that when the sanitary engineering profession long ago adopted disposal into surface waters rather than into soil, it was taking a wrong turn, since the soil system should be better than open water for handling sewage effluent.

The adsorptive capacity of soils is one of the most important inactivating mechanisms in man's environment, hence, soil is the "sink" for pollution from many sources. The action of soil microbes and oxygen, and sorption by fine particles, makes the layer above the water table effective in retention and bio-degradation of sewage effluents and alien chemicals. Due to variations in the texture and properties of soils, and other factors that influence the change of contaminating ground water and wells, soil experts and geologists can best interpret disposal potentials in specific cases. Poor drainage and associated waste disposal problems may be expected in areas of high water table, geologically indicated by the presence of peat, muck, marl, and other soft sediments that accumulated under water but are now above water level. County soil maps are detailed and accurately identify such deposits, which are also unstable for supporting foundations of houses.

Electric power, cars, and septic tanks have made possible extensive suburban housing developments. Bacterial action in septic tanks is largely anaerobic; the effluent should be led into tile lines or disposal field where degradation can be continued aerobically and the liquid seep into surrounding soil. Both impervious clays and overly permeable soils are unsuitable for this septic tank method of sewage disposal; the systems there are likely to deliver bacteriologically suspicious effluent to the ground surface or down to the water table, especially during rainy periods, so that intestinal bacteria may show up in water wells. Homes with septic tanks and wells should be placed on large lots so these facilities can be well separated and safely distant from neighbors; this consumes much space for residential use. Nitrates and phosphates are not well handled and eventually reach underground aquifers and surface streams. Along with nutrients from barnyard and pasture sources, they cause aging of lakes downstream. Compared with sewer lines leading to public secondary and tertiary treatment plants, the present methods of individual home disposal are inferior.

Garbage disposal landfills also can be effective or ineffective, depending on whether or not geologic conditions permit leaching water to carry away solutes and bacteria from the decaying material to discharge points or into the groundwater reservoir. Landfills are more likely to be safe where impermeable or slowly permeable earth materials separate them from the aquifer than where limestone or dolomite enables the leachate to reach underground water supplies rapidly. Regional plans should identify lands suitable for disposal by landfill. Such land near cities is running out. Proposals to fill in exhausted strip mine or gravel pit areas should be considered by competent geologists in each instance.

Subsoils in Cold Regions

One-fifth of the world's land surface is underlain by perennially frozen ground, or permafrost. Above it lies the "active layer," which thaws and freezes annually. Permafrost is a resource in the sense that as long as it remains unthawed it supports the natural ground surface with vegetation and animal life, or man-made structures.

Arctic tundra vegetation consists of perennial plants, chiefly low heath-shrubs, willows, sedges, grasses, other herbs, and lichens such as "reindeer moss." The northern permafrost belt extends also throughout the subarctic taiga, where tundra plants form the groundcover between conifer trees which root shallowly in the "active layer." Except on ground that is too well drained, or very unstable due to frost action, the lower vegetation forms a dense mat; beneath it a thick layer of peaty organic material decomposes very slowly. The living and dead cover insulates the underlying ground against thawing during the short warm season and keeps the active layer shallow, above the permafrost. When human activities disrupt the protective layers, rapid thawing of the ground beneath may penetrate deeply and erosion may produce incredible and spreading devastation. Roads and trails can turn into gullies or even canyons.

Gravel deposits where the top of the permafrost is deeply buried provide good construction sites. At normal shallow depths, however, the major engineering problem is the ecological one of minimizing disturbance of the ground cover and the permafrost. Buildings are erected on pillars to leave ample open space between the floor and the ground. Airfields, roads, water supply, and sewage disposal must be carefully located and specially designed.

Since 80 percent of Alaska is federal public land, the Interior Department received application from private oil companies for a 100 ft. right-of-way for a proposed $900 million, 800-mile, 48-inch oil pipeline from Prudhoe Bay to Valdez, an ice-free port in southern Alaska. Since the line would carry crude oil at 150° to 170° F., its underground portions would be expected to melt all permafrost within a radius of up to 40 feet, if constructed in usual pipeline fashion. The service road to parallel it would open up wilderness areas to many more tourists and hunters. In the most fragile of all American wildlands, the plan creates many ecological, engineering and construction problems never before encountered. Due to these uncertainties, and the 500,000 gallon per mile capacity, concern has arisen over possible serious effects of oil spills caused by landslides, earthquakes and permafrost melting. There are also matters of erosion, revegetation, impact of gravel removal from stream beds for both pipeline and road construction, and the line's blocking big game migration routes and salmon spawning migrations. Experiments with mock-up pipelines in permafrost have shown that some of these problems could be overcome, if the findings were actually applied in construction and maintenance. Demand for new United States sources of petroleum has become so insistent that it seems inevitable that north-slope Alaskan oil will soon be extracted one way or another, despite environmental risks.

The Land Ethic

After 70 years of American resource conservation, a striking change in popular thought on the subject brought about the popular environmental movement. The traditional conservation idea stressed techniques, focused independently on single problems, and was guided primarily by economics—usually rather short-term economics. The work of technicians, in the many applied sciences that contribute to good land use, will remain important. But more important is that broad ecological understanding, rather than detailed knowledge of techniques, permeate the general public. Such understanding is an essential basis of ethical responsibility for maintaining, or restoring when necessary, a human environment of high quality for our own and future generations. The "new conservation," based on integrated environmental science and a land ethic, goes beyond supplies of usable resources, essential to a high standard of (material) living, to the perpetuation of a productive-protective total environment capable of sustaining a high standard of life. For the latter, open space, natural beauty, wildlife, wilderness areas, opportunities for outdoor exploration and adventure are among the intangible amenities that most people now consider necessities of the good life. Land has great value beyond that for the production of commodities.

The land ethic was strongly developed in some American Indians. Tecumseh declared "My father is the Sun, my mother the Earth. Would we sell our mother to the white man?" He told Mad Anthony Wayne, the American general, "You are not at war with us, you are at war with the Earth." Recently, the Quinault tribe closed its beaches on the Washington coast to whites, as the only solution to trash littering and vandalism that was intolerable to these Indians though no worse than what is considered commonplace elsewhere.

The "new conservation" is not really new but is newly appreciated. In 1864 the remarkable statesman-ecologist-linguist-philosopher George Perkins Marsh published his prophetic book *Man and Nature*. This included the first scientific writing on soil conservation. The inspired books of John Muir, the father of the national parks, advanced public appreciation of wild nature. Liberty Hyde Bailey named the "biocentric concept," suggesting that our world view is too selfishly man-centered, that plants and animals have a rightful place in the world irrespective of their usefulness or even aesthetic appeal to man, and that knowledge about them is worth acquiring for its own sake.

Conservation based on ecology and ethics was urged in the writings of Paul Sears, William Vogt, and Fairfield Osborn in the 1930s and 1940s. Aldo Leopold, a president of the Ecological Society of America, in his 1949 book, *A Sand County Almanac*, called for acceptance of ethical obligations toward land and living things. Economic considerations should be tempered by those of ecology, especially human ecology broadly interpreted. Conservation is not primarily a matter of techniques, but lies in the realm of sanctions; not uncontrolled exploitation, but a measured restraint is required in meeting our responsibilities to the natural world and man's future. "The social conscience must be extended from people to land."

Estimates of future needs and benefit to cost formulations now loom large in public decision making, though certainly not as large as political advantage. The public tends to be unnecessarily overawed by published estimates and projections of need. At best, such projections are made by conscientious economists, based on trends and hopefully valid assumptions. At worst, they are sometimes self-serving promotional devices emanating from single-interest private groups or governmental construction agencies. Yet often such projections are uncritically accepted as "the given," that is, as solid, sacrosanct bases for action programs.

While the grosser forms of outdoor recreation may be expressed in man-hours and dollars, how can we program for the computer questions of intangible values and the land ethic? Educational, cultural, scientific research, esthetic, emotional, scenic, and perceptive-recreational uses of outstanding natural land parcels cannot be quantified, certainly not by nose-counts. In land-use decisions involving such natural areas, traditional benefit to cost ratios are quite unsatisfactory unless we are satisfied to assume that many alternative uses, social costs, and intangible values should be assigned the figure zero. The difficulty in trying to combine intangibles with technical tangibles into a single aggregate figure is that they are not commensurable, being different kinds of things.

The National Environmental Policy Act of 1969 may eventually mitigate the above criticisms considerably. It requires the preparation of environmental impact statements from federal agencies contemplating actions, so that potential adverse effects will at least come under discussion. With improvements in its administration (including more input by scientists), and with the federal agencies coming to take environmental considerations as seriously as Congress intended, this mechanism may bring about a real turning point in our country's history.

Rachel Carson and Stewart Udall brought about in the 1960s an acceptance of the ecological viewpoint by many thoughtful and influential people. Today this is being disseminated, with impressive help by the information media, to the general public. Conservation teaching in schools is gradually advancing beyond the level of the piecemeal resource conservation based on economics.

The crux of the land ethic has been stated by novelist Aldous Huxley:

> In the light of what we now know about the relationships of living things to one another and their inorganic environment, . . . it has now become abundantly clear that the Golden Rule applies not only to the dealings of human individuals and human societies with one another, but also to their dealings with other living creatures and the planet upon which we are all traveling through space and time. "Do as you would be done by." Would we like to be well treated by Nature? Then we must treat Nature well. Man's inhumanity to man has always been condemned; and, by some religions, so has man's inhumanity to Nature. . . . For the ecologist, man's inhumanity to Nature deserves almost as strong a condemnation as man's inhumanity to man. Not only is it profoundly wicked, and profoundly stupid, to treat animals as though they were things, it is also wicked and stupid to treat things as

though they were *mere* things. They should be treated as though they were component parts of a living planetary whole, within which human individuals and human societies are tissues and organs of a special kind.

It is no accident that most conservationists and preservationists started out as hunters or fishermen, "for each man kills the thing he loves" or, better, comes to love what he used to kill. The hunter is the cave-man reborn; he recapitulates, for instinctual satisfactions, this male activity that was once a life and death matter. Boys are, thus, natural hunters and fighters who, as they mature, usually find their interests gradually shifting from an exploitive, destructive approach to nature toward one that involves the mind and spirit more deeply. Hunting and fishing, although they call for skill and bring adventure, stand lower in the scale of outdoor values than many activities based on exploration, perception, and respect for life as expressed in private husbandry or sociopolitical trusteeship of the objects and systems of the natural world.

References

BESTON, HENRY (original ed., 1928), *The Outermost House: A Year of Life on the Great Beach of Cape Cod.* New York: Viking Press, 1961.

CARSON, RACHEL, *Silent Spring.* Boston: Houghton Mifflin Co., 1962.

Chemistry and Public Affairs Committee. *Cleaning Our Environment: The Chemical Basis for Action.* Washington, D.C.: American Chemical Society, 1969.

Council on Environmental Quality, *Environmental Quality.* Washington D.C.: U.S. Government Printing Office, 1970, 1971, 1972, 1973.

GRAHAM, EDWARD H., *Natural Principles of Land Use.* New York: Oxford University Press, 1944.

LEOPOLD, ALDO, *A Sand County Almanac and Sketches Here and There.* New York: Oxford University Press, 1970 reprint.

MEADOWS, DONELLA H., DENNIS L. MEADOWS, JORGEN RANDERS and WILLIAM W BEHRENS III, *The Limits to Growth.* New York: Universe Books, 1972.

ODUM, HOWARD T., *Environment, Power, and Society.* New York: Wiley, 1971.

RUDD, ROBERT L., *Pesticides and the Living Landscape.* Madison: University of Wisconsin Press, 1964.

RUSSELL, TERRY, and RENNY RUSSELL . *On the Loose.* San Francisco: Sierra Club, 1967.

11

Toward an Ecology of the Urban Environment

PAUL SANFORD SALTER
University of Miami

Paul Sanford Salter was born in Springfield, Massachusetts. He was educated at Massachusetts State College at Westfield (B.S.E. 1950), Indiana University (M.A. 1951), and the University of North Carolina at Chapel Hill (Ph.D. 1965). Presently he is Associate Dean of the College of Arts and Sciences and Associate Professor of Geography at the University of Miami, Coral Gables, Florida.

His particular geographic research interests are in urban geography, urban ecology, and environmental perception. Dr. Salter is the author of numerous research articles of which the latest is: "The Projected Impact of Cuban Settlement on Voting Patterns in Metropolitan Miami, Florida" (with R.C. Mings), *The Professional Geographer,* May 1972. Currently Dr. Salter is writing a college text on urban conservation for Holt, Rinehart and Winston, Inc.

Although the ecology issue is a popular one today and appears to be developing into a major political and social issue, particularly with the world's younger people, there is yet very little cause for optimism. The destruction and deterioration of our environment is concentrated in our urban regions. The pollution and destruction of the urban setting, despite repeated warnings, has been going on extensively and intensively for a great many years. Indeed, the great urbanologist, Lewis Mumford, claims it has been occurring for the past 300 years. Mumford points out that despite the warnings and, unfortunately, in some cases, outright disasters, it has been only in recent years that "any systematic effort has been made to determine what constitutes a balanced and self-renewing environment, containing all the ingredients necessary for man's biological prosperity, social cooperation and spiritual stimulation."

Today, most of us are aware of the tremendous population explosion the world is experiencing. Much of this population growth is concentrated in our urban centers. Patricia Hodge and Philip Hauser project that, in the United States alone, the metropolitan centers will grow from a population of 113 million in 1960 to some 178 million by the year 1985. This aggregate will be a total of 71 percent of all Americans as compared to 63 percent living in metropolitan areas in 1960 (Table 11.1). This growth, plus the realization of the serious nature of our urban problems, both physical and social, has promoted the scientific study of urban processes and resulting problems. Indeed, this concern with the urban phenomena has generated the creation of distinct, systematic segments of various disciplines for example, urban geography, urban sociology, city and regional planning, and urban ecology.

Ecology

The central theme of this chapter concerns urban ecology—in today's world a frequently overused and misunderstood term. Ecology is a word of rather recent origin. The first use of the term is attributed to the German biologist, Ernst Haeckel, in the late 1860s. The term "ecology" is derived from the Greek word "oikos," which means "house" or "place to live." However, since its introduction it has been given many meanings by various scholars. Originally the

Table 11.1 Metropolitan Population by Region 1960 and Projected 1985 (numbers in thousands)

Metropolitan Region	Population 1960	Population 1985	Change 1960–1985 Amount	Change 1960–1985 Percent
United States	179,323	252,185	72,862	40.6
Metropolitan	112,884	178,138	65,254	57.8
Northeast	44,678	58,517	13,839	31.0
Metropolitan	35,350	47,328	11,978	33.9
North Central	51,619	65,723	14,104	27.3
Metropolitan	30,963	44,642	13,679	44.2
South	54,973	78,910	23,937	43.5
Metropolitan	26,436	46,156	19,720	74.6
West	28,053	49,035	20,982	74.8
Metropolitan	20,135	40,012	19,877	98.7

SOURCE: Patricia Leavey Hodge and Philip M. Hauser, *The Challenge of America's Metropolitan Population. Outlook, 1960 to 1985* (New York: Frederick A. Praeger, 1968), p. 9.

term was applied to the biological sciences, particularly botany and zoology. In the early formative years, around the turn of this century, ecology was defined as the study of the relationships between plant and animal life and their physical environment. It was also defined as the relationship between organisms and their environment or the interrelationship between organisms and other organisms. The ecologist attempts to relate the organism to the total environment. Eugene Odum, the noted ecologist, writes: "In the long run the best definition for a broad subject (ecology) is probably the shortest and least technical one, as, for example, *The science of the living environment.*" This definition is essentially holistic or all inclusive. Ecology has also been defined as a state of mind, since it deals with not a particular segment of life, but total life, as a process. Naturally this must include an examination of the total environment—both natural and man-made.

Human Ecology

At the turn of this century a group of social scientists at the University of Chicago developed an innovative approach to the study of man and his environment. They attempted to apply the biological concepts of ecology to the study of human populations. This attempt to establish analogous biological concepts and theories to the study of man and his relationship to his environment developed into the field of human ecology. The philosophical and theoretical concepts of the newly created discipline were not limited by the Chicago scholars to urban regions, for they considered the new field applicable to the total man-land relationship. However, probably due to the location of the University of Chicago within the metropolitan region of Chicago, a greater proportion of the human ecologists' research concentrated on urban problems and conditions. The Chicago School, led by R.E. Park, was subjected to criticism by a number of other social scientists. Objections were raised to the newly developed field on the grounds that Park and his colleagues could not scientifically support the separation of the cultural from the physical (biotic) communities. Park had recognized this problem and attempted to explain away the fact that man, unlike other creatures whose lives are governed by the laws of survival, had a decision-making capacity based on innumerable human variables. To counter this, Park and his colleagues had developed a set of guiding principles based on the recognized dichotomy between the physical and cultural worlds. Park believed the physical community should be the principal interest of the human ecologist, while the study of the social or cultural community belonged to the field of social psychology. Park divided the community into the biotic and cultural realms on the basis of differing levels of human action. The inhabitants of the physical (biotic) community were considered to be functioning under the forces of competition. At this level people were considered something akin to human manikins, operating in isolation from social factors. The English geographer B.T. Robson summed it up most succinctly when he detailed Park's considerations of the inhabitants of the biotic community in his outstanding book, *Urban Analysis*, as being "subject to the same impulses and forces as were plants or animals in their struggle for existence and for the acquisition of the most favorable circumstances in which to live. The cultural level, on the other hand, gave rise to society and was based on the strictly social processes of communication and consensus in which people became 'persons' with social attributes. It was the operation of these cultural processes which distinguished man from other organic elements in nature."

The Chicago School drew on the analogy of plant competition for space with variations in land values. Land values reflect competition of people for the most desired and valuable locations. It was stressed by the Park group that people reacting to competition would be subjected to economic segregation and would be sorted out by their ability to pay for land. The inner-city ghetto area would be the location of minimum choice, while the wealthy suburb would be the maximum desired area. Park's followers also brought the botanical concept of plant dominance into consideration in explaining the patterns of development of the physical community. Land values, thus, determined the dominance of various levels and types of activities.

Botanical analogies of invasion and succession were also applied to the

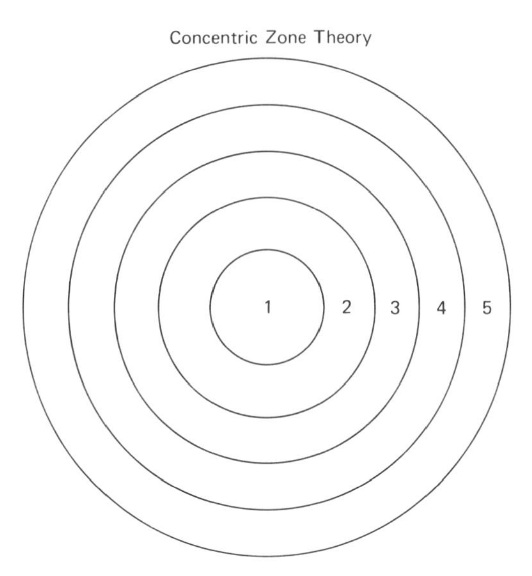

Figure 11.1 Concentric Zone Theory
SOURCE: Adapted from Burgess.

physical community—particularly in consideration of invasion and succession of particular ethnic and racial groups into residential neighborhoods and the intrusion of particular business activities into residential neighborhoods. These concepts later led to the concentric theory of urban growth as proposed by E.W. Burgess in the mid 1920s. Burgess argued that cities grow naturally into a five concentric ring pattern and that this growth was the natural ecological result of competition, dominance, and invasional succession (Fig. 11.1). Burgess pointed out that the highest valued land was consistently located in the center of the urban area—normally at the point of greatest accessibility. This area, the central business district (CBD), was surrounded by a zone of transition between those business enterprises in need of large floor space, for example, furniture showrooms, and the urban's oldest residential dwellings, generally in a state of advanced decay and disrepair. The area also was considered to be transitional, for if the CBD expanded outward, this area would bring very high land prices. The economics of land value competition turned this zone into densely packed multiple dwellings, occupied by those least able to afford decent housing and those racially or ethnically trapped by a prejudiced society. Land values in this transitional zone were based on potential, for the zone was, according to Burgess, in the first stages of invasion from the advancing business ring.

Next to this transitional ring would come the poorer, working class residential zone. Many of the inhabitants of this ring would be second generation citizens having moved from the transitional slum areas, thus illustrating the validity of the theory of succession. The housing in this ring was inexpensive, usually small frame houses and multiple storied tenements. The fourth ring comprised the better and more expensive homes of the upper-middle class and wealthy population. Housing in this area was expensive and almost exclusively single dwelling units, built on spacious lots. People who occupied this area were business people, white collar workers, and professional people. The outer and

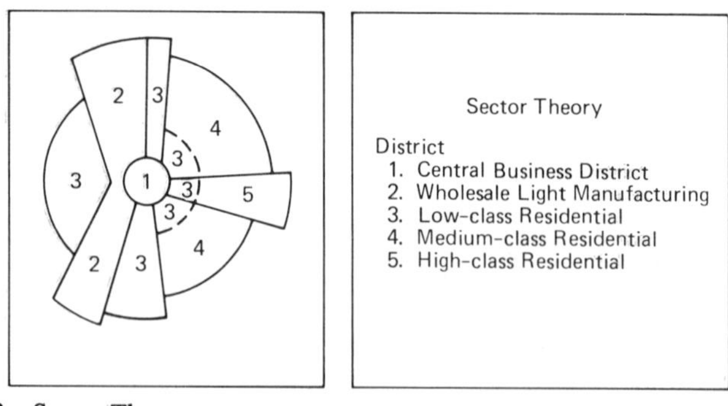

Figure 11.2 Sector Theory
SOURCE: Chauncy P. Harris and Edward L. Ullman, "The Nature of Cities," *Readings in Urban Geography* (Chicago: The University of Chicago Press, 1959), p. 281.

fifth ring was the commuter zone—the suburbs. Here people commuted distances, estimated to be at a maximum of an hour's duration one way to the center of the city. Since Burgess introduced his concept of concentric-ring urban development, based on invasion and succession, other scholars, notably Homer Hoyt and his sector theory and Chauncy Harris and Edward Ullman's multiple-nuclei concept, have made the best-known attempts to explain the spatial patterns of the western hermisphere industrialized urban centers.

Homer Hoyt, a noted land economist, published in 1939 the Federal Housing Administration Report, *The Structure and Growth of Residential Neighborhoods in American Cities.* On the basis of research involving rental data collected from a number of American cities, Hoyt countered the Burgess model, and in its place suggested the theory that cities develop outward from the CBD along the principal axial lines of transportation. He further theorized that these radial lines divide the city into sectors and that these sectors retain their residential character as they evolve outward from the city center. Most importantly, the wealthy rent sector is ultimately the most important element in the growth of a city, for it influences and pulls the entire city growth along with its line of development (Fig. 11.2).

There is no need to go into the particular criticisms leveled at the concentric and sector theories. Indeed, even though Burgess did not include industrial areas in the pattern and an ideal topographic arrangement must be assumed and even though Hoyt did not, as Walter Firey argued, take into consideration historical and human decisions, most cities, particularly in the United States and Canada, resemble to a degree these spatial patterns in their present city outlines.

A more recent theory concerning urban growth, and one that reflects the important influence of the automobile, is the multiple-nuclei theory proposed by Harris and Ullman (Fig. 11.3). This proposal theorized that numerous North American cities did not initially nor are now developing around a central

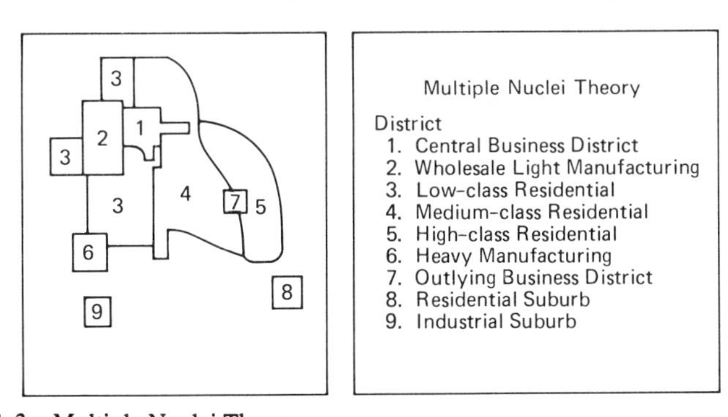

Figure 11.3 Multiple Nuclei Theory
SOURCE: Chauncey D. Harris and Edward L. Ullman, "The Nature of Cities," *Readings in Urban Geography* (Chicago: The University of Chicago Press, 1959), p. 281.

business district. Instead, they are expanding around several discrete nuclei. In several cases the nuclei have existed since the first settlement of the city; in other cases, the nuclei are newly developed, most likely the outgrowth of the acceptance of the automobile as a mode for commuting to work, coupled with the out-migration of people from the central city to the suburbs.

Two principal criticisms leveled against the human ecologists are related to their environmental deterministic philosophy and the distinction they placed between the physical (biotic) and social or cultural communities. The interrelationships between the physical and cultural communities are so interlocked and complex that it is impossible to separate the community into distinct and unrelated features of development. Contemporary human ecologists recognize that to attempt to study man and his relation to his environment, cultural factors must of necessity be included. A.W. Hawley was among the first to attempt this theoretical reconstruction and his writings are used as a base for modern human ecological theory. Contemporary scholars, including Duncan and Schore, point to the necessity of including the role of culture in man's adaptation to his environment. According to them, "The functional and analytic approach of human ecology involves a concern not with culture as an undifferentiated totality but with aspects of culture as they play into the process of adaptation. The basic flaw of the human ecologist school of thought lies in the analogy of the urban community to the biological one. The analogy is false, for the biotic community is not the result of conscious thought and its population is not free to make individual choices. As Constantinos Doxiadis so discusses it in his thought-provoking book, *Ekistics*, "The great difference between human settlements and natural organisms is that settlements are the product of both natural and conscious forces and thus their evolution can be guided, while natural organisms are the result of natural forces only and their evolution cannot be guided except within very natural limits."

In the preceding pages we have examined the human ecologist's effort to adopt the biological ecologist parlance and theory to fit his own needs and theories. This brings ecological terminology full circle, for originally, the biological ecologist adopted such everyday words as population and community to represent nonhuman groups and biotic environments. However, it must be recognized that whether we speak of ecology in a human or nonhuman frame of reference, we are discussing populations (human or nonhuman) occupying a given space (community). And to this concept we must add the interrelationships and interactions between the populations, their communities, and their environment, functioning together as an ecological system, or ecosystem.

The most important concept to keep in mind is that an ecosystem implies an interdependence of relationships. Odum states it this way: "As long as the major components are present and operate together to achieve some sort of functional stability, even if for only a short time, the entity may be considered an ecosystem." Odum considers the community concept as fundamental in the practice of ecology, "because as the community goes so goes the organism." If we accept this principle and apply it to man it becomes apparent that the ultimate future of man depends upon the nature of his communities and their associated ecosystems.

Conservation as an Ecological Concept

As we have discussed in an earlier section of this chapter, the term "ecology" has a variety of meanings depending on who or which discipline is applying the definition. It is hard to find a more precise definition of the word than the one proposed by Raymond Dasman, who considered ecology to be "the study of ecosystems, to determine their status and the ways in which they function." Ecology, also, has broad applications. Obviously, the application, like the definition, is dependent upon who or which group is doing the applying. Odum wrote in 1958, and the passage of 15 years has not altered the truth or his vision of it, that the most important application of ecology is in the area of conservation. F. Fraser Darling writes that "conservation as ecology in action is indeed looking outward from its traditional field of nature." In our case, we are primarily interested in the application of ecology and the various ecosystems to man and his communities. We are particularly interested in establishing relationships between man and his environment—in particular, the urban environment.

A disadvantage to be faced with the usage of the phrase "urban conservation" is the stereotype thinking associated with the term conservation. Odum remarks that the term suggests to many people a sort of "hoarding." Philip Hauser agrees and suggests the term "maintenance" might be more useful, and will help people to break away "from restrictive thinking which up to this point has been associated with conservation." It is apparent and logical that man must be included in any conservation consideration, and more importantly, he must be considered as a resource and a part of the environmental complex and not studied in isolation, or apart from his total environment. Conservation strives to maintain a balance between yields and renewals. It is concerned with relationships between man and man, and man and his environment. "The principle of the ecosystem, therefore, is the basic and most important principle underlying conservation."

The City and Conservation

Prior to World War II, the great conservation issues were concerned with saving our remaining forests and mineral reserves, fighting soil erosion, restoring our grasslands, and preventing future dustbowls. Though many of these conventional issues are still with us and are of important, some even critical, concern, the present conservation crisis is concentrated in the cities and their environs. Our cities are, in almost every instance, in desperate condition, and the situation is not encouraging, for if the past is any indicator, our future does indeed look bleak. The cities are not satisfactory, whether we examine them in human or physical terms. The underlying cause for many of the urban problems can be charged to the urban population explosion (really an implosion!). This phenomenon has not only aggravated and even accelerated our old urban difficulties, it has developed a whole new panoply of environmental problems. There is little doubt that if the more serious of these problems are not soon resolved they probably will spell the end of our urban civilization. Doxiadis struck the chord in *Ekistics* when he wrote, "If we deprive the society of its physical settlement, it is doubtful whether it can survive. Human societies need the settlement in order to organize themselves, and even to operate. How can ten million people of London operate as a society within an area of 40 miles square without all the existing networks of transportation, communications, and facilities?"

It is not just the sheer numbers of people that are laying waste to the urban environment, but it is also the attitude of man to the environment. The western world is a man-oriented society. Ian McHarg has written, "Show me a man-oriented society in which it is believed that reality exists only because man can perceive it, that the cosmos is a structure erected to support man on its pinnacle, that man exclusively is divine and given communion over all things, indeed that God is made in the image of man, and I will predict the nature of its cities and their landscapes." What is it that McHarg is getting at? He is saying that man with his egocentricism, his anthropomorphic and anthropocentric views does not choose to live with nature, but instead chooses to dominate it, often to degrade and despoil it. McHarg is correct, of course. One needs only to look around the urban centers to see and believe. What prompts man to feel this superiority? It is suggested that man's destructive attitude toward the environment has evolved from the Judeo-Christian religious observance that man is supreme. McHarg charges that this belief in the uniqueness of man stems from the concept of monotheism. Prior to the unfolding of the great Judeo-Christian religious revolution, man had faith in nature; man worshipped the natural environment and ascribed to it spiritual powers. However, with the evolution of monotheism and the belief that man is made in the image and likeness of God and that this God created the universe, including this earth and its environment, for man's benefit—that man should dominate and subdue nature, it is little wonder that man is despoiling his environment at such a ferocious pace. McHarg iterates, "For me the indictment of city, suburb, and countryside became comprehensible in terms of the attitudes to nature that society has and does expose." These are strong words, but the truth is there for all to see. Our environment is deteriorating rapidly and we may need to look in more than one direction for our salvation.

A new and rather startling tool for measuring the destruction of the environment has been proffered by W.H. Davis, of the University of Kentucky. Davis believes we should adopt the term "Indian Equivalents" to measure the deterioration of the environment. He defines his idea "as the average number of Indian citizens required to have the same detrimental effect on the land's ability to support human life as would the average American." Davis is saying that the average Indian is agrarian oriented, that he eats sparsely, either rice or wheat, that he frequently utilizes water drawn from a communal well, that he walks nearly everywhere and he burns animal dung to cook his food and heat his hut. In other words, he works at an environmentally slow destructive pace. The average American, on the other hand, through his great affluence, materialistic culture, and Keynesian economic concepts, is destroying his environment at an ever accelerating rate. For example, it is a well documented fact that each American citizen is responsible for helping to destroy the American environment by contributing to the annual disposal of some 7 million junked cars, 150 million tons of solid wastes, 20 million tons of paper, 26 million bottles and jars, and an astonishing 55 billion cans. In essence, even using the very conservative "Indian Equivalent" figure of 25 (Davis thinks that with our per capita gross national product some 38 times that of India, an "Indian Equivalent" figure of 500 would be more realistic), the damage a single American contributes is equal to the environmental deterioration caused by 25 Indians. If we accept this concept, the population of the United States, is, in ecological considerations, nearer to 5 billion people than the present 200 million.

Development of Urban Land-Use Patterns

In a chapter such as this it is only feasible to introduce the reader to urban ecology and to touch briefly on some of the more pressing urban problems facing cities everywhere. One major universal urban problem today is the uncontrolled spreading out of cities. As urban populations multiply, the city is not only developing vertically, but it is growing spatially. Uncontrolled spatial growth is referred to as urban sprawl. Frequently, as in the northeastern region of the United States, urban sprawl becomes so rapid that one city will merge into another. When this occurs we refer to it as an urban conurbation and when it occurs on a large spatial scale, for example, when cities not only merge, but urban regions overlap, we refer to it as a megalopolis. The brilliant French geographer, Jean Gottman, coined this term in describing the urban sprawl along the U.S. eastern seaboard from New England to Virginia.

Urban growth, whether controlled or uncontrolled, commonly develops along preset patterns formulated from the city's earlier years. It is true, of course, that all cities are different, and most have some particular identification; nevertheless cities do develop, spatially, along similar lines. For example, the spatial pattern of early American cities usually was one of compactness. This was due to the fact that site requirements, in those formative years, were quite restrictive. Settlement sites were almost always chosen for a topographic arrangement that would facilitate trade and transportational needs. Favored sites were well-protected harbors, deep anchorages, river crossings, the confluence of waterways, break-in-bulk points, and trail intersections.

A common pattern that early American cities developed was for the residential area to be tucked in close at hand to business and industrial districts. Due to lack of speedy local transportational facilities space was at a premium and people had to reside within walking or animal-drawn-vehicle distances of their places of employment. As Raymond Vernon poetically described the pattern: "Since shank's mare was the standard means of transportation in the pre-trolley age, the urban microcosm of the era economized carefully on space; the yards were small, the streets were narrow, and the houses stood cheek by jowl." Thus, the residential pattern of the early city was largely predetermined by inner-city transportation. The wealthy neighborhood usually developed near the business district, or close to the most scenic locations of the city, even if this meant these more fortunates had to travel some distance to and from the central core area.

The less wealthy settled farther out from the inner-city areas, but, nevertheless, within walking, or animal-drawn conveyance to the heart of the city. The poor had to manage for themselves, usually occupying those areas that were the least desirable, whether by aesthetic or health standards or, more probably, both.

Railroads also had a tremendous influence on the early spatial patterns of American cities. In the formative years of railroading, the tracks were laid where the traffic dictated. This, more often than not, meant close to the center of the urban center. Thus, many of our cities are plagued with freight and marshaling yards in their very hearts. Also, with the development of the spur lines, our cities became a morass of track, switches, smoke, noise and soot. The spur concept also permitted industry to choose locations away from the inner city. Of course, in terms of time-distance that we experience today, these newer industrial locations were still quite close to the center of the city. The important point is that these spur-industrial sites, with their noxious noises, smoke and fumes, are, today, seats of urban rot and decay.

Thus, the reader can begin to understand the cause of the congestion and transportation plight in present-day cities. The restricted urban residential development, coupled with the pre-automobile-railroad design of the cities, combined to bring us to the present complex transportation difficulties. It is true that some scholars, notably Professor Edward Banfield, have suggested that traffic and congestion are a normal part of the urban environment, and should be accepted (enjoyed?) as a penalty for urban living. However, it is difficult to imagine that many will agree with this idea.

Hence, urban sprawl, the potpourri of industrial locations and traffic congestion, can be laid at the feet of the evolution of railroads and highways. These links of transportation and communication, however unintentional, have formed the backbone for our present confused intraurban and interurban patterns. Regardless of the arguments presented against planning, such as limitation and loss of personal freedom, with the continued and predictable growth of urban centers, we must bring some order out of the present chaos. As Doxiadis has gloomily summed it up: "What we could do is to predict what is going to happen and to plan for it in time, but that would mean planning the growth of the urban areas, and such a thing has somehow been considered quite immoral. 'We don't want any larger cities,' we cry, 'these will mean our disaster.' And thus we let them 'happen' in the worst possible way. This is a problem that is becoming more critical every day. I know of no urban area in the world where the situation is under control."

Contemporary Urban Land-Use Patterns

There is no greater single factor that has conditioned the twentieth century U.S. urban land-use pattern than the development of the automobile and the common acceptance of it as a method of commuting to work. It is true that at the close of the last century new innovations in transporting people from their residences to places of work began the centrifugal urban pattern that continues through today. The development of high-speed electric public conveyances, combined with technical advances in bridge-building and tunnel construction, allowed the average citizen to move outward, away from the older, congested neighborhoods to newly created outlying residential districts. Some businesses and industries also took advantage of this centrifugal urban force to relocate outside the central business district and older fixed transportational sites. However, most central business districts continued to grow in size and importance. With the public acceptance of the automobile and the new, faster methods of public transport, new arterial systems were built from the center city to the expanding suburbs. As Jean Gottman and others have pointed out, the development of the skyscraper was also quite significant to the expanding growth of the city. These structures, built of steel and concrete and reaching hundreds of feet into the air, could house enough employees in one or two acres to populate a hamlet or village. Technical breakthroughs such as these continued to feed and strengthen the central business districts and with this economic growth, combined with the new transportation-residential patterns, cities began to reach outward into the rural areas.

It is true, of course, that in the earlier part of this century, prior to the automobile, but after the establishment of rapid mass-transit forms of commuting, cities tended to maintain well-fixed land-use patterns. That is to say, Burgess' concentric theory of urban residential development seems to have been quite accurate. The wealthiest neighborhoods were either located in restricted downtown areas, near the CBD, or in suburbs composed of large estates. Other economic, ethnic, and racial groups established themselves in particular sections of the city, forming what, today, is sometimes referred to as "solid, permanent neighborhoods."

As has been previously stated, the internal combustion engine radically changed this residential and business-industry pattern and made it much more complex. The automobile, with its speed and mobility, allowed people to travel greater linear distances in about the same time they had previously done utilizing older mass-transit methods. The basic land-use patterns of the urban centers began to break up and new ones were developed. The mobility of the automobile allowed people greater selectivity in choice of home locations. No longer were they bound to the fixed transportational arteries. The older, fixed neighborhoods became a scene of out-migration, particularly of the younger people, who wanted their own ranchhouses with their picture windows, in the open spaces of the suburbs.

The automobile and the movement away from the center city has also affected urban land-use patterns in the areas left behind. Today, the traffic jam is a way of life in our cities. However, this ever-present annoyance, whether acknowledged as normal or not, has led to a change in the inner city. The central business district, with its smart shops, theaters, hotels, and large department stores, is in a state of deterioration and decline in most of our cities. The time-consuming and nerve-racking driving problems to get to the center of the city and then the almost impossible parking problems have led to a mass exodus of business and professional people from the central city to the outlying districts. These people are moving back toward their places of residence, back to the new shopping plazas and industrial parks that now ring most of our urban centers.

What is happening to the inner city now that this movement out is taking place? The inner city has become the home of the poor and the segregated minorities. It is, in almost every instance, an area of decay, where school facilities are old and outdated, where empty stores and buildings are falling into disrepair, and where the once most valuable shopping location is being turned into an isolated shopping center for the poor and minority groups. The "flight to the suburbs" has had two significant effects: (1) The economic structure of our cities has been weakened. As the more affluent leave the city, the people least able to afford the cost of urban environmental maintenance are left behind to pay the bills. (2) The poorer residential sectors of the city, the slum, and the ghetto are expanding. White Americans are leaving the larger urban centers to black Americans and other depressed minority groups. The legacy left to these people is a sad one indeed. Many of our larger cities are on the verge of bankrupcy and nearly all are in a state of deterioration and decay. Even more alarming is the change taking place in the spatial nature of the inner city area. Due to the "flight to the suburb" the ghetto and slum areas are expanding. As more housing has become available the depressed areas have spread out and now cover more acreage than they did just a generation ago.

A significant proposition to consider in evaluating urban land use and the deteriorating environment is that, contrary to popular opinion, there is generally not a shortage of urban land. In the United States, for example, urban areas account for only approximately 2 percent of the nation's total land area—some

30,000 square miles. As William H. Whyte stresses in his thought-provoking book, *The Last Landscape,* nearly one-half of this urban land is unused. Whyte goes on with the stimulating argument that much of the urban land is either empty or underpopulated. And, aesthetically he asks, what is uglier to the human eye than urban lands either empty or partially used by, for example, gravel pits, empty lots, and junk yards?

Whyte believes that urban planners could settle more people in the metropolitan areas if they stopped treating the urban scene as a frontier. Whyte's argument has a great deal of credibility when we reflect on the typical ranch house centered on its own half acre. "We are using five acres to do the work of one, and the result is not only bad economics but bad aesthetics." You must keep in mind, of course, that when Whyte speaks of greater metropolitan densities, he is not referring to the inner city, which is, as we all know, too crowded.

Land-use patterns on the urban fringe (ecotone) have also been affected by urban sprawl. An ecotone can be defined as a transitional area between two ecosystems. In urban usage the ecotone is the transitional zone between the edge of the city and the rural landscape. This zone had deteriorated with the growth of unplanned suburbs and the location of aesthetically noxious, or perhaps even environmentally harmful, industries, who are seeking escape from restrictive zoning ordinances. Also, the farmer in the ecotone area is under heavy urban economic pressures. With the constant spreading our of our urban centers, rural lands have become increasingly more valuable as residential subdivisions. The pressure of the farmer to sell his lands is tremendous. Not only is the farmer tempted with over-inflated land prices, but his land is frequently unfairly taxed. The combination of these factors is rapidly changing traditional urban-rural patterns.

Pollution and Cities

It is important to realize that nearly all major pollution problems, air, water, and solid wastes originate in the urban centers. For example, in Los Angeles County alone, according to a 1969 report, approximately 13,500 tons of air pollutants are being emitted daily. This is due primarily to the exhaust emissions from some 4 million automobiles. In discussing air pollution, McHarg puts it in blunt terms: "The city creates the filthy air. Clean air comes from the countryside." The same can be said for most of the water and solid pollution problems. Sewage disposal, industrial wastes, thermal pollution of our surface waters, garbage, trash—all are associated urban environmental matters. It is with these problems the urban ecosystems can be understood and examined with the least difficulty, that is, cause and effect can be examined in minute detail. It is not difficult to understand the reaction of automobile exhaust and hydrocarbons and nitrogen oxides in the presence of sunlight and oxygen. The relationships are clear and the results forseeable. The same can be said for many other pollution problems.

However, what is difficult is just how to motivate the people to demand that the problems be resolved and the systems brought back into ecological balance. Of course, galvanizing the people to action is not the complete remedy. A serious obstacle to corrective action involves the economics of corrective action. The building of sewage disposal facilities and large, efficient incinerators requires large outlays of funds. On the other hand, some solutions can be inexpensive. McHarg offers us a fascinating examination of urban air pollution and a possible, inexpensive solution—one that stresses the conception of natural growth and metropolitan open space distribution. McHarg suggests the development of urban-rural air sheds, the size of which would assure an adequate refurbishment of clean air to the city. These sheds, many miles long and wide, although utilized as residential, business, and recreational zones would be, by law, free of any types of air polluting enterprises. The geographic location of the "sheds" would be predetermined for each city by examining the relationship between prevailing wind directions during temperature inversion conditions (the latter, the principle cause for most urban air pollution conditions).

A brief mention should also be made of the urban microclimatic changes that are occurring from the dirty air given off by our cities. Research in this phase of climatology is still in an early stage. However, it does seem probable that the pollutants in the urban atmosphere are altering the so-called "Greenhouse Effect." Briefly, this means the more pollutants in the atmosphere, the greater the gaseous blanket covering the city and the greater potential for capturing radiating heat from the earth's surfaces. Paradoxically, it also results in cutting off incoming solar electromagnetic short rays from the sun. Unquestionably, if the pollution goes on, climatic changes will be felt in some urban areas. La Porte, Indiana is reported to have had an increase in the last few years of thunderstorms and hail and this is being attributed to the city's geographic location, downwind from metropolitan Chicago.

Noise—The New Pollutant

Recent research relating to noise levels has indicated it to be growing in magnitude and increasingly dangerous to urban inhabitants. Recent findings indicate that needless noise in urban centers all over the world has placed millions of urbanites on the precipice of committing violent acts and mental breakdowns. A serious hindrance in gaining control over the problem is that noise has gradually gained a form of acceptance among people of the modern world. In the United States, noise levels are twice today what they were in 1955! The word "noise" is ironically derived from the Latin root word for "nausea." Scientists measure it in decibels—a word coined in honor of Alexander Graham Bell. Decibels measure the intensity of sound waves that batter against our eardrums. The shuffling of a newspaper in silent surroundings produces about 15 decibels; the rustling of leaves is measured at 20; normal restaurant conversation

Table 11.2 Noise Readings in the dbA* Scale at Distances at Which People Are Commonly Exposed

Threshold of hearing	1 dbA
Rustling leaves	20
Window air conditioner	55
Conversational speech	60
(Beginning of hearing damage if prolonged)	85
Heavy city traffic	90
Home lawn mower	98
150 cubic foot air compressor	100
Jet airliner (500 feet overhead)	115
(Human pain threshold)	120

*dbA scale gives less weight to low tones

SOURCE: "Turn Down the Noise," *Fortune Magazine*, Vol. 80, No. 5 (October 1969), p. 133.

is about 50 decibels; heavy city traffic about 90 decibels. Most humans can tolerate sounds up to 80 decibels. However, continuous exposure to a decibel rating of 85 or more not only makes us uncomfortable but can cause hearing damage. Table 11.2 list some of the more frequently encountered noise producers. Researchers in the field of noise have gathered much data that indicates that excessive exposure to such loud noises can affect the physical well-being of people. Continuous exposure to excessive noise can constrict the arteries, increase the heart rate, dilate the eyes, create vertigo, bring on hallucinations, induce paranoia, cause stomach ulcers, develop allergies, generate enuresis (involuntary urination), and give rise to hypertension.

The principal cause of urban noise is motor vehicle traffic. And as automobile numbers are increasing at an annual rate of 4 percent, it seems likely that the problem will become increasingly more serious. Vehicular noise levels emanate chiefly from direct radiation from the exhaust, motor noise, tires, transmissions, and surface vibrations. Three approaches have been suggested to reduce the traffic noise levels: (1) reduce the noise of the source (do away with the internal combustion engine); (2) eliminate the source through the use of an underground transportational network; and (3) reduce highway noise by depressing roadways or constructing sound-absorption walls along the highway edge.

Another principal source of urban noise is from construction within the city. Most experts agree this phase of the noise plight can be reduced considerably and without great cost, mostly by improved construction techniques, for example, muffling equipment, welding in place of riveting, steel wire mesh blankets for blasting. Also, time considerations can help modify the problem. That is, blasting, demolition, and similar sound-producing techniques can be done during off-hours, when most of the urban population have left the city to return to the city edge.

Noise levels around metropolitan airports present yet another serious problem. This affects the residential area more than the other urban zones. A large jet aircraft on take-off produces a sound level of approximately 85 decibels, and this, on a path one-half mile wide on either side of the take-off line of flight! Many residents living in proximity to major urban airports have left their homes because the noise levels became unbearable. There are today, in the courts of our land, lawsuits approaching some 20 million dollars in noise damage claims. Unfortunately, only a handful of people have been compensated for the financial losses incurred by the decline in their property value due to the noise pollution.

Operation "Bongo," conceived by the Federal Aviation Agency, was designed as a test to see what, if any, ramifications could be expected from a city exposed to a continuous pattern of sonic booms. Oklahoma City was chosen as the target city and starting on Feburary 3, 1964, jet sonic booms were patterned to break across the city eight times per day for a period of one-half year. At the end of the first three months a total of over 5000 personal and property claims had been filed with the FAA. By the end of the six months over 15,000 cases were on file. Of the people who were sampled in the experiment (three interviews over the six-month period), approximately one-quarter reported they could not tolerate the noise. It is true the majority of the citizens reported they could live with the sound; however, the FAA received over 15,000 complaints, and a 25 percent minority is difficult to ignore.

Strong federal and state laws have been introduced only recently to curb noise. In 1968, former President Johnson signed into law a measure assigning aircraft noise control to the FAA. Unfortunately, not much improvement in this area will come about until (1) quieter jet engines are developed, (2) new airports are constructed away from concentrated urban areas, and (3) other forms of highspeed transportation are developed that will reduce air traffic.

Motor vehicle noise laws have not been particularly effective nor widespread. New York has a law limiting automobile sounds to 88 decibels while traveling at 35 miles per hour, at a distance of 50 feet. California has a motorcycle and truck law on the books, limiting those vehicles to a limit of 92 decibels at 35 miles per hour. Unfortunately, violations of these vehicle laws are very difficult to enforce on the open road. A more practical approach would be to enforce the acceptable engine noise level on to the manufacturer and to require noise level inspections. One note of optimism that can be heard above the urban roar is that, although the problem is serious, it is one that can be attacked immediately, and at a comparatively low cost. Most importantly, tangible results will be recognized almost immediately. This, unhappily, cannot be said for most other urban conservation issues.

The Human City

A less easily understood aspect of the urban environment is the role played by man in the city. Man invented the city and conditioned its evolution and the city responded and has helped to condition man. If the present trend to urbanization continues, the portion of people living in cities of 100,000 or more by the year 1990 will be greater than 50 percent. The significance of this growth rate will be the associated increases in urban densities. These density patterns will force an increase in human contacts which create complex social situations never experienced by man previously. Kevin Lynch asks us to consider the appalling picture in the future in which the world's population growth, paired with the maturation of technology, has urbanized the entire world, turning it into one huge megalopolis. Lynch asks us to imagine being trapped in an endless city with rows upon rows of houses, apartments, tenaments, ranch houses, under the pressure of being in the constant presence of other people, mostly strangers, with no escape! The thought is terrifying. One envisions an endless surrealistic landscape. There would be no natural environment—all would be man-made: an environment of polluted air and water, unbearable noise, an unbelievable urban morass cluttered with neon signs, billboards and television antennas. As Lynch asks: "Where would one find a wilderness or start a revolution? Would there be anything to challenge or excite the human spirit? Would not this world, entirely man-made, be utterly alien to every man?"

Although such a global conurbation is quite unlikely to develop into reality, it is occurring now on a limited scale in various metropolitan centers around the world. For example, New York City and its environs had in 1950 an average population density of 9810 persons per square mile. It is projected that this same area will have in the year 2010 an average density of 24,000 persons per square mile! Understand that when we are referring to the New York City metropolitan region, we include areas of southwestern Connecticut, northern and northwestern New Jersey, and the lower Hudson River Valley. Today, much of this region is still quite sparsely settled, that is, rural. Interestingly, some world cities have already exceeded this density pattern. New York City proper has an average density of approximately 25,000 people per square mile already and other world cities also have density figures approaching this mark.

The Urban Sink

Such actual and projected density figures as these raise important questions. What effect does this densely populated urban environment have on the inhabitants? The warning signal flag is already flying over the crowded urban scene. A recent psychological research study on mental health conducted in New York City found that 25 percent of those interviewed were psychotic, another 43 percent suffered some type of neurotic disturbance, and only 18.5 percent were classified as "normal," or free from mental problems. It is even more shocking to note that this research project avoided the more improverished sections of the city and did not include persons in hospitals!

Another recent urban study concentrating on violence and the urban environment carried on over a two-year period was conducted by research psychologist Dr. Philip G. Zimbardo. Dr. Zimbardo has stated, "Americans are being transformed into potential assassins by the powerful social pressures of big city living." (See *New York Times,* April 20, 1969, p. 49.) Dr. Zimbardo reinforces his thesis on violence and urban pressures by citing the increase in urban homicides, the horrifying figure of 40,000 children beaten by their parents and relatives, the 230 city riots in the last five years, and the assassinations of public figures. As an example of urban madness, Dr. Zimbardo noted that in 1967 hoodlums smashed in New York City 202,712 school windows, damaged or destroyed 360,000 pay telephones, did $750,000 worth of damage to the city's parks and $100,000 of wanton destruction to the city's transportation systems.

A revealing psychological field experiment conducted by Dr. Zimbardo and his colleagues, on the relationship between urban vandalism and urban size, took place in the comparison between New York City (7,895,563) and Palo Alto, California (56,800). In these two cities, Dr. Zimbardo and his associates placed, in each city, in a similar type neighborhood across from a university campus, an automobile, hood raised, as if disabled by mechanical trouble. Both vehicles were left unattended for a period of a week. However, both were kept under surveillance by hidden observers with recording cameras. The Palo Alto vehicle was untouched after being left unattended for over one week; however, the car in the Bronx section of New York was completely demolished in a three-day period. But the really surprising part of the findings was that the vandals who destroyed the New York car were all well-dressed, clean-cut whites. "Under other circumstances they would be mistaken for mature, responsible citizens demanding more law and order." Professor Zimbardo believes that the sheer size of our cities causes a "deindividuation" of the inhabitants. They are overwhelmed and feel completely helpless and overpowered by the forces working within the big city system and, as Dr. Zimbardo points out, this raises serious questions about law enforcement and deterring crime in large urban centers through the presence of normally accepted deterrents of crime, for example, street lighting, concentrations of people, police patrols. None of the above deterred the vandals from "doing their thing" in a large crowded urban center in broad daylight!

The term "sink" used in the title of this section was first used by the noted ethnologist, Dr. John Calhoun, who used the word in the phrase "behavioral sink." Sink is an appropriate word for expresing the concept of declining value, or lowering a level. Webster's *New World Dictionary* defines the term "to become less in value or amount." Calhoun wrote in *Scientific American,* in 1962, of an experiment performed with rats in which he attempted to establish relationships between overcrowding, stress, and behavioral patterns. Calhoun found in a number of experiments that rats exposed to stress conditions resulting from overcrowding over an extended period of time reached a "sink condition." In the "sink condition" Calhoun found that the rats became totally disorganized. Sink mothers did not feed or care for their young, they suffered failure to bear offspring, the building and maintaining of nests ceased, resulting in the scattering of the young, and normal sexual mores were abandoned and some sadism and homosexual behavior developed while other rats withdrew from sexual activities altogether. Social and territorial behavioral patterns were also upset by increased numbers. The "sink rats" were unable to establish territories and their social classes (rats who share the same territory and exhibit similar behavioral patterns) were confused by the division and subdivision introduced by the overpopulation. Of course, we must take great care in interpreting rodent behavioral patterns and associating them with human response. Arthur Koestler calls this "ratamorphology"! However, the research evidence indicates that there is a probable relationship between stress, overcrowding, and behavioral patterns. As Paul Ehrlich remarked concerning the population densities, "The possibility that our cities could eventually deteriorate to the point of causing complete social breakdown is something to consider."

Besides knowing that incidences of emotional and mental problems are greater in large urban environments than in others (this does not mean they do not occur in other types of environments) we are also aware that certain other physical and social conditions are more prevalent in the urban environment. Lung cancer, for example, is more common to the city than rural areas; this is true of certain other cancers, too. Ian McHarg and his students at the University of Pennsylvania, in mapping the disease incidence patterns for the city of Philadelphia, found a higher rate of occurrence of heart disease, tuberculosis, diabetes, syphilis, cirrhosis of the liver, ameobic dysentery, bacillary dysentery, and salmonellosis in the center city. These diseases varied in incidence within the center city itself, but all diminished in frequency of occurrence from it. Likewise, McHarg reported a greater incidence of social problems, including suicide, murder, alcoholism, crime, infant mortality, and drug addiction in the center city, with a diminishing of these social problems to the peripheral areas of the city.

Recently released testimony presented to a House of Representatives Appropriations Subcommittee by Dr. Stanley Yolles, former Director of the National Institute of Mental Health, revealed a detailed house-to-house survey investigation of drugs in a 40-block area of Harlem. The survey area was composed of 22,000 adults, some 30,000 children and some 6000 individuals who lived in streets and alleys in this 40-block area. The conclusions of the investigation are shattering. The results show that in this 40-block area there were 10,000 addicted adults, 6000 addicted children between the ages of 16 and 21, and 2000 between the ages of 7 to 15, who were addicted. Of the 2000 children between the ages of 7 to 15, 90 percent lived by themselves without the presence of an adult in their immediate environment." (See *The Miami News,* June 12, 1970, p. 1.) There can be no doubt that this points out, at least for this center city, a narcotic epidemic. The causes of this horrendous situation are undoubtedly varied and complex, and probably are not all related to the urban environment, but surely these figures indicate the immediate necessity for in-depth research in this phase of the urban environment. We know the urban sink in our urban centers is concentrated in the inner city, and somehow, in some manner, this portion of the urban environment must be brought into ecological balance with the rest of the urban area. If we do not, it is quite probable it will destroy all of the city—the city will be uninhabitable! McHarg realized this fact when he wrote, "The pattern is very clear—the heart of the city is the heart of pathology and there is a great concentration of all types of pathology encircling it."

Urban Design—A Response

We must bring some order and meaning to the human urban environment. There have been attempts in the past. Many of the attempts have been rational and some functional. Many of these earlier urban designs laid the foundation for modern urban and regional planning. It is impossible in a few pages to review all of the earlier urban designs. One basic concept, nevertheless, does deserve particular mention for it forms the basis for much contemporary urban thinking. To defeat unplanned urban growth, particularly of large metropolitan centers, and to restore a balance between the urban and rural environment, a visionary Englishman, Ebenezer Howard, in 1898 proposed his "Garden City" concept. Howard put forward a plan to build new, smaller cities away from developed urban areas. The cities were to be separated by greenbelts, open spaces, and farmlands. They were to be interconnected by highways and rail systems. Howard's plan called for each city to have its own supporting economic base and to supply its fresh-food needs from the outlying rural areas.

Howard not only conceived his "garden cities" as combining the best of two environments, but he also conceived his plan as helping to limit the size of great urban centers such as London and as offering a method for its redesign. It was his thought that the garden cities would prove to be so attractive that people would migrate from the large metropolitan center to them, thus putting an end to the big-city growth. Also, he believed that with an out-migration from the metropolitan center, land values in the center would decline, thus open spaces resulting in fewer persons per square mile. These concepts of Howard's, featuring open space, dispersal of cities, and limited urban size, are still much in step with the thinking of city designers today. The post-World War II "New Towns" of England and new urban designs in the United States such as Reston, Virginia and Columbia, Maryland, all reflect, in some way, the conceptual thinking of Howard.

A modern visionary, Athelstan Spilhaus, calls for the dispersal and development of new cities. "I dream of a city where the dwelling units are simple and adequate but the services in education, sanitation, health, recreation, and all forms of culture are outstanding. To achieve this in realistic dimensions, I am convinced that we need, as a corrective, the development of a system of dispersed cities of controlled size, differing in many respects from conventional cities, and surrounded by ample areas of open land."

Spilhaus is developing plans for an experimental city, to be located somewhere in the Great Plains or in the state of Minnesota. It is to be located outside the range of existing cities and will be utilized to develop new urban ideas and systems, such as new underground high-speed tube transportation, the development of recycling waste methods, mass computer utilization procedures, and experimental low-cost housing. When completed (construction will begin in the mid 1970s) the city is planned to house some 250,000 citizens. Spilhaus rejects the idea of rebuilding our present urban centers. "Without exception our cities are bound by tradition, outmoded building codes, restrictive legislation, and the consequences of unplanned, unhealthy growth."

It is impossible for us, of course, to solve our urban plight by abandoning the old, and constructing a totally new set of cities. It would be a wonderful and a simplistic method—wipe the slate clean, so to speak, and start all over, but it cannot be done in the real world. To be sure, we will design and build some new cities, but we also are going to have to live and work with the old. One new exciting urban research tool is the concept of environmental perception. That is to say, how important, is it, that people view the urban scene around them? Open spaces, greenbelts, parks all are desirable concepts—indeed, needed. But the critical factor regarding space is not its size, but how it is perceived. The separation of neighborhoods, residential areas from industrial zones, does not always require a great deal of space. Parks do not have to be thousands, or even hundreds, of acres to be valuable. The most valuable quality concerning space is its accessibility. As a matter of fact, people perceive space on differing levels of perception. It is safe to generalize and say nearly all people view space on a quite restricted scale. William H. Whyte made an excellent point concerning space when he stated, "It does most of its work along the edges. This is the part that people use most often for recreation. This is the part people see the most and it is often the best part."

What we are discussing is scale. Doxiadis refers to it as "the human city." That portion of the city the citizen lives and works in must have meaning to him and his family. Doxiadis argues the city must be divided into cells (neighborhoods) with each cell recognizable to its inhabitants. He feels this will make them feel comfortable and give them a sense of belonging. Such cells must be designed on a human scale, "allowing people to walk or ride from one point to any other point—without the congestion of today's confusion between man and machine." (See Doxiadis, *Newsweek Global Report,* May 27, 1968.)

Kevin Lynch believes that one method of bringing the city into human scale is to accentuate the uniqueness of each urban center. He suggests that the diversity between urban centers should be emphasized and the differentiating character of centers—physical, historical, cultural—should be expanded. Lynch believes each urban center should have a positive identifying focal point, for example, a plaza, a fountain or body of water, a crossroads, a public room, that the urban inhabitants can relate to. Writing in *Scientific American,* he says his "goal would be a city-wide system of differentiated, compact centers, each reinforced by high-density housing and new educational or recreational institutions. The centers would be stable in location, giving the city continuity in time, but they would be changeable in form, reflecting the city's flux of activities and aspirations."

Urban Pricing—A Response

We have previously mentioned that nearly all large urban centers are in financial difficulties. As cities continue to grow and urban ecological problems become more pressing, the financial demands faced by the city increase at an ever-increasing rate. For too long the main burden of urban financing has been carried by the property tax payer. (In this context, renters are included as well as property-owners. Renters pay taxes in the form of their rents. If property taxes are increased, generally, so are rents.) It is clear that property taxes have reached the saturation point. It is unrealistic to expect the property owner to pay all the urban bills. In fact, much of our present urban dilemma is the result of the almost total reliance on revenue from the general (property) tax.

Urban pricing as a system of increasing urban revenue has been largely ignored by our cities. In most cities, the property tax is the only major tax that can be correctly classified as a "price." The property-price concept is that property taxes are the price that property owners pay for the receipt of urban services, for example, police and fire protection, trash and garbage collection. However, it is a sad paradox that this pricing system, now commonly accepted by most urban areas, works both to the disadvantage of the property owner and the urban center itself. For example, if the property owner physically improves his holdings, prices in the form of tax assessments are raised. Conversely, if property holdings are allowed to deteriorate, the property owner is rewarded by having his tax assessment reduced, or at the very least allowed to remain static, while tax assessments are increased on other urban properties.

A complete and realistic system of urban pricing would not only reduce the financial inequities imposed on the property-tax payer, but it would insure a new and larger source of revenue to meet and help solve some of the most pressing ecological issues. As an example: the tax on undeveloped urban land (ecotone) is generally far too low. Frequently these lands are taxed as vacant lands, or at the most, farm lands. Taxes should be raised on these lands to help bear the costs of their future development, such as building of roads, installation of water and sewer systems. Developers, if faced with a realistic pricing system, would be forced to consolidate neighborhoods and reduce house lot sizes to lessen installation costs.

Similarly, land developers who create urban sprawl by skipping over the more expensive contiguous urban lands to develop cheaper outlying districts could be discouraged from this practice by paying a higher property tax (price) for these outlying districts. Property taxes on the undeveloped contiguous lands could likewise be reduced to induce development.

Corrective pricing can also be of help in solving urban ecological problems such as that brought on by the automobile, for example, traffic congestion and air pollution. Generally, the public has assumed that gasoline taxes defray most of the public expenses created by the automobile. However, today we are aware that this is not true. For example, the costs of air pollution are borne by all of the citizens of the city. Likewise, noise pollution resulting from vehicular traffic and all of its ramifications is also borne by the general public. In this area the system of pricing could be most effective. Pricing (in the form of tolls or permits) could help reduce the volume of city traffic at peak rush hours by pricing those motorists who drive into and through the city at such times. Conversely, drivers who refrain from using highways during rush periods would be rewarded with reduced pricing. Generally, multilane expressways are used to their maximum only during the peak commuting hours. It seems obvious that the expense of building these rush-hour extra lanes should be borne by those motorists who drive at these times. It is discouraging to note that, presently, not only do most cities have inadequate vehicle pricing systems, but most urban highway systems have no price system whatsoever. Utilization rates are zero!

We have but briefly indicated the general lack of adequate and equitable urban pricing systems in operation in the urban centers. There are many other areas in the urban environment where the taxation system is completely distorted and out of balance. Municipal golf courses and city-operated marinas are outstanding examples.

Urban economist Professor Wilbur Thompson summarized it most pointedly when he wrote, "The central role of price is to allocate—across the board—scarce resources among competing ends to the point where the value of another unit of any good or service is equal to the incremental cost of producing that unit. Expressed loosely, in the long run we turn from using prices to dampen demand to fit the quantity demanded, at a price which reflects the production costs."

Conclusion

Many more approaches and systems have been proposed for solving our urban crises, but they cannot be presented here. Of this, however, we can be positive. If we are to save our urban world, action must be taken immediately. The urban picture, although bleak, is not hopeless. Technology, combined with the will of the people, can solve, or at the minimum reduce, the physical environmental ills. Of greater urgency and unquestionable intricacy, are the conditions affecting the human city. Methods are available to the people to make the city a more desirable place to live out their lives. But here is the heart of the matter. Somehow people must be made to want to return to the city. Housing and educational facilities must be made attractive and available to all. Money for this must come from a new allocation of tax revenues among the federal, state, and urban governments. Cities must be made more attractive aesthetically.

As our urban centers grow in size, open lands will become increasingly more difficult to retain. We must conduct new research on open space perception so that we can economize on open space and still make the city attractive. We must, as McHarg argues, design *with* nature. The physical properties of urban centers must be mapped. Social values must be assigned to these physical areas, and both must form the core for future city plans.

However, nothing we do will make our cities better places to live if we do not first reconstruct the human element and bring it into an ecological balance throughout the city. It is probably true, as we are told so often by the prophets of doom, that we will always have poor people and slum areas, but we have to bring all of the urban people together. The racial issue must be resolved first and foremost. Lewis Mumford said it all when he spoke these words before a U.S. Senate Subcommittee hearing: "Unless human needs and human interactions and human responses are the first consideration, the city, in any valid human sense, cannot be said to exist, for as Sophocles long ago said, 'The city is people.'"

References

CALHOUN, JOHN, "Population Density and Social Pathology," *Scientific American*. February 1962, **206** (2) pp. 139–148.

DARLING, F. FRASER, "A Wider Environment of Ecology and Conservation," *Daedalus*, Fall 1967, **96** (4), p. 1012.

DASMAN, RAYMOND F., *Environmental Conservation*. New York: John Wiley & Sons, 1968, p. 9.

DAVIS, W.H., "Overpopulated America," *The New Republic*, January 10, 1970, p. 14.

DOXIADIS, CONSTANTINOS, *Ekistics*. London: Hutchinson, 1968, p.42.

DOXIADIS, C., and T.B. DOUGLAS, *The New World of Urban Man*. Philadelphia: United Church Press, 1963, p. 22.

DUNCAN, O.D., "Human Ecology and Population Studies" in *The Study of Population*, P.M. Hauser and O.D. Duvean (eds.). Chicago: University of Chicago Press, 1959, p. 602.

EHRLICH, PAUL R., and ANNE H. EHRLICH, *Population and Resources Environment*. San Francisco: W.H. Freeman and Co., 1970, p. 143.

HAUSER, PHILIP M., *Population Perspectives*. New Brunswick, N.J.: Rutgers University Press, 1960, p. 146.

HOYT, HOMER, *The Structure and Growth of Residential Neighborhoods in American Cities*. Washington, D.C.: Federal Housing Administration, 1939, pp. 112–122.

LYNCH, KEVIN, "The City as Environment," *Scientific American*, September 1965, **213** (3), p. 209.
MCHARG, IAN L., *Design with Nature*. Garden City, N.Y.: The Natural History Press, 1969, p. vi.
MUMFORD, LEWIS, Statement, Senate Subcommittee on Executive Reorganization, April 27, 1967.
ODUM, EUGENE P., *Fundamentals of Ecology*. Philadelphia: W.B. Saunders Company, 1959, p. 4.
PARK, ROBERT E., ERNEST W. BURGESS, and RODERICK D. MCKENZIE, *The City*. Chicago & London: The University of Chicago Press, 1925.
ROBSON, B.T., *Urban Analysis*. Cambridge: Harvard University Press, 1969, p. 10.
SPILHAUS, A., "So That Man Might Live Better," *Bell Telephone Magazine*. September/October 1969, p. 21.
THOMPSON, WILBUR, "The City as a Distorted Price System," *Psychology Today*, August 1963, **2** (3), p. 30.
VERNON, RAYMOND, *The Myth and Reality of Our Urban Problems*. Cambridge, Mass.: Harvard University Press, 1966, p. 13.
WHYTE, WILLIAM H., *The Last Landscape*. Garden City, N.Y.: Doubleday & Co., Inc., 1968, p. 2.

The International Biological Program

W. FRANK BLAIR

University of Texas at Austin

W. Frank Blair was born in Dayton, Texas in 1912. He was educated at The University of Tulsa (B.S., 1934), The University of Florida (M.S., 1935), and The University of Michigan (Ph.D., 1938). He has been a member of the Department of Zoology at the University of Texas at Austin since 1946, serving as Assistant and Associate Professor, and Professor. He is former Chairman of the U.S. National Committee for the International Biological Program and is Vice-President for the Special Committee for the IBP for the International Council of Scientific Unions.

Research interests in recent years have been concerned with the evolution of the toad genus *Bufo*. He is editor of the book *Evolution in the Genus Bufo*, University of Texas Press, 1972. In addition to numerous technical papers he is senior author of the text *Vertebrates of the United States*, 2nd. ed., McGraw-Hill, 1968. He is currently Director of the U.S./IBP Integrated Research Program, the Origin and Structure of Ecosystems.

The International Biological Program (IBP) represents the first worldwide effort by biologists to pool their research in order to seek answers to environmental questions of great importance to man's continued survival on this planet. Fifty-eight countries are participating to greater or lesser degree in this massive, cooperative research effort. These range from small developing countries such as Panama and Malawi to the giants of technological development such as the United States and Russia. The stated objectives of the IBP are "integrated study throughout the world of the biological basis of productivity and human welfare."

The IBP has catalyzed the development of new

perspectives in the environmental sciences. The most important new development is that of multidisciplinary research by teams of scientists in seeking solutions to problems that are beyond the capacities of single scientists, no matter how competent and versatile they may be. This pattern of team research is having major effects on the fields of ecology and environmental science generally and on management of the world's environments.

The general concept is one of data sharing by scientists, each of whom works on his own aspect of a complex ecological problem such as the functioning of an ecosystem. Each scientist receives credit for the data which he contributes and is free to publish them independently, but his data are also freely available to the others who are participating in the integrated research program.

The IBP pattern of research is being eyed by agencies that will have programs extending far beyond the official life of the IBP. Thus, the sometimes painfully gestated plans for doing integrated multidisciplinary research on environmental problems will have effects far beyond the IBP itself.

Origin and Evolution of the IBP

First efforts toward an International Biological Program are generally attributed to Sir Rudolph Peters, then President of the International Council of Scientific Unions (ICSU), and Professor G. Montalenti, then President of the International Union of Biological Sciences (IUBS). The original concept of an IBP has been attributed to the late Lloyd V. Berkner and other internationally minded members of ICSU, who thought originally of an International Biological Year, following the precedent of the International Geophysical Year. Subsequently, the Ninth General Assembly of ICSU, at London in 1961, appointed a Planning Committee for the IBP with Professor Montalenti as Chairman.

It is of interest that one of the themes for the IBP that was originally considered was one of a year of human populational genetics. In a remarkable case of fortuitous serendipity, the emphasis turned to environment and to man's place in the environment. This emphasis was to be a precursor to the enormous concern for the quality of the environment that was to surface in the Congress, in the general public, and, ultimately, in the Executive Branch of the federal government during 1969.

In the United States, an *ad hoc* committee of six ecologists and five other members representing a range of specialties was appointed by Dr. Frederick Seitz, President of the National Academy of Sciences. The charge to this committee was to examine the desirability of U.S. participation in the IBP. After four meetings between December 1963 and November 1964, this *ad hoc* committee recommended that the United States participate in further planning and that a U.S. delegation be sent to the Paris planning meeting of ICSU in July 1964.

The IBP program that emerged from the Paris meeting is set out in the ICSU preamble for the IBP as follows:

> As a consequence of the rapid rate of increase in the numbers and needs of the human populations of the world and their demands on the natural environment, there is an urgent need for greatly increased biological research.
>
> It is proposed that there shall be an International Biological Program (IBP) entitled *The Biological Basis of Productivity and Human Welfare,* with the objectives of ensuring a world-wide study of (1) organic production of the land, in fresh waters, and in the seas, so that adequate estimates may be made of the potential yield of new as well as existing natural resources, and (2) human adaptability to changing conditions.
>
> In proposing such a program it is considered essential that it shall be limited to basic biological studies related to productivity and human welfare, which will benefit from international collaboration, and are urgent because of the rapid rate of the changes taking place in all environments throughout the World.

After the Paris meeting, the U.S. delegation recommended participation of the United States in the IBP even though there was some concern about the preoccupation with problems of increasing productivity rather than the overall problem of population numbers and the ability of the earth's resources to support a freely growing population.

Participation of the United States in the IBP was approved on November 27, 1964, when the Governing Board of the National Research Council met and formally approved of U.S. participation.

Subsequent to this decision, a permanent U.S. National Committee was formed with seven, and ultimately nine, subcommittees, as the United States added two subcommittees (never recognized by SCIBP) on Environmental Physiology and Systematics and Biogeography to the seven recognized by SCIBP.

Several meetings failed to produce a definitive U.S. program until a large meeting at Williamstown, Massachusetts was called. At this working session, the concept of Integrated Research Programs (IRPs) emerged. These would be multidisciplinary, integrated research programs aimed at the solution of major environmental problems that would be beyond the competence of single scientists. From this point on, the U.S. participation in the IBP was to break mostly new ground in environmental research and was to have a profound effect worldwide on environmental research.

The nine subcommittees of the U.S. National Committee were all charged with the development of projects within their respective purviews. However, by the beginning of 1968 it was obvious that a streamlining of the U.S. National Committee was a necessity. Consequently, the U.S. National Committee was reorganized into a seven-man Executive Committee, a Program Coordinating Committee (PROCOM) consisting of the approved directors of IRPs, and an International Coordinating Committee (INTCOM) consisting largely of the equivalents of the chairmen of the original nine subcommittees.

Internationally, the IBP is coordinated by the Special Committee for the IBP (SCIBP). This is a committee answerable to ICSU. A secretariat is maintained in London, where space has been provided by the Royal Society. Seven convenors of international sections, with their members, coordinate the IBP internationally through conferences and communication. These seven sections are: PT—Productivity, Terrestrial,PP—Production Processes, CT—Conservation, Terrestrial, PF—Productivity, Freshwater, PM—Productivity, Marine, HA—Human Adaptability, and UM—Use and Management.

How Is the IBP Coordinated?

As might be expected of an international program involving 58 national groups, the IBP has developed into a highly complex research effort. There are varying degrees of national and international coordination of the various research efforts. At the simplest level of national participation, the committee for the IBP in that particular country identifies on-going research projects that are pertinent to the objectives of the IBP. In this instance, the main contribution of the IBP is to identify and make known comparable research efforts in the various countries and hence facilitate interchange of information and ideas among workers on comparable research projects around the world. Of necessity, this has been the pattern of participation by many countries which have been unable to provide new research funds for the IBP.

An example of this kind of international program may be seen in the IBP program on nitrogen fixation as published in *IBP News*, No. 15. Individual projects dealing with some aspect of nitrogen fixation are listed for 35 countries. The degree of concern with this problem ranges from highs of 45 projects in the U.S.S.R., 25 in Australia, and 17 each in Japan and Yugoslavia, to only one in Colombia, Denmark, East Africa, Indonesia, Malaysia, Nigeria, Norway, Papua, New Guinea, and Rhodesia.

Another pattern of international coordination has involved the formation of regional coalitions to work as subunits of the IBP. The Scandinavian countries, for example, have formed such a regional group. The East African countries of Kenya, Tanzania, and Uganda have done likewise. The U.S. program has been closely coordinated with the programs of Canada and the Latin American countries, although no formal organization has been formed. The same has been true of some European and African nations.

Another pattern has been that of coordination of major national projects through the formation of international task forces. Thus the Grasslands IRP of the U.S. IBP, that of the Canadian IBP, and 15 other national grasslands studies identified as having the "ecosystem" approach, and 15 lesser national studies of some aspects of grasslands (mostly of primary productivity) are all coordinated through an international task force. As with other international themes, IBP handbooks have been prepared to standardize procedures, and coordination is also effected through international conferences.

Another pattern of cooperative research under the IBP is that of multinational cooperation by scientists who work at the same time and place on integrated research programs that are aimed at answering major questions. The United States is involved in several such programs, especially with polar nations and with Latin America.

The Emergence of Big Biology

The most significant accomplishment of the IBP seems certain to be its fostering of multidisciplinary, integrated research efforts to solve complex environmental problems. The concept of the Integrated Research Program (IRP) originated in the U.S. National Committee for the IBP. It has since spread to other IBP countries, and it has received recognition from the SCIBP panel on Production, Terrestrial (PT). In establishing a national program based primarily on the concept of the IRP, the U.S. National Committee established four criteria for acceptance of an IRP as a component of the U.S. program as follows:

1. Each IRP proposal should demonstrate that it is necessary to group a set of research projects in space and time in order to achieve the scientific goals of the program. That is, convincing arguments must be presented to show that support of individual, independent research projects would not achieve those goals. The whole, in this case, must be greater than the sum of its parts.
2. Substantive, operational questions of general ecological significance must be posed in order to give coherence and direction to the suite of research projects in the program. Without such questions the projects will tend to go their separate ways, thus placing the ultimate goal of the program in jeopardy. Management proposals must also clearly indicate a means of shifting direction and emphasis as new information and ideas dictate.
3. Each proposal must describe a system for rapid exchange of information and data among investigators so that each participant will profit from the efforts of others as the work proceeds. This is obviously essential for coordination. Without it, much of the advantage of simultaneous support of projects will be lost.
4. Each IRP proposal must provide a means of synthesizing the results of the several projects. Otherwise, the hoped for higher-order goals of the program will elude us.

U.S. Participation in IBP Research

The U.S. National Committee in 1968 adopted the theme "Man's Survival in a Changing World." It also shifted some of the SCIBP emphasis on the two primary themes of the IBP, "The Biological Basis of (1) Productivity and (2) Human Welfare" to ones of (1) environmental management and (2) human adaptability.

The main thrust of the U.S. program is found in the five subprograms that are grouped under the IRP on "Analysis of Ecosystems." These five subprograms (Biome Studies) represent coordinated efforts to understand the functioning of

Table 12.1 Total IBP Effort in Study of Ecosystems (for countries listing "ecosystem analysis" no other programs are counted)

	Ecosystem Analysis	Primary Production	Secondary Production	Nutrient Cycling	Decomposition and Soil Processes
Grasslands	17	14	1	1	2
Woodlands	18	14	4	—	3
Arid lands	4	5	2	1	1
Arctic–Subarctic	6	2	—	—	—
Wetlands	4	1	—	—	—

five major ecosystems. These efforts involve many breeds of scientists, ranging from ecologists, systematists, pedologists, hydrologists to specialists in remote sensing, data processing, systems analysis, and ecological modeling. The purpose is to develop an understanding of major ecosystems such that it is possible to predict the effects of specific human manipulation and alteration of these systems. The kind of information emerging from these studies will provide guidelines for environmental management of the type that we have yet to achieve, but of a type that we must achieve if we are to slow and, hopefully, reverse the rapidly accelerating deterioration of the earth's environments.

The five biomes, listed in order of their priority in the activation of the U.S. program are: (1) Grasslands, (2) Desert, (3) Deciduous Forest, (4) Coniferous Forest, (5) Tundra. The level of treatment of these biomes internationally is seen in Table 12.1, while the breakdown differs somewhat from that in the U.S. National Program. Deciduous, Coniferous, and Tropical Forests are combined by the international group. Wetlands are not separately recognized in the U.S. program.

The complexity of the biome studies is illustrated in Figure 12.1, which shows the management structure for the research on the Grasslands Biome. The research itself is divided into two main types: (1) research at the "intensive" site, which is located on the Pawnee Site of the Agricultural Research Service of the U.S. Department of Agriculture, and (2) a comprehensive network of research activities throughout the major types of grasslands that occur in the United States. Each of these two subdivisions has its own project director.

International coordination of the various national studies of grasslands is effected through the Canadian Matador Grasslands Project under the leadership of the director of that project.

The five biome studies have made possible the comparison of basic ecological processes among five quite different ecosystems. Eight interbiome specialists committees provide interbiome communication and comparisons. These committees are serviced by an Environmental Component Coordinating office located at the University of Texas.

Efforts to mount a biome study of the internationally important tropical rain forest have been unsuccessful because of the inability to obtain funding.

The IRP for "Origin and Structure of Ecosystems" (originally called

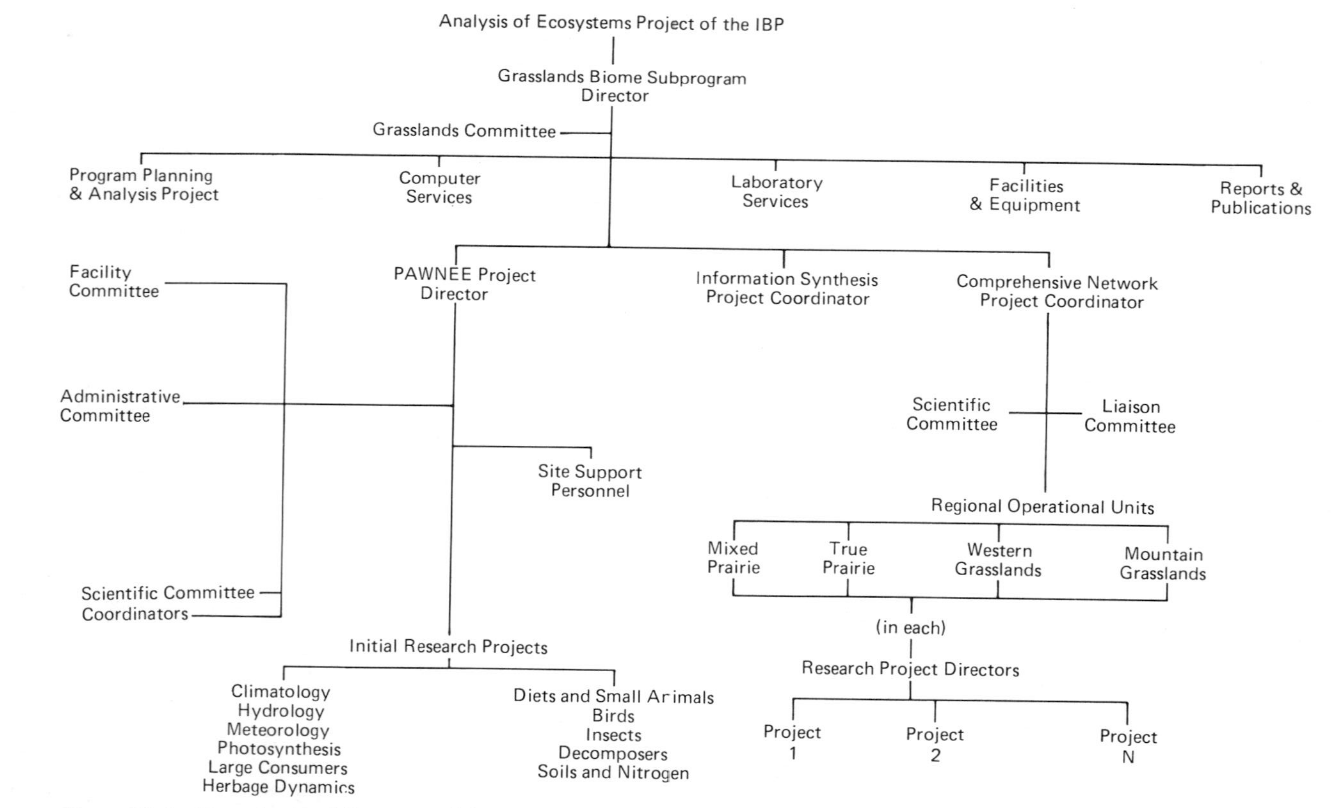

Figure 12.1 Organizational Structure of the Grasslands Biome Structure

"Convergent and Divergent Evolution") represents another example of integrated, cooperative research. The program as presently constructed consists of two subprograms, each of which stands as a relatively independent research unit. One of these subprograms is directed toward an understanding of the ecosystem structures that have evolved under similar physical environments in comparable areas of the southern and northern hemispheres. The degree of similarity imposed on ecosystems by their evolution in similar environments is of great importance to the whole ecosystem concept. Two sets of systems are being compared. One is the dry thorn scrub of the monte vegetation of northern Argentina versus the comparable Sonoran desert of North America. The other set is the Mediterranean Scrub as it occurs in western Chile and at comparable latitudes in California.

In these studies, vegetational structure and species diversity are being compared. Representative taxa of animals are being compared with respect to ecological niches occupied and with respect to the degree of similarity of these niches. Chilean, Argentinian, and U.S. scientists are working jointly on these studies, which should result in the formulation of general theory relative to the evolution of ecosystems.

The second subprogram is being carried out in Hawaii and is called "Island Ecosystems Stability and Evolution." This is directed mainly to an understanding of the ability of colonizing species to invade ecological systems relative to the age and degree of disturbance of the system. The generalizations that may come from this component are important because of the increasing alteration of the world's environments by man and because our modern methods of transportation have resulted in accelerated movement of weed species of plants and animals throughout the world.

Another IRP that involves multinational participation is one entitled "Upwelling Ecosystems." Areas on the west coasts of the continent where cold, nutrient-rich water is brought to the surface as the surface water is driven offshore by the trade winds support some of the world's most productive fisheries. The primary goal of this IRP is to obtain a quantitative understanding of the effects of enhanced nutrient supply. The results will be of potential value in the understanding and management of fisheries supported by upwelling.

A cruise of the research vessel *Thompson* in the summer of 1969 was directed to investigate the rich upwelling area off the coast of Peru. Scientists from the University of Washington, Oregon State University, the University of Alaska, Scripps Institution of Oceanography, Woods Hole Oceanographic Institution, U.S. Bureau of Commercial Fisheries, *Instituto de Investigaciones Pesqueras de Barcelona, Stations Marine d'Endoume de Marseille,* University of Athens, *Instituto del Mar de Peru,* and Peruvian Hydrographic Office participated in the cruise. A Mediterranean cruise took place in February 1970, and a major interdisciplinary cruise in 1972.

A somewhat different pattern of international cooperation is seen in the

CORE for "Aerobiology." In the United States program a CORE (Coordinated Research Program) differs from an IRP in that the objective is to coordinate a series of individual projects rather than to manage an integrated program. Internationally, aerobiology in the IBP has the main objective of information exchange and collation with respect to the dispersal of biologically significant materials in the atmosphere. Such materials would include such things as fungus spores, pollen grains, minute fragments of algae and other plants, bacteria, encysted protozoa, tiny insects, and other microfauna.

Here, emphasis is on better information exchange and on standardization rather than on cooperative research. The Program Director for this CORE of the U.S. program is also chairman of an international coordinating committee under SCIBP.

A relatively specialized but important IRP is that on "Integrated Pest Control." Control of agricultural pests through use of chemical biocides has created severe problems which may depend for their solution on great expansion of alternative methods for controlling agricultural pests. These major problems that have been created by agricultural pesticides include: (1) disbalanced ecosystems in which natural enemies that control agricultural pests have been destroyed, (2) evolution of pesticide-resistant strains of pests, (3) chemical pollution of the environment by the pesticides. Biological control projects are being carried out in 19 of the countries that participate in the IBP. Internationally, attention is being directed to five major groups of agricultural pests: (1) green peach aphids, (2) rice stem borers, (3) spider mites, (4) armored scale insects, (5) fruit flies.

The primary objective of this program is to establish the potential of and to apply a biological control approach (use of natural predators or parasites) to control the representative groups of pests just mentioned. The United States is an active participant in this program. International coordination is effected through an international committee.

One IBP project that is participated in by essentially all participant countries is called "Conservation, Terrestrial" by SCIBP ("Conservation of Ecosystems" in the U.S. program). The aim of this program is the establishment of the necessary scientific basis for a comprehensive world program of conservation and safeguarding areas of biological or physiographical importance for future world needs. The international coordinating group has established a vigorous and comprehensive system for recording precise data concerning the main relevant factors for all areas coming under consideration (*IBP News,* No. 16). A check sheet for IBP areas (questionnaire) has been established as an internationally approved basis for the collection of a wide range of information about sites throughout the world. The main accomplishment in the United States has been the preparation and publication of "Research Natural Areas," U.S. Government Printing office, Washington, D.C., 1968.

Other projects of the U.S. and/or international program are mostly peripheral to the present subject. These include "Conservation of Plant Genetic Materials," "Ecology of Nitrogen," "Phenology in Relation to Human Welfare," and the five projects that make up the "Human Adaptability" component of the U.S. participation in the IBP. The latter include: (1) "International Studies of Circumpolar Peoples," (2) "Biology of Human Populations at High Altitudes," (3) "Biosocial Adaptations of Migrant and Urban Populations," (4) "Nutritional Adaptation to the Environment," and (5) "Population Genetics of the American Indian."

Assessment of the IBP

Even with the modest financial support that it has been able to command, both nationally and internationally, the IBP has achieved an impressive series of accomplishments.

First, it has provided a prototype for the kind of multidisciplinary studies that are essential in providing baselines for environmental management. At the outset of the IBP, it was by no means certain that scientists of the many disciplines required for complete analysis and modeling of the functioning of an ecosystem could be induced to work in a team effort. Experience with the biome studies has shown that it is indeed possible to enlist and interrelate the efforts of a large group of scientists with diverse individual interests to develop an understanding of the functioning of a total ecosystem.

Second, the IBP in the United States and in other participant countries has provided a focus on environmental problems at a time when the attention of policy makers has been turning more and more to a consideration of environmental pollution and to means of restoring and maintaining environmental quality throughout the world. In the United States, the IBP organization has provided advisory service to the State Department and the World Bank. Internationally, the IBP organization has maintained liaison with FAO, WHO, and UNESCO of the UN agencies.

The IBP is a precursor of what promises to be a long-lasting, worldwide attack on problems of environmental quality. The "Man and the Biosphere" program of the UN, assigned to UNESCO for implementation, has been copied almost completely from the ongoing programs of the IBP, and it seems likely to succeed only if the ongoing IBP research is phased into the MAB program without interruption. The International Council of Scientific Unions (ICSU) has established a "Scientific Committee on Problems of the Environment" (SCOPE) with objectives very similar to those of the IBP. Some of the international projects of the IBP may be more appropriate for sponsorship by SCOPE than by UNESCO.

Finally, the IBP has spawned a major international effort to monitor the quality of the world's environments. At the IBP General Assembly at Varna, Bulgaria in March 1968, a three-man international committee was set up to study the feasibility of establishing a global network for environmental monitoring (GNEM). The Committee consisted of the author, Bengt Lundholm of Sweden, and N.N. Smirnov of the U.S.S.R. They all held work groups in their own countries, and one full committee meeting in London in September 1969. A full report of the Committee was presented to the IBP General Assembly in Rome in the fall of 1970. This report was subsequently transmitted to ICSU and through it to SCOPE, which formed a Monitoring Commission. The SCOPE Monitoring Commission built on the IBP report in preparing a report that contributed significantly to the planning of an "Earthwatch" program by the UN Stockholm Conference on the Human Environment in June 1972.

All in all, the IBP has been a success as judged at this point in its history. It has brought scientists of 58 participating countries into cooperative, coordinated effort to solve major problems of the world's environments. It has also brought together to considerable degree the efforts of those scientists who are "man oriented" in their outlook and those who are "environment oriented." Both groups have profited and should continue to profit from this development. However, its greatest contribution certainly has been in demonstrating the feasibility of a holistic approach to environmental research. This research will not solve the problems of environmental deterioration, but it can provide the basic information on which solutions must be based.

References

BLAIR, W. FRANKLIN (co-chairman), *Environment, the Quest for Quality.* U.S. National Committee for the International Biological Program and the Public Affairs Council, 1970. (A condensation of the proceedings of the February 18–20, 1970 conference mobilizing science, industry and government.)

"Research Natural Areas," U.S. Government Printing Office, Washington, D.C., 1968.

Government and Academia

Roles and Responsibilities in Man's Environmental Crisis

EDWARD J. KORMONDY*
The Evergreen State College

Edward J. Kormondy was born in Beacon, New York in 1926. He was educated at Tusculum College (B.S., 1950) and the University of Michigan (M.S., 1951, and Ph.D., 1955). He served as Instructor in Zoology at the University of Michigan, 1955–1957 and as Curator of Insects, University of Michigan Museum of Zoology, 1955–1957. From 1957 through 1968 he was a member of the Department of Biology at Oberlin College serving as Assistant and Associate Professor and Professor. During 1966–1967 he served as Acting Associate Dean of Oberlin's College of Arts and Science. During the summers of 1960–1962, he served as a summer staff member of the University of Pittsburgh Laboratory of Field Ecology. He held a National Institute of Health Post-Doctoral Fellowship in 1963–1964 at the Laboratory of Radiation Ecology on the Savannah River Atomic Energy Plant. He served as Director, Commission on Undergraduate Education in the Biological Sciences from 1968–1972 and Director, Office of Biological Education, American Institute of Biological Sciences from 1968–1971. He became Member of the Faculty at The Evergreen State College in Olympia, Washington in 1971 and currently is Vice President and Provost.

His research interests have ranged from the systematics and ecology of dragonflies to productivity, metabolism, and succession in beach ponds to decomposition of aquatic plants and to the use of radioisotopes in studying the nutrient cycle of zinc. Among his research papers are the following: The Systematics of *Tetragoneuria* Based on Ecological, Life History, and Morphological Evidence, *Miscellaneous Publications of the Museum of*

*Manuscript accepted in July, 1970.

Zoology, University of Michigan; "Territoriality and Dispersal in Odonata," *Journal of New York Entomological Society;* "Uptake and Loss of Zinc-65 in the Dragonfly *Plathemis lydia,*" *Limnology & Oceanography;* and "Ecological Succession in Sandspit Ponds," *American Midland Naturalist.* He is the author of five books, including *Readings in Ecology,* Prentice-Hall, 1965, and *Concepts of Ecology,* Prentice-Hall, 1969. He is editor of the BSCS/Pegasus Science & Society Series.

In his Inaugural Address, the charismatic and youthful President John F. Kennedy coined the aphorism "Ask not what your country can do for you, but what you can do for your country." Loyalty and devotion aside, I would like you to focus on the first part of this statement in a slightly different form. As you begin to look ahead, what do you want from life? What should you be able to expect from life? What responsibilities does the establishment, in the form of government and education, have in enabling you to reach these goals?

In asking this question, I am thinking not only about the immediate period of your college career: though terribly important and encompassing, it is of too short range. I am not thinking of the period immediately following college: though terribly personal and consequential, it, too, is of short range. I am asking you to think about the next 50 years, for that is the period over which most of your generation will live; this is the period of college, maybe graduate study, the first job, marriage, family, grandchildren and great grandchildren, vacations, travel, infirmity and senescence, and, for some, senility. What kind of quantity, what kind of quality do you want during that half-century span? What do you want for yourself? What do you want for society, or do you need to be concerned about society? How much are the wishes for self counter to those for society?

As I travel and talk with young people, listen to their speeches, and read their writings, I realize the answers to the kinds of questions I am asking are as diverse as is pluralistic society itself. I also realize the answers are largely personal. What many want are the basic amenities: a job with a future, marriage and family, home and car, that is, the "good things" of life, the advantages they perceive to be the deserved consequences of our competitively industrialized society.

But, there is a significant contingent of your peers, and particularly those in college, who question the contemporary values of our economic system; they see "more and bigger" as not necessarily better. They question the loss of privacy and aloneness which follows from densely crowded everything—cities, parks, movies, planes. They question hypocrisy in an officialdom which pleads for "law and order" while some of those very officials bribe and embezzle and otherwise flout the law. They question the contradictions in those government agencies which have responsibilities on the one hand for the development and promotion of a product or resource, typically in cooperation with business and industry, and, on the other hand, for regulation of those very businesses and industries to protect the citizen-consumer. They question the cumbersome, slow-moving machinations of democratic government when the need in their eyes is for fast, efficient, and comprehensive response. They question this same cumbersome and slowly responding mechanism of academia, too—in the colleges particularly, but increasingly in the high schools. They question whether there are national goals and priorities and assessments; they ask who can and should set them, and who should see to their achievement; they wonder whether political government can cope with these kinds of issues in nonpolitical ways, but they do not see the university rising to fill the void in leadership.

At the risk of being presumptuous, but with the idea of promoting your reflection on the matter, I would like to explore some aspects of the role and responsibility of both government and education in man's environmental crisis. My ideas are not unique, nor were they delivered on tablets from Mt. Sinai. They have developed out of personal experience and reading, but largely they have developed from numerous discussions with friends and associates in the university community and in various branches and levels of government. Prior to delving into this topic, and at the risk of perhaps repeating some things said elsewhere in this book, I should like to briefly encapsulate what I see as the nature and dimension of man's environmental problem. It will be with the intent of treating that malady that we can proceed with defined objectives to the main focus of this chapter.

The Nature and Dimension of Man's Environmental Problem

The basic dimension of man's environmental problem is that of too many with too little—too many people with too little room, too many people with too little resources. Most assuredly, not only will there be more people, but they will live in denser aggregations: by some estimates, 300 million people in the United States alone by the year 2000 living on the same total acreage and largely in the already densely populated cities, suburbs and megapoli; 6 to 7 billion people in the world by the year 2000 living on the same total acreage and largely in the already densely populated countries and cities. The New Yorks and the Hong Kongs will be only somewhat geographically larger but considerably larger in the number of inhabitants. More people means more food, more clothes, more necessities, and more amenities. More, more, more. But the more must be met from the ultimately limited resources spaceship Earth affords in land, water, power, and minerals, and by the ultimately limited technological capacity of man to engineer these resources to his use.

Food

Consider our present world of some 3 billion persons, wherein two-thirds are currently existing on a below-standard nutritional level. What consequences can one expect with a population twice that size, distributed approximately as now with a maximum potential food production that will not be sufficient? The best and most recent available estimates indicate a potential yield from the sea of only about twice the present production, or a maximum of about 100 million tons (Ryther, 1970), and about eight times the present production from land (under optimum conditions of high fertilization and pesticide applications, and with all potentially arable land under cultivation) (Committee on Resources and Man, 1969). This total food production "approaches a limit that seems to place the earth's ultimate carrying capacity at about 30 billion people, *at a level of chronic near starvation for the great majority*" (Committee on Resources and Man, 1969). Furthermore, one needs to keep in mind that although products of the sea are excellent sources of protein and would meet 30 percent of the world's protein need, the energy foods are those which come from the land.

The seemingly obvious solution to this impending crisis of demand and supply is to decrease the demand, that is, to decrease the growth (if not also the present size) of world population. However, as Ivan Bennett has noted (1969), this is a long range solution, not an immediate remedy. Food needs will increase markedly in developing countries even if no new children were to be added. The biological explanation of this lies in the fact that the "amount of food required for growth increases steadily from the time of birth to about 19 years of age. Half of those living in the developing countries are less than 15 years old. To maintain the Indian population at its *present level* of nutrition would require 20 percent more food in 1975 than in 1965 *if no new children were added during this 10 year period*" (Bennett, 1969). Elevation of the diet just to the minimal standard would require a 30 percent increase!

MINERAL RESOURCES

More people alone, or the same number of people desiring a higher standard of living, will run smack against the coldest stark reality of all—an ultimate limitation on mineral resources, not only because of the constraints of accessibility and recoverability, but fundamentally because they are nonrenewable. Long-term trends indicate there will be an increasing demand over the next 50 to 75 years for what I like to call "cultural nutrients" in contrast to the "biological nutrients"—those used by organisms in their metabolism. Man's cultural demand for coal and natural gas, mercury and tin, chromium and platinum, and similar resources has been growing at the rate of 6 percent a year for a decade or more. At that rate of increase the total to be used in the next 30 years will equal the total amount used in all previous time!

For our purposes here, the two sobering consequences of this increasing demand against a limited supply are that, according to projections of demand based upon current trends: (1) mercury, tin, tungsten, and helium reserves will be exhausted by the end of the century; petroleum and liquid gas supplies will last some fifty years and coal supplies some two to three centuries and (2) the

United States will be increasingly dependent on foreign supplies and will have to deal with countries of differing ideologies, largely communist, for therein are most of the reserves of tungsten and antimony, and a large part of the reserve of manganese, nickel, chromium, and platinum (Committee on Resources and Man, 1969).

Based on existing trends, the rising level of expectation for goods that would accompany the development of presently non-industrial nations, the rate of consumption will increase and the stockpile of resources will be decreased that much more quickly.

There is another aspect of man's cultural resources which deals with the by-products of our cultural *practices*, that is, the pollutants in our water, air, and soil. Because this matter has been treated in earlier chapters, I need not deal with it here. But, I do need to remind you that this is a significant, yes, *terribly* significant dimension of man's environmental crisis.

Energy Resources

More power to make more goods; more power to run more electrical devices; more power to make the machinery, fertilizers, and pesticides to produce more food; more people alone or the same number of people desiring a higher standard of living—all these will place ever-increasing demands on the world's energy resources. Yet, according to the assessment by the Committee on Resources and Man (1969), the present and future demand is such that, of the major current energy sources, coal supplies are sufficient to serve as a major source for industrial energy for only two or three centuries, and petroleum for only 70 to 80 years. Although the potential supply of water power is comparable to the present rate of energy consumption from fossil fuels, most of this potential occurs in the industrially undeveloped areas of Asia, South America, and Southwest Asia. Further, large-scale power production from solar energy is technologically unpromising, and geothermal and tidal energy sources are seen as supplying only a fraction of the world's present needs. Finally, although nuclear power is the only significant and practical source, present technology is far from that needed both in developing more efficient fission reactions—those in which the nucleus of an atom is split with the release of energy, such as occurs in the splitting of uranium in the atomic bomb—and in achieving controlled fusion reactions —those in which lightweight atoms join together to form a heavier atom and release energy, such as occurs on the sun with the fusion of hydrogen to form helium, and in the hydrogen or thermonuclear bomb.

Nature of the Problem

The preceding comments are not intended as doom-saying, but, rather as attempts to set forth a realistic assessment of the dimension of man's environmental problem: too many people and too many demands and expectations on the finite set of resources with which he is endowed. The nature of the problem and its basic causes lie in the nature of man himself and in the dilemmas which arise therefrom.

Applied science and technology can certainly go far toward developing solutions to some of the difficult problems I have listed, particularly in the areas of pollution and food and power production. Thus, we shall need to explore the role of government and academia in providing the circumstances in which and the resources with which this can be done. As you shall see, there are a number of factors of a nontechnical and nonscientific nature, but, rather, of a political and economic sort, which will complicate, if not make impossible, the achieving of those solutions.

What basic science and its applications will be unable to do, however, is to resolve the root causes of the problems and the dilemmas which arise from these causes. Science and technology will, at best, offer palliatives, alleviating the symptoms without excising the causes that are deeply ingrained in western, industrialized cultures of Judeo-Christian heritage (White, 1967). These are cultures embued with an ethic for the land and for the products of that land which exploits the former and aggrandizes the latter. The consequence of this ethic is the dilemma in choosing between the notions of quantity and quality: the quantity notion that more is better—more cars, more people, more power, more of everything, and the quality notion that more is not necessarily better—more cars is more noise, more people is more starvation, more power is more air pollution, more of everything is less of something. A widespread acceptance of unlimited growth is inimical to survival on a finite plant. The contenders are, indeed, growth versus quality of life (Wagar, 1970).

We need, then, to explore also the implications of this fundamental aspect of man in respect to government and academia. It must be immediately obvious that the scientific and technological considerations by government and academia are dwarfed and even potentially impotent in the light of the magnitude and pervasiveness of the problems posed by the nature of man himself.

Government and Environment

The Congress

> The Congress is the only institution having the scope to deal with the broad range of man's interactions with his physical-biological surroundings. We therefore believe that leadership toward a national environmental policy is our responsibility.

This profound and far-reaching statement appears in the letter of transmittal to the U.S. Congress of "A National Policy for the Environment," submitted October 1968 under the auspices of Congress' major environmental watchdogs, the Senate Committee on Interior and Insular Affairs and the House Committee on Science and Astronautics. The statement, based on a colloquium of the preceding summer which avoided committee jurisdiction and convened leaders of the Executive branch, industry, commerce, academia, and science, also contains the following policy statement:

> The ultimate responsibility for protecting the human-serving values of our environment rests jointly with the legislative, executive, and judicial branches of our Government. The Congress, as a full partner, has the obligation to provide comprehensive oversight of all environment-affecting programs of the executive branch, and also to participate in the overall design of national policy, thus serving both as architect of environmental management strategy and as the elaborator of goals and principles for guiding future legal actions.

I have been considerably impressed by the implications of these two statements for the future of man and by the sincere commitment with which several of the key Congressional leaders who were instrumental in the development of this statement of proposed policy faced their responsibility in this matter. I also have been considerably impressed by the continuing efforts of many members of the Congress to become informed on the basic underlying principles of man-environment relationships, to keep abreast of new developments, and to become aware of potential untoward consequences.

As a case in point, the Legislative Reference Service, an agency of the Congress in the Library of Congress that handles inquiries from any member of Congress or any committee (but honors no request from the Executive or Judiciary branches!) established an Environmental Policy Division in September 1969 (Carpenter, 1970). This division, like other divisions of the Service, is charged with determining what information will be useful to those making decisions, culling the answers from the literature, individuals, and institutions and organizing it in the "pro and con" mode. Like the successful debater who is as fully prepared to argue either side of a question, this technique not only sharpens the data search itself, it avoids the hazard of meeting only the polarized view with which the initial request might have been made. The fact that the Legislative Reference Service employs 325 persons with professionals from every field of knowledge and handles 155,000 requests a year (remember these are limited to the Congress only!) is considerable evidence not only of the desire for information, but of highly filtered, evaluated, documented, and articulated information.

A different mechanism of information is provided by the Environmental Clearinghouse, Incorporated, which was chartered in June 1968 for the purpose of providing staff assistance to the unofficial Congressional Ad Hoc Committee on the Environment, whose membership consists of some 120 Senators and Congressmen from both political parties (Ottinger, 1969). Through a 130-member Board of Advisors comprised of scientists, businessmen, and conservationists, information is provided to the Committee about the environmental consequences of decisions they are called upon to make. Its

Table 13.1 Principal Congressional Committees Concerned with Environmental Quality and Productivity

Senate	House of Representatives
Agriculture and Forestry	Agriculture
Commerce	Government Operations
Subcommittee on Energy, Natural Resources, and the Environment	Subcommittee on Conservation and Natural Resources
Government Operations	Interior and Insular Affairs
Subcommittee on Intergovernmental Relations	Interstate and Foreign Commerce
Interior and Insular Affairs	Merchant Marine and Fisheries
Labor and Public Welfare	Subcommittee on Fisheries and Wildlife Conservation
Subcommittee on Health	Subcommittee on Oceanography
Public Works	Public Works
Subcommittee on Air and Water Pollution	Subcommittee on Flood Control
	Subcommittee on Rivers and Harbors
	Science and Astronautics
	Subcommittee on Science, Research and Development
Joint Senate-House	
Atomic Energy	

SOURCE: Adapted from an article by R. A. Carpenter, printed in *Science*, **168**, pp. 1316–1322 (June 12, 1970). Copyright 1970 by the American Association for the Advancement of Science.

director, Frank Potter, a lawyer by profession, told me that the Clearinghouse "serves as a pipeline directly to the scientific community for any member of Congress, whether a member of the Clearinghouse or not."

As the old saw has it, "The road to hell is paved with good intentions," and we should be concerned about the extent to which the government, particularly the Congressional and Executive branches, can wear the mantle it has made for itself in "protecting the human-serving values of our environment." By historical accident, by back-scratching, and by design, the government has entered the 1970s woefully unprepared to do the type of job the system demands. It is not so much a matter of sending a child on a man's errand, but of neither knowing what a man is nor what the errand should be nor where to go to get whatever it is that no one will know when he gets it.

In the instance of Congress, varying aspects of environmental management reside within the jealously guarded jurisdictions of 14 principal committees (Table 13.1). These committee jurisdictions overlap, however, so that a matter like control of water pollution may be considered in the Senate by the Committee on Interior and Insular Affairs as well as by the Committee on Public Works or the House Committee on Science and Astronautics. Conflicting jurisdictions exist in the matter of pesticides such that, although hearings, quite critical ones at that, have been held by Government Operations, Merchant Marine and Fisheries, and Commerce Subcommittees, tradition and precedent has it that regulatory legislation is the sole responsibility of the Agriculture Committees in both houses (Carpenter, 1970).

In the 89th Congress (1966–1968) alone, moreover, significant environmental measures were introduced in 8 of the 16 standing committees of the Senate (Agriculture and Forestry, Commerce, Finance, Foreign Relations, Government Operations, Interior and Insular Affairs, Labor and Public Welfare, Public Works) and in 11 of the 21 standing committees of the House (Agriculture, Banking and Currency, Government Operations, Interior and Insular Affairs, Interstate and Foreign Commerce, Judiciary, Merchant Marine and Fisheries, Public Works, Rules, Science and Astronautics, and Ways and Means). And as if this were not enough of an organizational web—a web defying successful travel even for a schizophrenic spider—the best legislative intentions can go to naught before the purse holders, the powerful Senate Committee on Appropriations, and the House Committee on Appropriations.

That everyone in Congress wants and has a ladle in the stew underscores the classic understatement of the aforementioned white paper, "The proper development of such a far-reaching body of policy raised many difficult organizational, economic and legal problems." In suggesting ways in which Congress might be influential on environmental policy, the Congressional authors, it seems to me, have some tired, merely rhetorical remedies: (1) resolutions declaring the interest of Congress in establishing a national environmental policy or calling for an amendment to the Constitution on "environmental rights" (this was subsequently introduced); (2) joint committees on environmental management; (3) an environmental surveillance unit, or a task force, or a commission. There is a saying in Washington that "Commissions come and go, but the problem still remains." The Commission route, at least as viewed by an alumnus of the several hundred formed in the last twenty years and of the 40 established by Richard Nixon in the first 17 months of his presidency sees it this way: "If the subject is a hot potato, if you don't know what to do, appoint a Commission" (Haynes Johnson, "The Study Commission Syndrome," *Washington Post*, June 28, 1970).

Not all Congressional well-saying is fruitless, nor, as I shall point out later, is all commission and/or council-type activity merely talk and report. Things do happen, and out of the Congressional white paper did come the National Environmental Policy Act of 1969 which was signed by President Nixon on January 1, 1970. This Act heralds something quite different for the 1970s and beyond. The fundamental elements of this policy are found in the following excerpt from the earlier white paper:

> It is the policy of the United States that:
>
> Environmental quality and productivity shall be considered in a worldwide context, extending in time from the present to the long-term future.

Purposeful, intelligent management to recognize and accommodate the conflicting uses of the environment shall be a national responsibility.

Information required for systematic management shall be provided in a complete and timely manner.

Education shall develop a basis of individual citizen understanding and appreciation of environmental relationships and participation in decision making on these issues.

Science and technology shall provide management with increased options and capabilities for enhanced productivity and constructive use of the environment.

One of the immediate implementational aspects of this Act is a requirement that "every recommendation or report on proposals for legislation and other major Federal actions significantly affecting the quality of the human environment" must be accompanied by a detailed statement on

(i) The environmental impact of the proposed action,

(ii) any adverse environmental effects which cannot be avoided should the proposal be implemented,

(iii) alternatives to the proposed action,

(iv) the relationship between local short-term uses of man's environment and the maintenance and enhancement of long-term productivity,

(v) any irreversible and irretrievable commitments of resources which would be involved in the proposed action should it be implemented.

The Act also established in the White House the Council on Environmental Quality whose responsibility it is to transmit to the Congress an annual report on status and trends in environmental affairs and to suggest remedies for deficiencies. President Nixon has referred to the Council as "the keeper of our environmental conscience," inasmuch as every department or bureau of government is to send their plans to the Council to be cleared for environmental impact. Others have thus gone so far as to compare it with the well-established Council of Economic Advisors which clears proposals on economic impact. However, it remains to be seen how effective the environmental Council can be considering the relatively small budget and staff with which it has been provided in its first several years.

It is nonetheless apparent that the Congress is well intentioned, if not politically motivated, on this matter and has taken steps to assure a flow of environmental information upon which appropriate legislative response might be made. The major information sources established by the National Environmental Policy Act, coupled with the services performed by the impartial evaluations of the Environmental Policy Division of the Library of Congress' Legislative Reference Service and of the Environmental Clearinghouse, and the witness of representatives of learned and technical professional societies, when balanced against the pleas of civic-oriented groups and the lobbying of industrial and commercial interests, should go some distance toward an enlightened definition and implementation of a national policy for the environment by the Congress.

Table 13.2 A Partial List of the Federal Agencies and Bureaus Dealing in Varying Degrees with Environmental Matters

Departments

- Agriculture
  - Agricultural Research Service
  - Forest Service
  - Soil Conservation Service
- Commerce
  - Business and Defense Services Administration
  - Environmental Science Service Administration
- Defense
  - Army Corps of Engineers
  - Organization of the Joint Chiefs of Staff
- Health Education and Welfare
  - National Air Pollution Control Administration
  - National Institutes of Health
  - Office of Education
  - Public Health Service
    - Bureau of Radiological Health
    - Bureau of Solid Waste Management
    - Environmental Control Administration
  - Environmental Health Service
  - Food and Drug Administration
- Housing and Urban Development
  - Federal Housing Administration
  - Model Cities Administration
  - Renewal Assistance Administration
- Interior
  - Bureau of Commercial Fisheries
  - Bureau of Land Management
  - Bureau of Mines
  - Bureau of Reclamation
  - Bureau of Sport Fisheries and Wildlife
  - Federal Water Pollution Control Administration
  - Office of Marine Resources
  - National Park Services
- Justice
  - Land and Resources Division
- State
  - Agency for International Development
  - International Scientific and Technological Affairs Bureau
  - Office of Environmental Affairs
- Transportation
  - Coast Guard
  - Federal Aviation Administration
  - Federal Highway Administration
  - Urban Mass Transport Administration

Executive Office of the President

- Bureau of the Budget
- Citizens' Advisory Committee on Recreation and Natural Beauty
- Council of Economic Advisors
- Council on Environmental Quality
- National Council on Marine Resources and Engineering Development
- Office of Science and Technology

Independent Agencies

- Atomic Energy Commission
- Federal Power Commission

Table 13.2 (Continued)

National Science Foundation
National Aeronautics and Space Administration
Smithsonian Institution
Tennessee Valley Authority

Quasi-Official Organizations

Environmental Studies Board
National Academy of Engineering
National Academy of Sciences
National Research Council

Selected Boards

Federal Radiation Council
Migratory Bird Conservation Commission
Mississippi River Commission
National Park Foundation
National Water Commission
Water Resources Council

The EXECUTIVE BRANCH

If the labyrinth of Congressional jurisdictions in environmental matters seemed incomprehensible to you, imagine being a mouse in the maze which exists on environmental concerns in the Executive branch. According to a Library of Congress report (Boswell, 1970) some 100 separate agencies, boards, committees, and commissions play in the federal environmental game (Table 13.2). In turn, many of the agencies are hydra-headed. For example, the Environmental Science Services Administration of the Department of Commerce has five components: Environmental Data Service, Weather Bureau, Research Laboratories, Coast and Geodetic Survey, and National Environmental Satellite Center.

Like Topsy, the agencies "just growed"—spawned by a national crisis (for example, the Bureau of Mines in response to the annual death of 3000 miners), a pet dream of a key leader (for example, the Coast Guard and Alexander Hamilton, or the Soil Conservation Service and Franklin D. Roosevelt), a response to scientific insight (for example, the Bureau of Radiological Health), or even a headliner for an astute politician (for example, almost any "popular" issue). As Theodore White put it, "All agencies reflect that hoary first principle of American government: when something itches, scratch it" (*Life*, June 26, 1970). And there has certainly been a lot of "you scratch my back and I'll scratch yours."

It is true that the problem with which these agencies deal is complex beyond measure, and it is not surprising therefore that a complex pattern of agency activity has resulted. This is one of the enrichments gained by not being a planned dictatorship as well as one of its hazards. The redundancy of responsibility, in fact, should give some assurance of meeting the fundamental objective of managing man's environment.However, more often than not, as the redundancy and ambiguity of agency titles suggest, the situation is one of conflict and contradiction, overlap and overkill. Charged with keeping the forests in preservative perpetuity, the Department of Interior's National Park Service runs headlong into Agriculture's Forest Service whose job it is to serve lumbermen and cattlemen; Agriculture's long defense of DDT's role in high crop yields counters the human health responsibility resident in Health, Education, and Welfare; rat control involves at least a dozen agencies including some seemingly unlikely custodians such as Fish and Wildlife Service, Agricultural Research Services, National Institutes of Mental Health, and the Office of Economic Opportunity.

Some agencies suffer from an internal identity crisis. The Bureau of Mines, for example, was created to protect miners but during World War II became a prospecting agency, right after the war a marketing agent for mining interests, and most recently it has engaged in mineral reclaiming projects. Yet other agencies suffer from a different kind of malaise. For example, under an 1899 law dealing with dumping of wastes in navigable waters, fines of up to $2500 and one year in jail can be imposed. Yet in our own nation's Capital, the first charges brought under the law occurred in May 1970! And, suffering from something more than malaise, the Army Corps of Engineers whose job includes keeping waterways navigable, and hence, whose responsibility it is to enforce the 1899 Refuse Act—a responsibility it has unashamedly failed to carry out—has itself been consistently one of the worst offenders, dredging one place and dumping in another. Only within the current year has there been some overtures from the Corps of change on this matter, and then only as a result of increased Congressional and public pressure, particularly through actions by civic groups. This privilege of pressure is one of the inherent rights and responsibilities of our type of government; it needs to be exercised if that government is to discharge its responsibilities.

Far worse than the inter- and intra-agency pot-shots, feuds, and outright skirmishes, or an occasional detente or even inaction, are those situations for which no responsibility exists or in which no agency believes it has the jurisdiction to proceed. The prime (or most sorry) example of this vacuum is Lake Erie whose sure fate at the hands of abundant city sewage, industrial acid wastes, agricultural fertilizers, and silt was predicted by numerous Cassandras in scientific circles both in and out of government in the late 1940s. But, strange as it may seem, no overriding responsibility for the Lake existed, or, at this writing, exists.

Admittedly, Lake Erie's problems are compounded by the international factor of Canadian jurisdiction of half the system; however, the situation of a lack of comprehensive overviewing authority on the part of the United States still stands. On the other hand, to think in terms of a "Lake Erie Authority" would cause one to ask, "Why not a Hudson River Authority or a Chesapeake Bay Authority?" Vertical and horizontal organizations both have their shortcomings, to be sure.

As the Congress has attempted to make some order out of mayhem, so, too, has the Executive branch. The Congress has gone the route of information, the Executive the route of reorganization. Based on the recommendations of his Advisory Council on Executive Reorganization, President Nixon reshuffled the bureaucracy in the summer of 1970 by creating two new super-agencies: the Environmental Protection Administration (EPA) and the National Oceanographic and Atmospheric Administration (NOAA)—pronounced NOAH, and causing some joker to ask where the ark was to be located.

EPA, an independent agency like NASA, has responsibility for all programs of air and water pollution and solid wastes; its genesis was in the frustration at the slow progress of pollution control programs which, as noted above, were largely in departments with dual responsibility for both regulating and promoting activities that cause pollution. Among other shifts, into EPA and out of Interior came the Federal Water Pollution Control Administration, from Health, Education and Welfare came the clean-air and solid-waste disposal programs and the pesticide research program, from Agriculture came pesticide regulations, and so on. Unlike its diffusely organized forebearers, the intent of EPA is to functionally integrate and consolidate the various pollution programs, separating the setting of standards and regulations from development activities. Given wise, but strong, leadership, EPA may go far toward implementing the many fundamental recommendations made earlier by the Environmental Pollution Panel of the President's Science Advisory Committee (1965).

The National Oceanographic and Atmospheric Administration (NOAA) is a kind of "wet NASA." Not only does it reflect a new and greater national emphasis on the science and technology of the oceans and atmosphere, it also reflects a long- and short-term concern over these integral compartments of man's environment. Its function is to explore and predict what is happening to our oceans and climate, and to take appropriate remedial steps. Housed in the Department of Commerce, NOAA brings together the Environmental Science and Service Administration, already in the Commerce Department and involving the Weather Bureau and the Coast and Geodetic Survey, Interior's Bureau of Commercial Fisheries, the Lake Survey Office of the Army Corps of Engineers, the Sea Grant Program of the National Science Foundation, the marine technology program of the Bureau of Mines, and yet other ocean- or atmosphere-oriented agencies.

Reshuffling and reorganizing the federal bureaucracy is both a Presidential prerogative and responsibility, reflecting new perceptions and insights, new problems and opportunities. President Nixon's reshuffling of the existing environmental agencies into EPA and NOAA were taking place as this essay was being written; unlike many other of his actions, this was widely applauded and supported as a definitely wise and needed move in the right direction. You will be reading this sometime after these new agencies have been in operation; how effective have they been? How much have they tripped over their own internal loop holes? How much have they been unable to do because yet other critical environmental matters are still diffusely "agencied"—things like energy resources which are monitored by both the Atomic Energy Commission and the Federal Power Commission among others, things like land resources which are the province of the Army's Corps of Engineers, the Forest Service and the National Park Service, and so on? How much have they been curtailed by the political process—by lobbyists for particular industrial or commercial interests, by the failure of the Congress to appropriate sufficient monies to execute programs, by the delaying tactics possible under our judicial system, by the failure of the populace to pressure?

Finally, but most significantly, how effective have these agencies been or can they be expected to be with the real can of worms that describes the local and regional jurisdictions and overlords—the towns and cities, counties and states, the regional and metropolitan authorities? In Long Island Sound alone, there are more than 50 local jurisdictional groups having regulatory powers on sewage disposal!

More sobering yet is a look in the other direction—to national jurisdictions, prerogatives and jealousies and the associated nose-thumbing at more than one international agreement on more than one major issue. But, as former Ambassador to Russia George F. Kennan (1970) has noted, "The entire ecology of the planet is not arranged in national compartments." Kennan argues, and I concur, that there is a need for an International Environmental Agency to review the work of existing organizations primarily to identify the gaps and voids, to advise governments and influence public opinion. He sees this as an "organizational personality—part conscience, part voice—which has at heart the interests . . . of mankind."

Without belaboring the point further, it should be clear that in man's environmental crisis responsibility and opportunity extend vertically as well as horizontally through the hallowed halls of the federal as well as local, regional, state, and national governments in monitoring, standard setting, regulating, and enforcing, research and development, prediction and assessment, informing and communicating. Some of these roles can be carried out by government *per se*, and only by the government. With many issues, however, the government may be unable to cope in an apolitical way. The political pressures are so very severe that the kind of clear thinking that must ensue in the establishment of national goals and priorities, and in the dispassionate assessment of the consequences of an advancing technology, cannot be realistically expected. John Platt (1969) has called for a mobilization of scientists as the only way to solve our crises and problems. It is my feeling that whatever the mechanism, the discourse leading to resolution can occur only in the freedom of the university community, although that community will have to alter much of its self-image and traditionally defined role to do so.

Academia and Environment

To you who have just joined the university[1] community, it may seem needless to comment on the nature and role of higher education in contemporary society. Although the literature abounds with lengthy discourses on this particular topic, we can perhaps agree that the university has perceived its mission in two major roles: that of transmitting from the past, and that of investigating what is in the present.

As a custodian of the past, it is a library of history, an encyclopedia of man's accomplishment and understanding. Rather than like a cradle springing forth it is, in this sense, much like a wheelchair tenaciously holding only the carcass of the past. While I subscribe to the role of the university in passing on what has been learned and that each ingenue at the university need not discover for himself all that man has already discovered, the reverential looking-back is singularly inappropriate in contemporary times. In its other role, that of investigating what is, the university is concerned with the acquisition of knowledge and typically in compartments referred to by the names of the respected disciplines. Seldom is it done in the context of the totality of man or of man with respect to his environment.

As the university entered the decade of the 1970s, under pressures largely from students, and from the professional voices that had been crying in vain in the wilderness for some time previously, came new currents and new dimensions, new roles and responsibilities. Not the least of these was the recognition that the university needs both to look ahead and to be concerned with the transmission of value judgments. Perhaps the clearest statement reflecting this newer purpose and function of the university was that prepared by the Interdisciplinary Studies Committee on the Future of Man at the University of Wisconsin (Potter, 1970). In this report we find the following:

> The university is one of the institutions that has a major responsibility for the survival and improvement of life for civilized man. We agreed . . . that an important purpose of the university is "to provide society with objective information and with imaginative approaches to the solutions of problems which can serve as the bases for sound decision-making in all areas."
>
> Government and industry can and will be primarily responsible for solutions to problems of the present, though they will in many instances draw upon university resources. On the other hand, the university by its very nature must be future-oriented because it is responsible for the joint effort by which faculty and student provide knowledge, skills, and social values for much of the leadership for the next generation.

In its scholarly tradition of upholding impartial judgment in the weighing of evidence accrued from various sources, and in its ofttimes abstract orientation in

[1]The term "university" as used here involves all institutions of higher education, two and four year colleges, as well as those actual graduate-degree granting institutions usually referred to as universities.

the "search for truth," the university has too often left its youth in an abyss of meaninglessness and purposelessness, without a rationale or a philosophy. A survey by pollster Louis Harris in 1970 found that 78 percent of the college students believe "the real trouble with U.S. society is that it lacks a sense of values." In the rightful attempt to have each student achieve his own self-identity and his own set of values, the university inadvertently left a leadership vacuum which, in the late 1960s and thus far in the 1970s was being filled by forces antithetical to the university as well as to the democratic system in which the American university survives.

Although catalog after catalog purports to the contrary, the university actually has largely failed to instill those time-honored values that constitute the warp and woof of American society. This I see as largely a consequence of compartmentalization of knowledge and specialization of discipline. The result of the pursuit of finer discriminations, resolutions, and particularizations is counter to a seeking for the broad brush strokes of man in his total being. Although individual faculty members, and even particular faculties and colleges, have been able to avoid this fault, the generalization holds that the university community in general has been remiss in discharging this responsibility of conveying an holistic view of man.

Not all is lost however; a few, and now in increasing numbers, members of the university community perceive the need for a shift in their functions. For example, the previously referred to University of Wisconsin Committee states elsewhere in its report:

> We believe that the university has an obligation to examine and preserve the value judgment that can elevate the condition of the society on which it depends. It can serve this function by a search for truth that is future-oriented and that explicitly recognizes the need to transmit not only knowledge but also meaningful value judgments to succeeding generations.

The university then must not only be concerned with the past and the present in terms of knowledge to be discovered and transmitted, examined and to be preserved, but also with that knowledge and that wisdom and those values that will work toward both the survival of man and improvement in the quality of his life.

In implementing these new directions, there may be as often as not, the need for expansion and redirection of existing functions; in other instances, new mantles will have to be donned. Let's consider some of these roles.

Generation of Information

Fully recognized as part of university activity is the generation of information. In the matter before us, however, there is need for a shift to the generating of information on the character of environmental change. This includes an assessment of short- and long-term effects of additions and depletions to the environment, of resource exhaustion, utilization and recycling, and the like. It includes the need for the generation of models of ecological systems, not merely of forests and fields, but of cities like Los Angeles, and of major aquatic systems like Lake Michigan. Some universities have such programs in process at this writing.

This systems approach will facilitate an understanding of the network impact of environmental alteration, of the fundamental nature of ecological interconnection. For the fact is that the depletion of one resource leads inevitably to the depletion of another, that one additive has effects on other components of the system removed from it both in time and space. None of the major resource or pollution crises can be solved in isolation, hence, the need for models which can be manipulated and used as a basis for prediction.

There is no question then but that this kind of information generation will require individuals trained in specific disciplines—engineers, biologists, physicists, and so on. However, since ecological machinations are interconnected, the disciplines must be interconnected. The popular solution for this is to have an interdisciplinary study. I much prefer to achieve an interdisciplinary attitude and understanding, to prepare a specialist who is a generalist in outlook and who is able to work effectively and cooperatively with other individuals on a particular problem.

By looking at the kinds of information needs which were spelled out in the aforementioned Congressional white paper on a National Policy for the Environment, you can see why only individuals carefully and explicitly trained can, in fact, bring some of these matters to solution. For example, the university must prepare its graduates to assume at least the following tasks:

> Alteration and use of the environment must be planned and controlled. . . . Alternatives must be actively generated and widely discussed. Technological development, introduction of new factors affecting the environment, and modifications of the landscape must be planned to maintain the diversity of plants and animals. Furthermore, such activities should proceed only after an ecological analysis and projection of probable effects.
>
> Manufacturing, processing, and use of natural resources must approach the goal of total recycle to minimize waste control and to sustain materials availability. Renewable resources of air and water must be maintained and enhanced in quality for continued use.
>
> A broad base of technologic, economic, and ecologic information will be necessary. . . . Ways must be found to add to cost-benefit analyses nonquantifiable, subjective values for environmental amenities.
>
> Ecological knowledge (data and theories) must be greatly expanded and organized for use in management decisions. Criteria must be established which relate cause and effect in conditions of the environment.
>
> Indicators for all aspects of environmental productivity and quality must be developed and continuously measured to provide a feedback to management.

These are tough demands which can be met only by those strongly disciplined in specialized fields. However, stress on the need for strongly discipline-trained individuals seems to run counter to my earlier plea for a comprehensive and encompassing outlook on man. It does not run counter, however, for what is needed is not untrained individuals but individuals trained to look beyond the narrow confines of the discipline in which they are becoming specialized. In no small measure, this can be achieved by a focus on the kinds of problems which man faces, rather than on the problem of the discipline *per se*. It is true, indeed, that ofttimes the problems which a discipline faces can, by a twist of the screw, have much to say about the problems faced by man.

Dissemination of Information

A second aspect of the university's obligation is its responsibility as a disseminator of information on environmental matters. Now, this is not particularly a new role for the university; tradition has it that transmission of information is one of its major functions. However, that information has been transmitted typically through the medium of the learned and professional journals, hardly a competitor for reading time of the public.

The university community has not, by and large, assumed an overall responsibility for public education to the degree required to offset the problems at hand. To be sure, some academic systems, particularly the land grant universities through their extension services, have been involved in providing public education within the confines of the university door, and in fewer instances those doors have been opened such that members of the community have gone forth to transmit information. What is needed now are much more aggressive and innovative steps. A public which has become well-intentioned as a result of the environmental crisis blast from the mass communication media, is still quite poorly informed regarding the very fundamental problems that underlie those crises.

It behooves those in academia to assist in making the public aware of the relationships of environmental quality to human welfare. This is a call for education at all levels, one which includes an appreciation of mankind's intimate relationship with his environment—if you will, a call for a literacy on environmental matters. Why! Ultimately the responsibility for change rests with the individual citizen. It is not the industrial-military complex which litters the streets, which uses nonreturnable bottles, which is profligate with resources; it is the individual citizen. It is the individual citizen who must vote on the bond issues for new sewage treatment plants. It is the individual citizen who must pay higher costs for goods produced under conditions in which higher costs are engendered by pollution-controlling devices and passed on to the consumer.

At this writing, before the Congress are bills in the Senate and House entitled "Environmental Quality Education Act." These bills would (1) grant aid to colleges and universities to develop materials dealing with the whole range of ecological concerns; (2) provide teacher-training programs on environmental studies; (3) enable the preparation and distribution of materials suitable for use by the mass media. Most important is the fact that these bills would provide grants not only to colleges and universities, the traditional recipients of federal largess for educational matters, but also to elementary and secondary schools, to other public and private organizations, and most significantly, to community groups—business, industrial, and civic leaders.

At this writing, the bills would appear to have an excellent chance of passage. However, as I noted above, what happens at the appropriations level may forfeit very considerably the implementation of these much-needed programs.

Involvement in the Democratic Process

In many ways, the most elusive of the major responsibilities which the university must assume in ensuring survival of man with an improved quality of life deals with goals and priorities. Although, as we have seen, the federal government has gone far toward defining basic national goals, these have not been hammered out by design but largely by default and political opportunism. Once the goals are established, the priorities will follow. But, their insinuation into everyday life involves a changing attitude and a changing life style, one of the most difficult of all objectives to be achieved. At the least, this involves the best from leaders in government and industry, leaders who themselves are prepared in the hallowed halls of academia.

The academic community has a responsibility to become involved and to participate in the democratic process, not only on local and national levels, but internationally as well. As Harvey Wheeler (1970) put it:

> The characteristic political problems of the present arise from disorders of the entire ecological order. Their solutions are to be found, not through the traditional interaction of local interests and pressure-group politics, but through a new politics of the whole—politics considered architectonically, as the ancients called it. This requires a politics that is more speculative and less mechanistic, it requires us to do our lobbying in the realm of thought as well as in the corridors of power.

It is this "lobbying in the realm of thought" in which the universities will play a significant role. For example, in arriving at goals and priorities, it will be important, as Wheeler has pointed out, "first to figure out how to preserve general ecological balances, and, second, how to calculate hidden social costs so as to determine how much is really being spent on side effects. . . . Complete ecological harmony is impossible to achieve, but the "trade-offs" necessary to approach it as closely as possible must become known."

It is in this context that the university can play a significant role, in a new kind of participatory democracy whose direction is the seeking of solutions to ecological problems. The technological-related problems as well as the science-related issues we face know no territorial bounds—they are not limited to the confines of the university, nor to the confines of City Hall. The present system is proving itself incapable of dealing with it since not all problems are brought together within an overall ecological framework. A new politics will be required—one designed to cope with science and technology, one which deals with both in the realm of thought. This will be an uncomfortably difficult role for the university to play, one in which many halting steps will be taken.

Implementing Schemes and Operating Assumptions

This kind of far-reaching demand on the academic community implies the disrobing of some vested interests, and the espousing and building of bridges among disciplines. I have elsewhere argued that this can be achieved only by "focusing on the problems of profound significance to survival and well-being, and by eradicating the notion that action programs are hostile to academic life" (Kormondy, 1970).

What we need for an ultimate resolution is not an applied, environmental science in all its dimensions, but the totality of man in all of his dimensions: his economics, his sociology, his psychology, his philosophy, his aesthetics, his theology—as well as his science, basic and applied. Man's ecology must be witnessed and examined in total perspective, not in isolated fragments—particularized segments that are neither conceived nor construed holistically. Because of its unlimited outlook, because of its lack of constraints, our educational system can do this; *it must do this*.

Bold and daring new directions and approaches must be enabled by moral commitment backed up with fiscal support: the university has the moral, the public must provide the fiscal. The innovations must be inclusive and comprehensive, far-reaching and penetrating; in a word, there must be a revitalization and revolutionizing of existing educational programs, and this must extend from pre-school through adulthood. In this context, no longer is science to be seen as "Science" and literature as "English" but as something untitled, each witnessed in a new context, a new relationship, and conveyed as a significant human endeavor, as a key to deepened human understanding, as a route to human survival of richer bounds. The university is the only part of the establishment which can bring this about.

How likely is it that our academic system can respond in this way? What evidence is there that in fact it has done so? The evidence is encouraging.

In 1958, the Natural Resources Committee of the Conservation Foundation held a Conference on Resources Training in Berkeley, California. To the conference were invited all those institutions which were developing multidisciplinary programs focused on the environment; only 20 were so identified. However, in 1969, the number was an order of magnitude greater. This information developed out of a study supervised by the Environmental Policy Division (see above) of the Library of Congress' Legislative Reference Service (1969). Of 2000 accredited colleges and universities surveyed on the question of whether a multidisciplinary unit of any kind had been created or was being planned to deal with numerous and varied fields of environmental education and/or research, nearly 500 responded with an indication that something was going on. Of these, 121, selected as having a pertinent and viable program in operation or on the drawing boards, were analyzed in some detail.

The common denominator of these centers of environmental science appears to be a nonuniformity of models: no single structural pattern predominated. The variety of approaches indicates that there is no single answer to what Congressman Emilio Daddario referred to, in the study, as ''the mismatch of historical institutional organizations and emerging social problems.''

Although the various centers dealt with a fantastic array of problems, it is an unhappy note that few of the programs involved the social sciences and virtually none included the humanities. Further, at the time of the survey, most of the centers concentrated on graduate and post-graduate instruction and research, and expressed little intent to include undergraduate work.

Several of these environmental science programs were subjected to further analysis in a study conducted under the auspices of the Office of Science and Technology (1969) by Dr. John Steinhart, of the University of Wisconsin, and a student, Miss Stacie Cherniak. The major call of this report was for a problem-focused educational program involving all disciplines in a common thrust.

The interim between the Carpenter and Steinhart studies and this writing has witnessed a tremendous response and searching on the part of the academic community to intelligently respond to the environmental issue. Some of these programs have been hastily and ill-conceived; others, however, have had good genes and a good environment and will generate a good organism (Aldrich and Kormondy, 1972).

The need for an education which is problem- and action-oriented, as well as multidisciplinary, stretching from philosophy and design through the natural and behavioral sciences to the professional fields of law, medicine, engineering, and business administration, places an array of demands on the university community largely unknown in a previous day. But, the synergism which will doubtless come from faculty and administrative initiative and involvement with a strong student participation will go far toward achieving the kind of participatory democracy which I talked about above.

In the National Academy of Science's study discussed earlier *(Resources and Man)*, there is also a recognition of this need in a calling for

> continuing systematic programs and structures to be organized to promote more pervasive interaction among the environmental sciences, and between them and the behavioral sciences, technology, and the strictly physical sciences. We need more schools and institutes of environmental studies where ecologists, hydrologists, meteorologists, oceanographers, geographers, and geologists will work closely together with scholars and practitioners from other fields. Such organizations might serve as the cores of new ''urban grant'' universities intended to nucleate new urban centers. . . . More interaction among governmental agencies concerned with different parts of the environment should also be generated, as well as among them and other part parts of the scientific and governmental communities.

In the final analysis, what will be required is a much greater capability for flexibility and self-direction in matters of curriculum. The Steinhart-Cherniak report called for "freedom to be innovative in introducing course materials, educational programs, work study programs, and curriculum requirements for degrees." The experimental trends in many colleges and universities, known by a variety of names, could well be a step in the "right" direction in achieving this particular goal. However, such programs must be well conceived, but in no sense hide-bound by tradition. They cannot be "fly-by-night" in their conception or execution. They must have a mission; they must be oriented to achieve that mission, an orientation that will in large measure be enhanced by focusing on a regional problem.

By taking a regional problem-focus, not only is the problem tangible and real, it is immediate and its consequences can thus be read in the local community. It is easy enough to talk about defoliation in Vietnam; in relevant education, it is more important to talk about the effects of the herbicide being used along the highway approaching the town in which the community exists. It is easy to deplore the "death" of Lake Erie when sitting next to a town with its own major water pollution problem. It is easy to smirk at the smog problems of Los Angeles while particulate matter spews from the local industrial smokestack. As one begins to explore the regional problem, the scientific and technological components, I surmise, will become dwarfed by the socio-economic-behavioral consequences of the alterations which science or technology suggests. Finally, a relevance and realism is to be found in the legal and community actions that can ensue, in the litigatory action rather than in liturgical rhetoric to which we are so heavily subjected.

Not all university communities will be able to do all things, nor should they attempt to do so; nor should all students be expected to do all these things. Some universities well may concentrate their talent on aspects of applied research; some well may utilize their resources in establishing environmental centers which deal with developing a capacity for assessing the consequences of technological development; others may have such a wealth of resources as to concentrate on forecasting and applying their findings to public policy, others on the refinement of legal techniques, yet others in assuming a greater public responsibility in the education of leaders of industry, government, and academia, and still others dealing with the deeply rooted social, political, and economic bases—the human behavior of nontechnological problems.

In a word, the university, in its very broad sense, must assume a role which, for me, was well described by Eldridge Cleaver in "Convalescence" in his *Soul on Ice*:

> It was as if a driverless vehicle was speeding through the American night down an unlighted street toward a stone wall and was boarded on the fly by a stealthy ghost with a drooling leer on his face, who, at the last detour before chaos and disaster, careened the vehicle down a smooth highway that leads to the future and life.

I would like to think that the university might be that stealthy ghost who safely steers the vehicle—man and his environment—down a smoother highway. On the other hand, a colleague of mine has serious reservations about being saved by a leering and drooling stealthy spook!

Epilogue

The very considerable kinds of efforts on the part of government and academia which we have explored will go far in prompting a rethinking of man's relation to his environment, or what Leopold (1949) referred to as a new ethic for the land. The arrogance with which man has raped his environment can be atoned if we are successful in developing an ecological conscience, a respect and reverence for the land on which we walk, the air we breathe, the water in which we bathe. Can man achieve such a new ethic? In reality, is it not the same stewardship and guardianship which was called for in the Old Testament but which became perverted into a pioneer ethic?

Can man come to see that the environment is no longer an adversary to be conquered, a servant to be exploited, a property of rightful and eminent domain, a possession of unlimited capacity? Can he reorder and subsume his divine status into that of a natural and integral part of this Earth spaceship? So long as man continues to conquer and exploit others of his own kind, what confidence can one have that his environmental ways will change, that a new ethic for the land will develop? Not much, I would argue. But, can government and academia working together effect that change? What is your role as both a citizen and as a member of the academic community in effecting this change? If there is cause to be hopeful, it is because there is government and academia—and mostly, you.

References

ALDRICH, J.L. AND E.J. KORMONDY. *Environmental Education: Academia's Response*, Washington, D.C.: The Commission on Undergraduate Education in The Biological Sciences, Publ. 35. 1972.

BENNETT, I.L., JR., "Problems of World Food Supply. XI International Botanical Congress. Allis Chalmers, 1969.

BOSWELL, E.M., "Federal Programs Related to Environment." Legislative Reference Service, Library of Congress, Washington, D.C. (TP 450, 70–30 EP), 1970.

CARPENTER, R., "Information for Decisions in Environmental Policy." *Science*, 1970, **168**: pp. 1316–1322.

Committee on Resources and Man, National Academy of Sciences and National Research Council, *Resources and Man*. San Francisco: W.H. Freeman, 1969.

KENNAN, G.F., "To Prevent a World Wasteland." *Foreign Affairs*, 1970, **48**: pp. 401–413.

KORMONDY, E., "Ecology and the Environment of Man." *BioScience*, 1970, **20**: pp. 751–754.

LEOPOLD, A., "The Land Ethic," in *A Sand County Almanac.* New York: Oxford University Press, 1949.

Office of Science and Technology, Executive Office of the President, "The Universities and Environmental Quality—Commitment to Problem Focused Education," *A Report to the President's Environmental Quality Council,* by John S. Steinhart and Stacie Cherniak. Washington, D.C.: U.S. Government Printing Office, 1969.

OTTINGER, R.L., "The Congress and Environmental Deterioration: Immovable Object/Irresistible Force?" *BioScience,* 1969, **19**: p. 554.

PLATT, J., "What We Must Do." *Science,* 1969, **166**: pp. 1115–1121.

POTTER, V., *et al.*, "Purpose and Function of the University." *Science,* 1970, **167**: pp. 1590–1593.

President's Science Advisory Committee—Environmental Pollution Panel, *Restoring the Quality of our Environment.* Washington, D.C.: U.S. Government Printing Office, 1965.

RYTHER, J., "Photosynthesis and Fish Production in the Sea." *Science,* 1970, **166**: pp. 72–76.

United States Congress. House of Representatives, Subcommittee on Science, Research and Development, "Environmental Science Centers at Institutions of Higher Education. A Survey." 91st Congress, 1st Session, Washington, D.C.: U.S. Government Printing Office, 1969.

United States Congress. Senate. Committee on Interior and Insular Affairs and House Committee on Science and Astronautics, "A National Policy for the Environment. Congressional White Paper." 90th Congress, Second Session, Washington, D.C.: U.S. Government Printing Office, 1968.

WAGAR, J.A., "Growth versus Quality of Life." *Science,* 1970, **168**: pp. 1179–1184.

WHEELER, H., "The Politics of Ecology." *Saturday Review,* March 7, 1970, pp. 51–52 and 62–64.

WHITE, L. "The Historical Roots of Our Ecological Crisis." *Science,* 1967, **155**: pp. 1203–1207.

American Institutions and the Ecological Ideal

Scientific and Literary Views of Our Expansionary Life Style Are Converging*

LEO MARX
Amherst College

Leo Marx was born in New York in 1919. He was educated at Harvard (B.S., 1941; Ph.D., 1950) and served four years (1941–1945) in the U.S. Navy. From 1949 to 1958 he was a member of the English Department and an Associate of the American Studies Program at the University of Minnesota. Since 1958 he has been Professor of English and American Studies at Amherst College. He has held two Fulbright lectureships, one at the University of Nottingham in England (1956–1957) and one at the University of Rennes in France (1965–1966) and two Guggenheim Fellowships.

His research interests center on the interplay between consciousness and society in America—but especially on the relations between literature and the environment, both primary and sociological. His study of "technology and the pastoral ideal in America," *The Machine in the Garden*, appeared in 1964. He has edited a number of texts in American literature, and has published many essays about literature and society in America. Of particular interest to students of the environment is "Pastoral Ideals and City Troubles," which appears in the Smithsonian Annual II, *The Fitness of Man's Environment*, New York, Harper Colophon Books, 1968.

*This chapter is based on a talk given to the general symposium on "Human Settlements and Environmental Design" at the annual meeting of the A.A.A.S., Boston, Massachusetts, December 29, 1969. It was printed in *Science*, **170,** pp. 945–952 (November 27, 1970).

Anyone familiar with the work of the classic American writers (I am thinking of men like Cooper, Emerson, Thoreau, Melville, Whitman, Twain) is likely to have developed an interest in what we recently have learned to call ecology. One of the first things we associate with each of the names just mentioned is a distinctive, vividly particularized setting (or landscape) inseparable from the writer's conception of man. Partly because of the special geographic and political circumstances of American experience, and partly because they were influenced by the romantic vision of man's relations with nature, all of these writers possessed a heightened sense of place. Yet words like *place, landscape,* or *setting* scarcely can do justice to the significance they imparted to external nature in their work. They took for granted a thorough and delicate interpenetration of consciousness and environment. In fact it now seems evident that these gifted writers had begun, more than a century ago, to measure the quality of American life against something like an ecological ideal.

By "ecological ideal" I mean, quite simply, the maintenance of a healthy life-enhancing interaction between man and the environment. This is layman's language for the proposition that every organism, in order to avoid extinction or expulsion from its ecosystem, must conform to certain minimal requirements of that system. What makes the concept of the ecosystem difficult to grasp, admittedly, is that the boundaries between systems are always somewhat indistinct, and our technology is making them less distinct all the time. Since an ecosystem not only includes all living organisms (plants and animals), but also the inorganic (physical and chemical) components of the environment, it has become extremely difficult, in the thermonuclear age, to credit even the relatively limited autonomy of local or regional systems. If a decision taken in Moscow or Washington can effect a catastrophic change in the chemical composition of the entire biosphere, then the idea of a San Francisco or a Bay Area or a California or even a North American ecosystem loses much of its clarity and force. Similar difficulties arise when we contemplate the global rate of human population growth, which is only to say that the case for world government on ecological grounds is beyond argument. Meanwhile, we have no choice but to use the nation-states as political instruments for coping with the rapid deterioration of the physical world we inhabit.

The chief question before us, then, is this: What are the prospects, given the character of America's dominant institutions, for the fulfillment of the ecological ideal? But first, what is the significance of the current "environmental crusade"? Why should we be skeptical about its efficacy? How shall we account for the curious response of the scientific community? To answer these questions I will attempt to characterize certain of our key institutions from an ecological perspective. I want to suggest the striking convergence of the scientific and literary criticism of our national lifestyle. In conclusion I will make explicit a few responses to the ecological crisis indicated by that scientific-literary critique.

The Limits of Conservationist Thought

In this country, until recently, ecological thinking had been obscured by the more popular, if limited, conservationist viewpoint. Because our government seldom accorded a high priority to the protection of the environment, much of the responsibility for keeping that end in view fell upon a few voluntary organizations known as the "conservation movement." From the beginning, the movement attracted people with enough time and money to enjoy the outdoor life: sportsmen, naturalists (both amateur and professional), and, of course, property owners anxious to protect the sanctity of their rural or wilderness retreats. As a result, the conservationist cause came to be identified with the special interests of a few private citizens. It seldom if ever had been made to seem pertinent to the welfare of the poor, the nonwhite population, or, for that matter, the great majority of urban Americans. The environment that mattered most to conservationists was the environment beyond the city limits. Witness the aura of genteel rusticity that clings to the names of such leading organizations as the Sierra Club, the National Wildlife Foundation, the Audubon Society, and the Izaak Walton League. This concern with the interests of a privileged minority is consonant with the shallow conception of nature, largely confined to notions of natural resources and scenery, that has characterized the mentality of conservationists. In their view, nature is a world of useful and pretty objects placed "out there" for the benefit of mankind.

The ecological perspective is quite different. Its philosophic root is the secular idea that man (including his works—the secondary, or man-made, environment) is wholly and ineluctably embedded in the tissue of natural process. The interconnections are delicate, infinitely complex, never to be severed. If this organic (or holistic) view of nature has not been popular, it is partly because, as we shall see, it calls into question many presuppositions of American culture. Even today an excessive interest in this idea of nature carries, as it did in Emerson's and Jefferson's time, a strong hint of irregularity and possible subversion. (Nowadays it is associated with the anti-bourgeois defense of the environment expounded by the "long-haired cop-outs" of the youth movement.) Partly in order to counteract these dangerously idealistic notions, American conservationists often have made a point of seeming hard-headed, which is to say, "realistic" or practical. When their aims have been incorporated in national political programs, notably during the two Roosevelt administrations, the emphasis has been upon the efficient use of resources under the supervision of well-trained technicians.[1] Whatever the achievements of such programs, as implemented by the admirable, if narrowly defined, work of such agencies as the National Park Service, the United States Forest Service, or the Soil Conservation Service, they did not raise the kind of questions about our overall capacity for

[1]S.P. Hays, *Conservation and the Gospel of Efficiency* (Cambridge, Mass.: Harvard University Press, 1959).

survival brought into view by ecology. In this sense, conservationist thought is pragmatic and meliorist in tenor, whereas ecology is, in the purest meaning of the word, radical.

The relative popularity of the conservation movement helps to explain why troubled scientists, many of whom foresaw the scope and gravity of the environmental crisis a long while ago, have had such a difficult time arousing their countrymen. As early as 1864, George Perkins Marsh, sometimes said to be the father of American ecology, warned that the earth was "fast becoming an unfit home for its noblest inhabitant," and that unless men changed their ways it would be reduced "to such a condition of impoverished productiveness, of shattered surface, of climatic excess, as to threaten the depravation, barbarism, and perhaps even extinction of the species."[2] No one was listening to Marsh in 1864, and some 80 years later, according to a distinguished naturalist who tried to convey a similar warning, most Americans still were not listening. "It is amazing," wrote Fairfield Osborn in 1948, "how far one has to travel to find a person, even among the widely informed, who is aware of the processes of mounting destruction that we are inflicting upon our life sources."[3]

The Environment Crusade, c. 1969

But that was 1948, and as we all know, the situation now is wholly changed. Toward the end of the 1960s there was a sudden upsurge of public interest in the subject. The devastation of the environment and the threat of overpopulation became too obvious to be ignored. A sense of anxiety close to panic seized many people, including politicians and leaders of the communications industry. The mass media began to spread the alarm. Television gave prime coverage to a series of relatively minor yet visually sensational ecological disasters. Once again, as in the coverage of the Vietnam War, the close-up power of the medium was demonstrated. The sight of lovely beaches covered with crude oil, hundreds of dead and dying birds trapped in the viscous stuff, had an incalculable effect upon a mass audience. After years of indifference, the press suddenly decided that the jeremiads of naturalists might be important news, and a whole new idiom (*environment, ecology, balance of nature, population explosion,* and so on) entered common speech. Meanwhile, the language of reputable scientists was escalating to a pitch of apocalyptic excitement comparable only with that of the most fervent young radicals. Barry Commoner, for example, gave a widely reported speech describing the deadly pollution of California water reserves as a result of the excessive use of nitrates as fertilizer. This method of increasing agricultural productivity, he said, is so disruptive of the chemical balance of soil and water that within a generation it could be relied upon to poison irreparably

[2]David Lowenthal (ed.), *Man and Nature* (Cambridge, Mass.: Harvard University Press, 1965), p. 43.

[3]*Our Plundered Planet* (Boston, Mass.: Little, Brown, n.d.), p. 194.

the water supply of the whole area. The *New York Times* ran the story under the headline: "ECOLOGIST SEES U.S. ON SUICIDAL COURSE."[4] But the demographers and population biologists, worried about behavior even less susceptible to regulatory action, used the most portentous rhetoric. "We must realize that unless we are extremely lucky," Paul Ehrlich told an audience in the summer of 1969, "everybody will disappear in a cloud of blue steam in 20 years."[5]

To a layman who assumes that responsible scientists choose their words with care, this kind of talk is bewildering. How seriously should he take it? He realizes, of course, that he has no way, on his own, to evaluate the factual or scientific bases for these fearful predictions. But the organized scientific community, to which he naturally turns, is not much help. While most scientists calmly go about their business, activists like Commoner and Ehrlich dominate the headlines. (One could cite the almost equally gloomy forecasts of Harrison Brown, George Wald, René Dubos, and a dozen other distinguished scholars.) When Anthony Lewis asked a "leading European biologist" the same question—how seriously should one take this idea of the imminent extinction of the race?—the scholar smiled, Lewis reports, and said, "I suppose we have between 35 and 100 years before the end of life on earth."[6] What is bewildering is the disparity between words and action, between the all-too-credible prophecy of disaster and the response, or, rather, the non-response, of the organized scientific community. From a layman's viewpoint, the professional scientific organizations would seem to have an obligation here—since nothing less than human survival is in question—either to endorse or to correct the pronouncements of their distinguished colleagues. If a large number of scientists do indeed endorse the judgment of the more vociferous ecologists, then the inescapable question is: *What are they doing about it?* Why do they hesitate to use the concerted prestige and force of their profession to effect radical changes in national policy and behavior? How is it that most scientists, in the face of this awful knowledge, are able to carry on business more or less as usual? One might have expected them to raise their voices, activate their professional organizations, petition the Congress, send delegations to the President, speak out to the people and the government. Why, in short, are they not mounting a campaign of education and political action?

Why Are Most Scientists Undisturbed?

The most plausible answer seems to be that many scientists, like many of their fellow citizens, are ready to believe that such a campaign already has begun. And

[4]*New York Times*, November 19, 1969.
[5]*New York Times*, August 6, 1969.
[6]*New York Times*, December 15, 1969.

if, indeed, one accepts the version of political reality disseminated by the communications industry, they are correct: the campaign *has* begun. By the summer of 1969 it had become evident that the media were preparing to give the ecological crisis the kind of saturation treatment accorded to the Civil Rights movement in the early sixties and to the anti-Vietnam War protest after that. (Observers made this comparison from the beginning.) Much of the tone and substance of the campaign was set by the advertising business. Thus, a leading teenage fashion magazine, *Seventeen*, took a full page ad in the *New York Times* to announce, beneath a picture of a handsome collegiate couple strolling meditatively through autumn leaves, "The environment crusade emphasizes the fervent concerns of the young with our nation's 'quality of life.' Their voices impel us to act now on the mushrooming problems of conservation and ecology."[7] A more skeptical voice might impel us to think about the Madison Avenue strategists who had recognized a direct new path into the lucrative youth market. The "crusade," as they envisaged it, was to be a bland, well-mannered, clean-up campaign, conducted in the spirit of an adolescent love affair and nicely timed to deflect student attention from the disruptive political issues of the sixties. A national survey of college students confirmed this hope. "ENVIRONMENT MAY ECLIPSE VIETNAM AS COLLEGE ISSUE," they reported, and one young man's comment seemed to sum up their findings: "A lot of people are becoming disenchanted with the anti-war movement," he said. "People who are frustrated and disillusioned are starting to turn to ecology."[8] On New Year's Day, 1970, the President of the United States joined the crusade. Adapting the doomsday rhetoric of the environmentalists to his own purposes, he announced that "The nineteen-seventies absolutely must be the years when America pays it debt to the past by reclaiming the purity of its air, its waters and our living environment. It is literally now or never."[9]

Under the circumstances, it is understandable that most scientists, like most other people (except the disaffected minority of college students), have been largely unresponsive to the alarmist rhetoric of the more panicky environmentalists. The campaign to save the environment no longer seems to need their help. Not only have the media been awakened, and with them a large segment of the population, but the President himself, along with many government officials, has been enlisted in the cause. On February 10, 1970 President Nixon sent a special message to the Congress outlining a comprehensive 37-point program of action against pollution. Is it any wonder that the mood at recent meetings of conservationists has become almost cheerful—as if the movement, at long last, really had begun to move? After all, the grim forecasts of the ecologists necessarily have been couched in conditional language, thus: *if* California

[7]*New York Times*, December 5, 1969.
[8]*New York Times*, November 30, 1969.
[9]*New York Times*, January 2, 1969.

farmers continue their excessive use of nitrates, *then* the water supply will be irreparably poisoned. But now that the facts have been revealed, and with so much government activity in prospect, may we not assume that disaster will be averted? No need, therefore, to take the alarmists seriously. Which is only to say that most scientists still have confidence in the capacity of our political leaders, and of our institutions, to cope with the crisis.

But is that confidence warranted by the current "crusade"? Many observers have noted that the President's message was strong in visionary language and weak in substance. He recommended no significant increase in funds needed to implement the program. Coming from a politician with a well-known respect for strategies based on advertising and public relations, this high-sounding talk should make us wary. Is it designed to protect the environment or to assuage anxiety or to distract the anti-war movement or to provide the cohesive force necessary for national unity behind the Republican administration? How can we distinguish the illusion of activity fostered by the media—and the President—from auguries of genuine action? On this score, the frequently invoked parallel of the Civil Rights and the anti-war movements should give us pause. For while each succeeded in focusing attention upon a dangerous situation, it is doubtful whether either got us very far along toward the elimination of the danger. At first each movement won spectacular victories, but now, in retrospect, they too look more like ideological than substantive gains. In many ways the situation of blacks in America has not changed fundamentally since 1960. Nor can it be said that the war in Southeast Asia was stopped by the peace movement. This is not to imply that the strenuous efforts to end the war or to eradicate racism have been bootless. Some day the whole picture may well look quite different; we may look back on the sixties as the time when a generation was prepared for a vital transformation of American society.

Nevertheless, scientists would do well to contemplate the example of these recent protest movements. They would be compelled to recognize, for one thing, that while public awareness may be indispensable to effecting changes in national policy, it hardly guarantees results. In retrospect, indeed, the whole tenor of the Civil Rights and anti-war campaigns now seems much too optimistic. Neither program took sufficient account of the deeply entrenched, institutionalized character of the collective behavior it aimed to change. If leaders of the campaign to save the environment were to make the same kind of error, it would not be surprising. A certain innocent trust in the efficacy of words, propaganda, and rational persuasion always has characterized the conservation movement in this country. Besides, there is a popular notion that ecological problems are in essence technological, not political, and therefore easier to solve than the problems of racism, war, or imperialism. To indicate why this view is mistaken, why in fact it would be folly to discount the urgency of the environmental crisis on these grounds, I now want to consider the fitness of certain dominant American institutions for the fulfillment of the ecological ideal.

THE DYNAMISM OF AMERICA

From an ecological perspective, a salient characteristic of American society is its astonishing dynamism. Ever since the first European settlements were established on the Atlantic seaboard, our history has been one of virtually uninterrupted expansion. How many decades, if any, have there been since 1607 when this society failed to expand its population, territory, and economic power? When foreigners speak of Americanization they invariably have in mind this dynamic, expansionary, unrestrained behavior. "No sooner do you set foot upon American ground," wrote Tocqueville, "than you are stunned by a kind of tumult; a confused clamor is heard on every side, and a thousand simultaneous voices demand the satisfaction of their social wants. Everything is in motion around you. . . ."[10] To be sure, a majority of these clamorous people were of European origin, and their most effective instrument for the transformation of the wilderness, their science and technology, was a product of Western culture. But the unspoiled terrain of North America gave European dynamism a peculiar effervescence. The seemingly unlimited natural resources, and the relative absence of cultural or institutional restraints, made possible what surely has been the fastest developing, most mobile, relentlessly innovative society in world history. By now that dynamism inheres in every aspect of our lives from the dominant national ethos to the structure of our economic institutions down to the deportment of individuals.

The ideological counterpart to the nation's physical expansion has been its celebration of quantity. What has been valued most in American popular culture is growth, development, size (bigness), and, by extension, change, novelty, innovation, wealth, and power. This tendency was noted a long while ago, especially by foreign travelers, but only recently have historians begun to appreciate the special contribution of Christianity to this quantitative, expansionary ethos. The crux here is the aggressive, man-centered attitude toward the environment fostered by Judeo-Christian thought: everything in nature, living or inorganic, exists to serve man. For only man can hope (by joining God) to transcend nature. According to one historian of science, Lynn White, the dynamic thrust of Western science and technology derives in large measure from this Christian emphasis, unique among the great world religions, upon the separation of man from nature.[11] But one need not endorse White's entire argument to recognize that Americans, from the beginning, found in the Bible a divine sanction for their violent assault upon the physical environment. To the Puritans of New England, the New World landscape was Satan's territory, a hideous wilderness inhabited by the unredeemed and fit only for conquest. What moral precept could have served their purpose better than the Lord's

[10]Alexis de Tocqueville, *Democracy in America*, Phillips Bradley, ed. (New York: Alfred A. Knopf, 1946, 2 vols.), vol. 1, p. 249.

[11]"The Historical Roots of Our Ecological Crisis," *Science*, **155,** p. 1203 (1967).

injunction to be fruitful and multiply and subdue the earth and exercise dominion over every living creature? Then, too, the millennial cast of evangelical Protestantism made even more dramatic the notion that this earth and everything upon it is an expendable support-system for a voyage to eternity. Later, as industrialization gained momentum, the emphasis shifted from the idea of nature as the devil's country to the idea of nature as commodity. When the millennial hope was secularized, and salvation was replaced by the goal of economic and social progress, it became possible to quantify the rate of human improvement. In our time this quantifying bent reached its logical end with the enshrinement of the GNP—one all-encompassing index of the state of the union itself. Perhaps the most striking thing about this expansionary ethos, from an ecological viewpoint, has been its capacity to supplant a whole range of commonsense notions about man's relations with nature which are recognized by some preliterate peoples and are implicit in the behavior of certain animal species. These include the idea that natural resources are exhaustible, that the unchecked growth of a species will eventually lead to its extinction, and that other organisms may have a claim to life worthy of respect.

The Expansionary System

The record of American business, incomparably successful according to quantitative economic measures like the GNP, looks quite different from an ecological perspective. Whereas the environmental ideal affirms the need for each organism to observe limits set by its ecosystem, the whole thrust of industrial capitalism has been in the opposite direction: it has placed the highest premium upon ingenious methods for circumventing those limits. After comparing the treatment that various nations have accorded their respective portions of earth, Fairfield Osborn said this of the United States: "The story of our nation in the last century as regards the use of forests, grasslands, wildlife and water sources is the most violent and the most destructive in the long history of civilization."[12] If that estimate is just, a large part of the credit must be given to an economic system unmatched in calling forth man's profit-making energies. By the same token, it is a system that does pitifully little to encourage or reward those constraints necessary for the long-term ecological well-being of society. Consider, for example, the fate of prime agricultural lands on the borders of our burgeoning cities. What happens when a landowner is offered a small fortune by a developer? What agency protects the public interest from the irretrievable loss of topsoil that requires centuries to produce? Who sees to it that housing, factories, highways, and shopping centers are situated on the far more plentiful sites where nothing edible ever will grow? The answer is that no such agencies exist, and the market principle is allowed to rule. Since World War II approximately one-fifth of California's invaluable farm land has been lost in this

[12] *Our Plundered Planet* (Boston, Mass.: Little, Brown, n.d.), p. 175.

way. Here, as in many cases of air and water pollution, the dominant motive of our business system, private profit, leads to the violation of ecological standards.

Early in the industrial era one might reasonably have expected, as Thorstein Veblen did, that the scientific and technological professions, with their strong bent toward rationality and efficiency, would help to control the ravening economic appetites whetted by America's natural abundance. Veblen assumed that well-trained technicians, engineers, and scientists would be repelled by the wastefulness of the business system. He therefore looked to them for leadership in shaping alternatives to a culture obsessed with "conspicuous consumption." But until now that leadership has not appeared. On the contrary, this new technical élite, with its commitment to highly specialized, value-free research, has enthusiastically placed its skill in the service of business and military enterprise. Which is one reason, incidentally, why today's rebellious students are unimpressed by the claim that the higher learning entails a commitment to rationality. They see our best educated, most "rational" élite serving what strikes them as a higher irrationality. So far from providing a counterforce to the business system, in fact, the scientific and technological professions have strengthened the ideology of American corporate capitalism, including its large armaments sector, by bringing to it their high-minded faith in the benign consequences of the most rapid possible rate of technological innovation.

But not only are we collectively committed, as a nation, to the idea of continuing growth; each subordinate unit of the society holds itself to a similar standard of success. Each state, city, village, and neighborhood; each corporation, independent merchant, and voluntary organization; each ethnic group, family, and child; each person, ideally speaking, should strive for growth. Translated into ecological terms, this popular measure of success—becoming bigger, richer, more powerful—means gaining control over more and more of the available resources. When resources were thought to be inexhaustible, as they were through most of our national history, the release of these unbounded entrepreneurial energies was considered an aspect of individual liberation. And so it was, at least for large segments of the population. But today, when that assumption no longer makes sense, those energies are still being generated. It is as if a miniaturized version of the nation's expansionary ethos had been implanted in every citizen—not excluding the technicians and scientists. And when we consider the extremes to which the specialization of function has been carried in the sciences, each expert working his own miniscule sector of the knowledge industry, it is easier to account for the unresponsiveness of the scientific community to the urgent warnings of alarmed ecologists. If most scientists and engineers do not seem to be listening, much less acting, it is because these highly skilled men are so busy doing what every good American is supposed to do.

On the other hand, it is not surprising that a gifted novelist like Norman Mailer, or a popular interpreter of science like Rachel Carson, or an imaginative medical researcher like Dr. Alan Gregg, each found it illuminating in recent years to compare the unchecked growth of American society, with all the resulting disorder, to the haphazard spread of cancer cells in a living organism.[13] There is nothing new, of course, about the analogy between the social order and the human body; the conceit has a long history in literature. Since the early sixties, however, Mailer has been invoking the more specific idea of America as a carcinogenic environment. Like any good poetic figure, this one has a basis in fact. Not only does it call to mind the radioactive matter we have deposited in the earth and the sea, or the work of such allegedly cancer-producing enterprises as the tobacco and automobile industries, or some of the new drugs administered by doctors in recent years, but even more subtly, it reminds us of the parallel between cancer and our expansionary national ethos which, like a powerful ideological hormone, stimulates the reckless, uncontrolled growth of each cell in the social organism.

In the interests of historical accuracy and comprehensiveness, needless to say, all of these sweeping generalizations would have to be extensively qualified. The record is rich in accounts of determined, troubled Americans who have criticized and actively resisted the nation's expansionary abandon. A large part of our governmental apparatus was created in order to keep these acquisitive, self-aggrandizing energies within tolerable limits. And of course the full story would acknowledge the obvious benefits, especially the individual freedom and prosperity, many Americans owe to the very dynamism that now threatens our survival. But in this brief compass my aim is to emphasize that conception of man's relation to nature which, so far as we can trace its consequences, issued in the *dominant* forms of national behavior. And that is a largely one-sided story. It is a story, moreover, to which our classic American writers, to their inestimable credit, have borne eloquent witness. If there is a single native institution which has consistently criticized American life from a vantage like that of ecology, it is the institution of letters.

America's Pastoral Literature

A notable fact about imaginative literature in America, when viewed from an ecological perspective, is the number of our most admired works written in obedience to a pastoral impulse.[14] By "pastoral impulse" I mean the urge, in the face of society's increasing power and complexity, to retreat in the direction of nature. The most obvious form taken by this withdrawal from the world of established institutions is a movement in space. The writer or narrator describes, or a character enacts, a move away from a relatively sophisticated to a simpler,

[13]N. Mailer, Introduction and *passim., Cannibals and Christians* (New York: Dial, 1966); R. Carson, *Silent Spring* (Boston: Houghton Mifflin, 1962); A. Gregg, "A Medical Aspect of the Population Problem," *Science*, **121**, p. 681 (1955).

[14]L. Marx, *The Machine in the Garden, Technology and the Pastoral Ideal in America* (New York: Oxford University Press, 1964).

more "natural" environment. Whether this new setting is an unspoiled wilderness, like Cooper's forests and plains, Melville's remote Pacific, Faulkner's Big Woods, or Hemingway's Africa; or whether it is as tame as Emerson's New England village common, Thoreau's Walden Pond, or Robert Frost's pasture, its significance derives from the plain fact that it is "closer" to nature: it is a landscape that bears fewer marks of human intervention.

This symbolic action, which reenacts the initial transit of Europeans to North America, may be understood in several ways, and no one of them can do it justice. To begin with, there is an undeniable element of escapism about this familiar, perhaps universal, desire to get away from the imperatives of a complicated social life. No one has conveyed this feeling with greater economy or simplicity than Robert Frost, in the first line of his poem, "Directive": "Back out of all this now too much for us." Needless to say, if our literary pastoralism lent expression only to this escapist impulse, we would be compelled to call it self-indulgent, puerile, or regressive.

But fortunately this is not the case. In most American pastorals the movement toward nature also may be understood as a serious criticism, explicit or implied, of the established social order. It calls into question a society dominated by a mechanistic system of value, keyed to perfecting the routine means of existence, yet oblivious to its meaning or purpose. We recall Thoreau's description, early in *Walden*, of the lives of quiet desperation led by his Concord neighbors, or the first pages of Melville's *Moby-Dick*, with Ishmael's account of his moods of suicidal depression as he contemplates the meaningless work required of the inhabitants of Manhattan Island. At one time this attitude toward the workday world was commonly dismissed as aristocratic or elitist. We said that it made sense only to a fortunate leisure class for whom deprivation was no problem. But today, in a society with the technological capacity to supply everyone with an adequate standard of living, that objection has lost a good deal of its force. The necessary conditions for giving a decent livelihood to every citizen no longer include harder work, increased productivity, or technological progress. But, of course, a program of this kind would require a more equitable distribution of wealth, and the substitution of economic sufficiency for the goal of an endlessly "rising" standard of living. The mere fact that such a possibility exists explains why our literary pastorals, which blur distinctions between the economic, moral, and aesthetic flaws of society, now seem more acceptable. In the nineteenth century, many pastoralists, like today's radical ecologists, saw the system as potentially destructive in its innermost essence. Their dominant figure for industrial society, with its patent confusion about ends and means, was the social machine. Our economy is the kind of system, said Thoreau, where men become the tools of their tools.

Of course there is nothing particularly American about this pessimistic literary response to industrialism. Since the Romantic Movement it has been a dominant theme of all Western literature. Most gifted writers have expended a large share of their energy in an effort to discover, or, more precisely, to imagine, alternatives to the way of life that emerged with the industrial revolution. The difference is that in Europe there was a range of other possible life styles which had no counterpart in this country. There were enclaves of preindustrial culture (provincial, aesthetic, religious, aristocratic) which retained their vitality long after the bourgeois revolutions, and there also was a new, revolutionary, urban working class. This difference, along with the presence in America of a vast, rich, unspoiled landscape, helps to explain the exceptionally strong hold of the pastoral motive upon the native imagination. If our writers conceived of life from something like an ecological perspective, it is largely because of their heightened sensitivity to the unspoiled environment, and man's relation to it, as the basis for an alternative way of life.

What, then, can we learn about possible alternatives from our pastoral literature? The difficulty here lies in the improbability which surrounds the affirmative content of the pastoral retreat. In the typical American fable the high point of the withdrawal toward nature is an idyllic interlude which gains a large measure of its significance from the sharp contrast with the everyday, "real," world. This is an evanescent moment of peace and contentment when the writer (or narrator, or protagonist) enjoys a sense of integration with the surrounding environment that approaches ecstatic fulfillment. It is often a kind of visionary experience, couched in a language of such intense, extreme, even mystical feelings that it is difficult for many readers (though not, significantly, adherents of today's youth culture) to take it seriously. But it is important to keep in view some of the reasons for this literary extravagance. In a commercial, optimistic, self-satisfied culture, it was not easy for writers to make an alternate mode of experience credible. Their problem was to endow an ideal vision—some would call it utopian—with enough sensual authenticity to carry readers beyond the usual, conventionally accepted limits of commonsense reality. Nevertheless, the pastoral interlude, rightly understood, does have a bearing upon the choices open to a post-industrial society. It must be taken, not as representing a program to be copied, but as a symbolic action which embodies alternative values, attitudes, modes of thought and feeling to those which characterize the dynamic, expansionary life style of modern America.

The focus of our literary pastoralism, accordingly, is upon a contrast between two environments representing virtually all aspects of man's relation to nature. In place of the aggressive thrust of nineteenth century capitalism, the pastoral interlude exemplifies a far more restrained, accommodating kind of behavior. The chief goal is not, as Alexander Hamilton argued, to enhance the nation's corporate wealth and power, but rather the Jeffersonian "pursuit of happiness." In economic terms, then, pastoralism entails a distinction between a commitment to unending growth and the concept of material sufficiency. The aim of the pastoral economy is enough—enough production and consumption to insure a decent quality of life. Jefferson's dislike of industrialization was based on this standard; he was bent on the subordination of quantitative to qualitative "standards of living."

From a psychological viewpoint, the pastoral retreat affirmed the possibility of maintaining man's mental equilibrium by renewed emphasis upon his inner needs. The psychic equivalent of the balance of nature (in effect, the balance of *human* nature) is a more or less equal capacity to cope with external and internal sources of anxiety. In a less developed landscape, according to these fables, behavior can be more free, spontaneous, authentic, or, in a word, more natural. The natural in psychic experience refers to activities of mind which are inborn or somehow primary. Whether we call them intuitive, unconscious, or preconscious, the significant fact is that they do not have to be learned or deliberately acquired. By contrast, then, the expansionary society is figured forth as dangerously imbalanced on the side of those rational faculties conducive to the manipulation of the physical environment. We think of Melville's Ahab, in whom the specialization of function induces a peculiar kind of power-obsessed, if technically competent, mentality. "My means are sane," he says, "my motive and my object mad."

This suspicion of the technical, highly trained intellect comports with the emphasis in our pastoral literature upon those aspects of life common to all men. Whereas the industrial society encourages and rewards the habit of mind which analyzes, separates, categorizes, and makes distinctions, the felicity enjoyed during the pastoral interlude is a tacit tribute to the opposite. This kind of pleasure derives from the connection-making, analogizing, and poetic imagination that aspires to a unified conception of reality. At the highest or metaphysical level of abstraction, then, romantic pastoralism is holistic. During the more intense pastoral interludes, an awareness of the entire environment, extending to the outer reaches of the cosmos, affects the perception of each separate thing, idea, and event. In place of the technologically efficient but limited concept of nature as a body of discrete manipulable objects, our pastoral literature presents an organic conception of man's relation to his environment.

What I am trying to suggest is the striking convergence of the literary and the ecological views of America's dominant institutions. Our literature contains a deep intuition of the gathering environmental crisis and its causes. To be sure, the matter-of-fact idiom of scientific ecology may not be poetic or inspiring. Instead of conveying Wordsworthian impulses from the vernal wood, it reports the rate at which monoxide poisoning is killing the trees. Nevertheless, the findings of ecologists confirm the indictment of the self-aggrandizing way of life that our leading writers have been building up for almost two centuries. In essence it is an indictment of the destructive, power-oriented uses to which we put scientific and technological knowledge. The philosophic source of this dangerous behavior is an arrogant conception of man and, above all, of human consciousness as wholly unique—as an entity distinct from, and potentially independent of, the rest of nature.

As for the alternative implied by the pastoral retreat, it also anticipated certain current insights of ecology. Throughout this body of imaginative writing, the turn back to nature is represented as a means of gaining access to governing values, meanings, and purposes. In the past, to be sure, many readers found the escapist, sentimental overtones of this motive embarrassing. As a teacher, I can testify that until recently many pragmatically inclined students were put off by the obscurely metaphysical, occultish notions surrounding the idea of harmony with nature. It lacked specificity. But now all that is changing. The current environmental crisis has in a sense put a literal, factual, often quantifiable base under this poetic idea. Nature as a transmitter of signals and a dictator of choices now is present to us in the quite literal sense that the imbalance of an ecosystem, when scientifically understood, defines certain precise limits to human behavior. We are told, for example, that if we continue contaminating Lake Michigan at the present rate, the lake will be "dead" in roughly ten years. Shall we save the lake or continue allowing the cities and industries which pollute it to reduce expenses and maximize profits? As such choices become more frequent, man's relations with nature will in effect be seen to set the limits of various economic, social, and political activities. And the concept of harmonious relations between man and the physical environment, instead of seeming to be a vague projection of human wishes, must come to be respected as a necessary, realistic, limiting goal. This convergence of literary and scientific insight reinforces the naturalistic idea that man, to paraphrase Melville, must eventually lower his conceit of attainable felicity, locating it not in power or transcendence, but in a prior duty to sustain life itself.

A Proposal and Some Conclusions

Assuming that this sketch of America's dominant institutions as seen from a pastoral-ecological vantage is not grossly inaccurate, what inferences can be drawn from it? What bearing does it have upon our current effort to cope with the deterioration of the environment? What special significance does it have for concerned scientists and technologists? I shall draw several conclusions, beginning with a specific recommendation for action by the American Association for the Advancement of Science.

First, then, let me propose that the Association establish a panel of the best-qualified scientists, representing as many as possible of the disciplines involved, to serve as a national review board for ecological information. This board would take the responsibility for locating and defining the crucial problems (they presumably would recruit special task forces for specific assignments) and making public recommendations whenever feasible. To be sure, some scientists will be doing a similar job for the government, but if an informed electorate is to evaluate the government's program, it must have an independent source of knowledge. One probable objection is that scientists often disagree, and feel reluctant to disagree in public. But is this a healthy condition for a democracy? Perhaps the time has come to lift the dangerous veil of omniscience from the world of science and technology. If the experts cannot agree, let them issue minority reports. If our survival is at stake, we should be allowed to know what the problems and the choices are. The point here is not that we laymen look to scientists for *the* answer, or that we expect them to save us. But we do ask for their active involvement in solving problems about which they are the best-informed citizens. Not only should such a top flight panel of scientists be set up on a national basis, but, perhaps more important, similar committees should help to make the best scientific judgment available to the citizens of every state, city, and local community.

But, there will also be those who object on the ground that an organization as august as the Association for the Advancement of Science must not be drawn into politics. The answer, of course, is that American scientists and technologists are now and have always been involved in politics. A profession whose members place their services at the disposal of the government, the military, and the private corporations can hardly claim immunity now. Scientific and technological knowledge unavoidably is used for political purposes. But it also is a national resource. The real question in a democratic society, therefore, is whether that knowledge can be made as available to ordinary voters as it is to those, like the Defense Department or General Electric, who can most easily buy it. If scientists are worried about becoming partisans, then their best defense is to speak with their own disinterested public voice. To allow the burden of alerting and educating the people to fall upon a few volunteers is a scandal. Scientists, as represented by their professional organizations, have a responsibility to make sure that their skills are used to fulfill as well as to violate the ecological ideal. And who knows?—if things get bad enough, the scientific community may take steps to discourage its members from serving the violators.

There is another, perhaps more compelling reason why scientists and technologists, as an organized professional group, must become more actively involved. It was scientists, after all, who first sounded the alarm. What action we take as a society, *and how quickly we take it*, depends in large measure upon the credibility of the alarmists. Who is to say, if organized science does not, which alarms we should take seriously? What group has anything like the competence of scientists and technologists to evaluate the evidence? Or, to put it negatively, what group can do more, by mere complacency and inaction, to insure an inadequate response to the environmental crisis? It is a well-known fact that Americans hold the scientific profession in the highest esteem. As long as most scientists go about their business as usual, as long as they seem unperturbed by the urgent appeals of their own colleagues, it is likely that most laymen, including our political representatives, will remain skeptical.

The arguments for the more active involvement of the scientific community in public debate illustrate the all-encompassing and essentially political character of the environmental crisis. If the literary-ecological perspective affords an accurate view, we must eventually take into account the deep-seated, institutional causes of our distress. No cosmetic program, no clean-up-the-landscape activity, no degree of protection for the wilderness, no anti-pollution laws can be more than the merest beginning. Of course such measures are worthwhile, but in undertaking them we should acknowledge their superficiality. The devastation of the environment is at bottom a result of the kind of society we have built. It follows, therefore, that environmentalists should join forces, wherever common aims can be found, with other groups concerned to change basic institutions. To arrest the deterioration of the environment it will be necessary to control many of the same forces which have prevented us from ending the war in Indochina or giving justice to black Americans. In other words, it will be necessary for ecologists to determine where the destructive power of our society lies and how to cope with it. Knowledge of that kind, needless to say, is political. But then it seems obvious, on reflection, that the study of human ecology will be incomplete until it incorporates a sophisticated mode of political analysis.

Meanwhile, it would be folly, given the character of American institutions, to discount the urgency of our situation, either on the ground that technology will provide the solutions, or that adequate programs already are in view. We cannot rely on technology because the essential problem is not technological. It inheres in all of the ways in which this dynamic society generates and uses its power. It calls into question the controlling purposes of all the major institutions which actually determine the nation's impact upon the environment: the great business corporations, the military establishment, the universities, the scientific and technological élites, and the exhilarating expansionary ethos by which we all live. Throughout our brief history, a passion for personal and collective aggrandizement has been the American way. One can only guess at the extent to which forebodings of ecological doom have contributed to the revulsion that so many intelligent young people feel, these days, for the idea of "success" as a kind of limitless ingestion. In any case, most of the talk about the environmental crisis that turns on the word "pollution," as if we face a cosmic-scale problem of sanitation, is grossly misleading. What confronts us is an extreme imbalance between society's hunger—the rapidly growing sum of human wants—and the limited capacities of the earth.

15

Summary

WILLIS H. JOHNSON
Wabash College

Willis Hugh Johnson was born in Parkersburg, Indiana in 1902. He attended Wabash College where he was awarded the A.B. degree in 1925. His graduate training was obtained at the University of Chicago; M.S., 1929 and Ph.D., 1932. Except for the period of his graduate training he was a member of the Department of Zoology at Wabash College from 1925 to 1935. In 1935 he joined the staff in biology at Stanford University where he served as Assistant Professor, Associate Professor, and then Professor until 1946. He then returned to Wabash College and served as Chairman of the Science Division and of the Department of Biology until 1968. Since then he has served as Treves Professor of Biology.

Professor Johnson has long been interested in the teaching of introductory biology at the college level. He served as Chairman of the Biology Examination Committee and Chief Reader in Biology in the Advanced Placement Program from its inception until 1960. He served as Chairman of the Conference on Undergraduate Curricula in the Biological Sciences held in 1957 under NRC auspices and is coauthor of the report of that conference "Recommendations on Undergraduate Curricula in the Biological Sciences," NAC-NRC publication 578. He was a member of CUEBS, 1963–1966. He is a past president of the Indiana Academy of Science and of the Midwest Conference of College Biology Teachers.

Professor Johnson is the author of numerous research articles, among which are: "A Purine and Pyrimidine Requirement for *Paramecium multimicronucleatum*" (with C.A. Miller), *Journal of Protozoology,* 1957; "Nutrition of *Paramecium:* A Fatty Acid Requirement" (with C.A. Miller), *Journal of Protozoology,* 1960, "Induced Loss of Pigment in Planarians" (with C.A. Miller and J.H. Brumbaugh), *Physiological Zoology,* 1962. He is also senior coauthor of the following texts: *Biology,* 4th ed. (1972), *Essentials of Biology* (1969), and *Principles of Zoology* (1969), all published by Holt, Rinehart and Winston, Inc. He was coeditor, with W.C. Steere, of *This Is Life* (1962), Holt, Rinehart and Winston, Inc.

As stated in the Introduction, the purpose of this volume is to call attention to the crisis facing civilization as a result of the human population explosion and the pollution of the environment, to point out some of the things that are being done to counteract these forces, and to suggest other things that must be done.

Environmental pollution, to the degree present in the world today, must be associated with our rapid technological developments and the much greater use of energy than in times past. But technological developments alone are not the sole causes of environmental pollution. The rapid increase in numbers of humans who use the products of technology is a very integral part of the picture. No longer can the interrelationship between expanding human population and expanding technological developments as major factors in our pollution problems be ignored. Closely linked with these two problems is the problem of the rapid depletion of many of our essential natural resources, as is pointed out in several of the chapters.

The first five chapters present a brief background against which both the population problem and the pollution problems must be viewed.

Background

Our *biosphere*, a thin layer of the earth's surface with its surrounding envelope of water and air and all the living organisms found there, is a very complex system and is the product of a very long evolution. And for all that we know at present, it may be unique in the universe.

Ecology is the study of the interrelations that exist in the biosphere. The basic functional ecological unit is the *ecosystem*. All of the plants and animals living in a given habitat constitute a *biotic community*. The plants and animals in a lake, a deciduous forest, a prairie would, in each case, constitute a biotic community. However, the broader, more inclusive term, *ecosystem*, is defined as an interacting community of organisms together with their abiotic (non-living) environment. Larger ecosystems that occur in similar climates and share a similar character and arrangement of vegetation are called *biomes*. Examples are tropical rain forests, northern coniferous forests, and deserts.

There are four components to a complete ecosystem: (1) *abiotic substances*, the non-living components of the environment; (2) *producers*, green plants (aquatic algae or land plants) that utilize light energy, CO_2 and H_2O to synthesize carbohydrates and in turn use some of them to produce more complex organic compounds; (3) *consumers*, mainly animals which consume other organisms or parts of organisms (there may be two or three levels of consumers, rarely more because of the basic energy limitations); (4) *decomposers* or *reducers*, mainly bacteria and fungi, which break down organic material into simpler compounds, utilizing some of the material themselves, and releasing the rest into the environment in a form which can be reused by producers. Every ecosystem is organized around these four components.

In Chapter 2, George Masters Woodwell discusses the structure and function of ecosystems. The flow of energy is often used in the analysis of ecosystems. All of the energy for the maintenance, growth, and reproduction of organisms comes either directly or indirectly from the sun. This energy is circulated through ecosystems by *food chains* (an animal eats a producer, another eats the first, and so on). The energy that flows through ecosystems in food chains is the chemical energy in carbon compounds. In many ecosystems the food interrelationships are so complex that the concept of *food web* better expresses the relationship. This flow of energy follows two basic routes—a grazing food chain and a decay chain, which often is differentiated into aerobic and anaerobic components.

As energy flows through an ecosystem there is a decrease in the amount of usable energy. Some energy in the form of heat is dissipated into the environment every time organic compounds are passed from one organism to another. In general, according to Woodwell, although there are many variations it seems reasonable to assume that 10 to 20 percent of the energy entering any trophic level is available for transfer to the next higher trophic level in a grazing food chain. This means that 10 to 20 percent of the energy fixed by plants is available to transfer to herbivores directly, and 10 to 20 percent of that energy is available to support carnivores that prey on herbivores. Woodwell points out that this does not account for the transfer of energy to the decay routes nor for the flux back into the grazing chain. The magnitude of these under most circumstances is less than 30 to 40 percent of the primary production.

In addition to the energy flow, ecosystems provide for a cycling of the essential nutrients or elements found in living organisms. There are cycles of carbon, nitrogen, phosphorus, calcium, and all of the essential elements for life. This permits the use of the same elements over and over.

Woodwell indicates that as ecosystems are progressively degraded, the amount of energy following the decay route increases. It is here that we have real cause for concern. With man-made changes in the structure of natural ecosystems there is a shift in the flow of energy from the grazing food chains, which are the principal chains tapped by man, into the decay chains.

This intricate web of nutritional interrelationships found among living things is one of the marvels of the evolutionary process, and associated with this is the great diversity of kinds or species of organisms as is described by Ernst Mayr *et al.* in Chapter 3. This great diversity provides for the maximum utilization of the energy input in a given region and is the product of a very long evolution. A mature ecosystem is a diversified one; it is a stable one; and it has a great capacity for trapping and holding nutrients in it for recycling. Such a system is better able to adjust to changes than is a simpler system.

Each species, as a result of its evolution, has a genetic composition that limits how far it can go in adjusting to sudden changes in its environment. Changes in ecosystems are occurring all the time. In the evolution of our biosphere tremendous changes occurred from time to time with many extinctions resulting.

The ecosystems adjusted to these changes but millions of years were required. Today we are concerned with the drastic changes in the environment that man has made and is making in a relatively short time. Man has been upsetting the intricate balance in our ecosystems—the product of millions of years of evolution—by draining swamps; plowing up grasslands; cutting off forests; placing poisons in the air, water, and land; and in many other ways. This not only results in extinctions of animal and plant life, it has tragic effects on man himself. The present human civilization cannot wait for the slow evolutionary adjustments to such changes. One of our great challenges today is not to make further drastic changes in our environment until we have some understanding of the possible consequences.

In addition to the energy that flows through ecosystems in the form of chemical energy, there are other important energy relations in the biosphere. These are the purely physical aspects of energy discussed by David Gates in Chapter 4. In addition to the energy utilized by green plants in photosynthesis, solar energy is the ultimate source of most of the heat energy that makes it possible for life to exist. Each species has evolved to live within certain temperature ranges. Temperature conditions cannot be altered very much without drastic effects on many kinds of organisms. Also solar heating drives the circulation of air and water, evaporates water, and, in general, is the primary force in determining the climates found in the different ecosystems.

Gates calls our attention to two of man's activities which affect the energy relationships of the atmosphere: (1) dumping into the skies of large amounts of particulates (dust, dirt, and aerosols that result from certain agricultural practices and from industry, automobiles, and home heating units); and (2) increasing the CO_2 content of the atmosphere through the burning of fossil fuels (coal, oil, and gas) as a source of power for industry and other operations.

Increased CO_2 concentration in the atmosphere causes increased absorption of terrestrial radiation and an increased amount of radiation emitted back toward the ground. The result is one of more energy at the ground surface and a slight warming of the surface and of the atmosphere. This has been called the "greenhouse effect." Today the average concentration of carbon dioxide in the atmosphere is about 320 ppm (parts per million); in 1870 the concentration was about 283 ppm. Between 1880 and 1940 the increase in mean global temperature was 0.5°C. As Gates points out, there has been an increase of 13 percent in the CO_2 concentration in a period of 100 years. A 20 percent increase would produce about a 1°C rise in the mean temperature of the atmosphere. A 3°C rise in mean temperature could result in sufficient warming to melt the ice caps of the world and to flood major coastal urban areas.

On the other hand the increase of particulates in the atmosphere has the opposite effect. This happens in two ways: (1) an increase of dustiness in the atmosphere reflects more sunlight into space, and (2) if the dust is black, such as carbon particles, then instead of reflecting light it will absorb more radiation. In the past 30 years the average temperature has dropped about 0.3°C. Gates points out that this may have happened due to natural causes that we know nothing about. On the other hand, he says that the massive growth of technology since 1940 may be associated with it.

So, under one set of conditions—the increase of CO_2 in the atmosphere—we must worry about the "greenhouse effect." Under another set of conditions—an increasing amount of particulates in the atmosphere—we must worry about the effects of a continued lowering of the global temperature.

Populations of different species make up the living parts of ecosystems. A population is defined by the ecologist as a group of similar organisms (same species) that occurs in a defined space during a specified time interval. A species (sexually reproducing one) is defined as a group of freely interbreeding organisms that produce fertile and viable offspring and do not reproduce with members of other populations. In attempting to understand what goes on in ecosystems, one of the first essential steps is to study what goes on in the various populations. In Chapter 5, Peter W. Frank discusses the regulation of populations—how the numbers of a population are determined or limited. He points out that the general processes that lead to population change are birth, death, immigration, and migration. He uses two models to represent the extremes of situations encountered in nature. One is a test tube culture of bacteria. In this culture in the process of population growth the food material is used up, metabolites are produced, and conditions change so that no further increase in numbers occurs—the population increases to a peak at which births equal deaths. Thereafter, it declines to extinction, unless provision is made for emigration by subculture. This model corresponds to the situation in nature when some factor becomes limiting for a particular population. The other model is provided by a bacterial population in a continuous culture chamber, the chemostat. Here a population is introduced and supplied with a constant, continuing input of energy and materials. Under the proper conditions of input and flow, the population will increase to a finite steady state. Frank points out that such continuous cultures are comparable to climax communities in nature where materials are recycled and a continuous, constant input of energy exists.

Given a favorable environment with an unlimited source of materials and energy, a population will increase exponentially. But this does not go on very long in natural populations; something stops the exponential growth. Frank shows that attempts to find the causes of changes in numbers of natural populations indicate that many different things in addition to energy and materials may be involved, not only in comparisons among different species but also within a given species at different times. Also, each change may not be due to a single cause, but to several. Among the factors involved in such changes are: density effects *per se*, extremes of physical factors, habitat selection, behavior, predators, parasites, absence of natural enemies, disease vectors, loss of virulence by disease producers, and immunity, to name some.

Although many species show only small variations in numbers from year to year, others may show great variations. In spite of this, the generalization made by Charles Darwin that the average number of each species remains relatively constant over long periods of time still seems to be valid. The exploding human population seems to be an exception. This we examine next.

Human Population Control

In Chapter 6, Garrett Hardin discusses the problems of human population and the possibilities of its control. Human population growth, like that of all populations, must ultimately level off. It is more complicated than that of other populations because of the social and political aspects. Hardin is not a prophet of doom but he does indicate that the problem of human population along with nuclear war are, in his opinion, the two most important problems facing mankind and that their solutions must be faced now.

Hardin points out that the controversy over human population started in 1798 with the publication of Malthus' *Essay on Population*. He clearly describes the thinking of Malthus and, in doing so, provides the historical perspective that every student of human population should have. He discusses the nature of exponential growth and points out that most of the time the average growth rate of any population is, and must be, zero. Yet for the last three centuries this has not been true for man; the human population has been increasing. However, this fact was not recognized until less than 200 years ago when human censuses were started. Since that time men have been interested in trying to predict human population size at some future date. At first these were called predictions; now they are called projections. Man really cannot accurately predict the human population density at some future date. He can only show what the present trend of population growth will produce assuming that no change in the trend takes place. Hardin points out that projections can be used in the process of engineering human consent—consent to planning for a world in which the average rate of population growth is zero. This is one of the revolutionary messages that human ecologists are trying to get across.

In general, until recent years, people have not been aware of the rate of population increase. If a population of one million in 50,000 B.C. is assumed and if we take the figure of 3500 million for today, only 0.016 percent per year rate of growth on the average was required. People do not notice this rate of growth. Even now in many towns and small cities in the United States, when you talk about the population explosion, people will say we have none here—our population has not increased in the last decade. What they forget is that, although they are reproducing at a rate as high as the nation's average, most of the young people are moving to the large cities.

In a graph of the human population for the last million years, the slope of the curve would be almost zero for most of its duration. There are three eras, however, when the curve moves up sharply to a higher level. These eras are: (1)

the tool-making revolution, (2) the agricultural revolution, and (3) the scientific-industrial revolution, and this last upward curve is still advancing. We are now faced with projections of 300 million people in the United States and 6 to 7 billion in the world by the year 2000.

Some people have said that we should solve our population problem by sending the excess to live on planets in outer space. We have good indications now that none of the planets in our system is suitable for human habitation. Hardin asks if it would be feasible to think of sending people to a planet of Alpha Centauri, the nearest star in another solar system, which is 4.3 light years away. By Hardin's calculations such a suggestion is preposterous; it is not possible either in terms of cost or in terms of the time required.

There are many people who are asking the question, "What is the maximum population that the earth can support?" Hardin points out that throughout most of human history food has been a limiting factor and in two-thirds of the world it is still a factor today. Improvements in agriculture have greatly increased the food supply. Recently, many people who do not want to face the population problem have been enthusiastic about the prospects of the "Green Revolution." It is interesting, in this connection, that Norman Borlaug, the man who in 1970 was awarded a Nobel Prize for his part in developing new strains of grains that figure in the so-called Green Revolution, said when he received his award that the effects of these new developments were only temporary—that unless the tremendous power of human reproduction could be slowed, there would be no long-run benefits.

There are those who are saying that we should by-pass the rather inefficient green plants and produce food for man using solar or nuclear energy directly in the synthesis. Hardin quotes the physicist, Fremlin, who has calculated that if this could be done, the earth could support a maximum of 10^{18} people. Hardin calls such a situation "one great Black Hole of Calcutta"; I would call it a situation where people exist like maggots in a wound.

Hardin then says "surely we can agree that it is not the *quantity* of human life that we want to maximize, but its *quality*."

There are many indications that we have already passed the level of optimum populations. Hardin calls attention to these—information and communication overload, great crowding of our national parks and forests, great increases in hunting and fishing licenses, and our many kinds of pollution and environmental disruption. Although improved technology can ameliorate pollution at a cost, the ultimate solution to pollution is not to be found in technology but in population control.

How can population control be effected? Not by birth control measures (although they are desirable) because it has been found in most societies that the number of children desired per family is greater than the number required for population equilibrium.

Hardin considers that population control, like pollution control, must be accomplished ultimately by political change. He says that the population problem should be put in the general framework that includes all the principal systems of exploiting the environment for man's use. A comparison of the systems found today in the United States and Russia shows that the United States is not wholly a free enterprise system and that Russia is not wholly socialistic. Furthermore, both have aspects of a third system, "the tragedy of the commons" system. This system is given its name from the time when men grazed their herds of cattle on a common pasture. Each herdsman was out to maximize his own gain and eventually this resulted in ruin for all herdsmen. It is Hardin's point that our attitude about common usage, whether it be of air, land, water, or the resources for human existence, must be rethought. Russian paper mills dump wastes into Lake Baikal and call it "socialism"; American pulp mills dump mercury into the Great Lakes and call it "free enterprise." The same situation exists in regard to population. Most nations, regardless of political system, have in recent years greatly increased their philanthropic activities. It is considered a disgrace to the community or state when a child starves to death, for example. Hardin says that few will argue against this, but we should notice its political significance. If parents are free to have as many children as they want because they know that they can dip into the commons, then population control becomes difficult if not impossible. "Unrestricted freedom to breed will, in the long run, produce ruin for all."

Hardin's logic seems clear. If we are to keep such freedoms as freedom of speech, freedom of assembly, freedom of association, freedom in the choice of residence, freedom in work, and freedom to travel, we will have to restrict the freedom to breed.

He concludes by saying that it is not clear at the moment just how this can be accomplished but that we do not have long to find acceptable answers. There are many obstacles to be overcome and the most difficult ones involve *attitudes* of people. Hardin emphasizes one of these. Beginning with the time of Condorcet and the idea of "Progress," population growth has been regarded as normal and this idea has remained quite firmly embodied in the thinking of most people. In this country we have racial and ethnic groups who cry about genocide and the deprivation of one of man's basic rights when population control is mentioned. And in certain religious groups the control of human population is frowned upon. In 1967 some 30 nations of the United Nations agreed to the following: "The Universal Declaration of Human Rights describes the family as the natural and fundamental unit of society. It follows that any choice and decision with regard to the size of the family must irrevocably rest with the family itself, and cannot be made by anyone else."[1]

[1]Since this chapter was written the President's Commission on Population Growth and the American Future reported to the President (May 5, 1972) on their two-year study. This was a broadly based study and the report was carefully worded. It said, "Recognizing that our population cannot grow indefinitely and appreciating the advantages of moving now toward the stabilization of population, the Commission recommends that the nation welcome and plan for a stabilized population." (See *Population Bulletin*, April 1972.)

Pollution

Various aspects of environmental pollution and disruption are described in Chapters 7, 8, 9, and 10. When different individuals write on the pollution of air, water, and land, the same pollutant may be referred to by each of them, and this is as it should be. The same pollutant often affects the air, the water, and the land.

AIR

In Chapter 7, James A. Oliver discusses air pollution. This is not a new problem. Man has been polluting the air ever since he started using fire. As far back as the twelfth century, London experienced heavy palls of smoke from thousands of soft-coal fired hearths. In 1881 the cities of Chicago and Cincinnati passed legislation for smoke control. Since that time many local and state governments have passed additional laws. The United States Clean Air Act was passed in 1963 and since then a number of other legislative measures have become law. International regulations are being discussed in the United Nations, and pacts among European nations are being effected. Several major disasters from air pollution have occurred in such widespread industrial areas as the Meuse Valley of Belgium in 1930, Donora, Pennsylvania in 1948, London in 1952 and 1962, New York in 1953, 1962, and 1966, and Tokyo in 1967. Los Angeles and the surrounding area has been plagued with smog problems for a long time.

There are many kinds of man-made air pollutants, but the major ones are carbon monoxide, sulfur oxides, hydrocarbons, particulate matter, and nitrogen oxides. The major sources of air pollution are transportation, industry, generation of electricity, space heating, and refuse disposal. The single greatest source of air pollution is the gasoline engine. The more than 80 million automobiles in the United States in 1967 produced 61 million tons of nitrogen oxides and 210,000 tons of lead. Coal burning operations produce most of the sulfur oxide and particulates.

The effects of air pollution are numerous. Some of the major effects are: harmful effects on human health and vitality, damage to and loss of animal and plant life, damage to property and materials with heavy economic loss, destruction of natural beauty, great deterioration of the quality of life in heavily affected areas, and changes in climatic and weather patterns, which ultimately could be very serious. Oliver documents all of this. Although to date it has not been possible to link specific pollutants to causes of specific human ailments, several ailments or diseases are associated with air pollution such as throat irritation, bronchial asthma, bronchitis, emphysema, and lung cancer.

Besides the major sources of air pollution, a number of other air pollutants are now causing much concern. Particulate asbestos, a highly toxic substance, is produced by the building industry and by auto brake linings, and it has become increasingly present in the urban atmosphere. A number of municipalities have passed specific ordinances to regulate the use of asbestos. Lead is a very toxic material and lead poisoning has become a problem, particularly in slum areas, because children often ingest flakes of paint containing lead. In the summer of 1971 two primates died in a New York zoo from lead poisoning. They had ingested neither paint, nor food, nor water containing lead. The obvious source of the lead seemed to be the air which contains lead from the exhaust of automobiles that burn leaded gasoline. Lead has been found in vegetation growing along highways in concentrations high enough to abort cows subsisting on the greenery. Cadmium is another heavy metal that may pollute the air in some places. A startling case of cadmium poisoning was reported in Japan early in 1971. In 1969 a Japanese woman, Takako Nakamura, committed suicide at the age of 28. She started to work at 18 as a lathe operator in a cadmium smelter. After 2 years she had to quit because of severe pains. Doctor's diagnosed her ailment as intestinal ulcers. Her pains continued after she became a clerical worker. By the time Takako committed suicide a number of other Japanese were complaining of the same mysterious pains. Seeking clues health officials exhumed Takako's body and performed an autoposy. By Japanese standards, one part per million of cadmium is harmful to humans. The autopsy showed 4540 ppm. in her liver and 22,400 ppm. in her kidneys! Still another heavy metal that must be added to the list of dangerous air pollutants is nickel carbonyl. It comes from diesel oil and has been reported as a cause of lung cancer in animals and exposed workers.

Oliver also discusses radioactive pollution—which he says is not properly considered as a form of air pollution since it is a form of energy, but it may be transmitted through the air and may cause human health hazards. There has been great concern about the pollution of our environment by radioactive materials since World War II. The explosion of nuclear bombs in the war and the testing of many new bombs in the years that followed released many forms of radioactive materials into the environment. One of the great difficulties associated with radioactive materials is the very long half-life of many of them. Strontium 90 has been studied as much as any other form of radioactive pollutant. It has been identified in human bones and in both cow and human milk. It has been reported to cause cancer. The effects of radiation on the genetic material have been studied extensively and have been found to cause mutations.

Although we now have a nuclear test-ban treaty, we have not been freed entirely from the dangers of radioactive materials. More and more nuclear power plants are being constructed and these pose dangers to the environment because of the problems associated with the dispersal of the wastes from such plants and because of the possibilities of leakage from them. As Oliver points out, because of widespread fear by the public of these hazards, no one wants to have a nuclear powered electrical plant located near him. This creates serious problems for our expanding economy and ways are being sought to meet the growing demands for power with a minimum disruption of the environment.

Oliver discusses air pollution control and indicates the many difficulties that have been, and still are, encountered. It has been very difficult, for example, to set emission standards that are both safe and feasible. He cited the progress that has been made in California, London, and New York in reducing air pollution. The United States Congress passed the 1970 National Air Quality Standards Act which is the stiffest antipollution law ever passed. It gave a firm mandate to the automobile industry to produce a pollution-free internal combustion engine; it also indicated to the electric power producers, the paper and steel industries, and all others who have been employing inadequate technology that they must make urgently needed changes. The new Environmental Protection Agency is in charge of enforcement of the law. It remains to be seen how the agency will fare in setting up the standards and, then, enforcing them.

WATER

In Chapter 8, Marion T. Jackson begins by pointing out that life itself, as well as the development and maintenance of human civilizations, is absolutely dependent upon water. He then describes the chemical and physical nature of water and the many properties of water that make it so important to life. He concludes this discussion with the statement: "The physical properties of water seem to have been designed purposfully to make life on earth possible. It obviously was not water that was shaped to meet the needs of life—life originated and was shaped by millions of years of evolution to become almost perfectly adopted to the nature of water."

He discusses the amount of water on the earth and the hydrologic cycle (the evaporation of water from the land and sea, the formation of clouds, and the precipitation of rain). This never-ending cycle purifies water in the process and sends it back to earth for man and other organisms to use.

There is so much water on earth (326 million cubic miles) that it has been customary for people to assume that we will not have to worry about water supply on a global scale. However, Jackson points out that 97.2 percent of earth's supply is in the oceans and cannot be used for most purposes without desalinization. Of the remaining 2.8 percent that comprises the fresh-water supply of the world, 2.15 percent is immobilized in ice caps and glaciers, leaving 0.65 percent of potential use to man. When the ground water, the soil moisture, and the water in saline lakes is subtracted, only 0.0091 percent, or 30,300 cubic miles, held in fresh-water lakes and streams, is potentially available for man's use at present. This is a very slender thread when viewed in the light of present-day water demands with human population growing at its present rate and with an ever-expanding technology. (It can be pointed out here that the recent report of the President's Commission on Population Growth and the American Future says: "Sooner or later we will have to deal with water as a scarce resource.") (See *Population Bulletin*, April 1972.)

Jackson discusses the increasing demands for water in our society: for drinking and washing, growing crops, transportation, sewage, generating electricity, cooling houses, making steel, recreation, fighting fires—to mention a few. Some of his figures are very illuminating, for example, "the total daily water requirement to produce food for one person in America is about 2500 gallons."

The problems of water pollution are many and great. Jackson defines water pollution as a situation where there is the presence of undesirable foreign matter in such quantities that the quality of the water or it usefulness for beneficial purposes is dimished. He discusses various tests and criteria used in testing for water pollution.

Jackson points out that the major general categories of water pollution are municipal wastes, industrial wastes, and agricultural wastes. Many specific pollutants are discussed: sewage; chemicals and heavy metals such as mercury, lead, and cadmium; pesticides; fertilizers; particulate matter from soil erosion; radioactive materials; petroleum; acids from strip mines; and detergents, to name some. Thermal pollution resulting from the dumping of hot water into streams and lakes by electric generating companies and certain industries is a growing kind of water pollution. He considers the effects of our water pollution in terms of human health, eutrophication and the modification of the ecosystem, the destruction of species, recreational uses, and added costs to society.

Jackson considers the status of sewage disposal in this country at the present time. Much needs to be done. He describes the various types of treatment—primary, secondary, and tertiary—and points out the great need to move to tertiary treatment in all communities. The cost for this will be enormous. Jackson states that "if water rates were tripled, water would still be a bargain in terms of price increases on almost all commodities in recent years." However, fixing up our sewage plants will not, as many people presently assume, solve our water pollution problems. Many other things must be done, including more research on various aspects of the problem, the control of industrial pollution, the control of agricultural run-off, the control of feed lots, the elimination of phosphate detergents, the control of mine wastes, and the control of soil erosion.

The Environmental Protection Agency has started to work in this direction but is hampered by the lack of an up-to-date and comprehensive water pollution law. Its major accomplishments to date in the control of industrial pollution has been in the use of the Refuse Act of 1899!

Jackson discusses the role of the Army Corps of Engineers in building large reservoirs to control floods and takes a very critical stand on their operations. He also discusses the oceans, pointing out that they are the ultimate sump for all the water borne waste from all the continents. Only recently has man realized that oceans, too, are being dangerously polluted.

In the final section Jackson asks the question: What are some of the proposed solutions to our water problems? He then examines three proposals: (1) that we augment our water needs by the desalinization of sea water; (2) that we move icebergs to our coast and use the melted ice; and (3) that we arrange to bring in fresh water in large conduits from Canada and Alaska. He concludes that a fourth option is more feasible—something he emphasizes throughout his discussion —purify the water that is available and reuse it over and over.

LAND

In Chapter 9 Alton A. Lindsey discusses the land resource base. Lindsey classifies the land under three major types of biotic systems. These are: (1) independent biotic systems, (2) dependent biotic systems, and (3) compromise systems. Independent biotic systems include natural areas, nature preserves, national parks, monuments and landmarks, wildlife sanctuaries, and wilderness areas. Such systems are mature ecosystems with a steady state. They help to balance other systems. These must be preserved for uses in hiking, riding, canoeing, fishing, adventure, and creative loafing, as well as for their uses and values of an aesthetic, scientific, educational, spiritual, artistic, and literary nature. The dependent biotic systems include cities, towns and suburbs, croplands, farm pastures and woodlots, monoculture tree plantations, orchards, and reservoirs. These systems are less stable, are vulnerable and poorly buffered, and tend to unbalance the world ecosystem. Intensive conservation techniques must be applied in these systems for protection, maintenance, pollution control, material recycling and beautification, while restraining abusive technologies and population increases. Compromise systems are intermediate between the other two. They include public and private forests in multiple-use, rangeland, most wetlands and recreation areas, such as state parks and public hunting grounds. Conservation in these systems depends on both use and management being light enough to take advantage of nature's protective buffering and restorative powers.

In Chapter 10 Lindsey considers many kinds of present land pollutants and their dangers to society. Included are fertilizer pollution, fallout pollution, pesticides, and herbicides. He points out that new methods for the control of pests are urgently needed and suggests that biological controls and the use of pest-resistant strains of plants may be part of the answer.

He comments on the undesigned urban sprawl which is rapidly changing the landscape and points to the urgent need for better planning and design for land use.

One of the most obvious pollutants of the land is solid waste. He considers the methods of handling solid wastes and the urgent need for recycling many of the solid waste materials. It is heartening that a number of industries are now working with community groups in the recycling of glass, metals, and paper, but not nearly enough is being done in this area. The natural beauty of many roadsides is destroyed not only by the trash and litter dumped along them but also by the many billboards that have been erected.

Lindsey discusses the problems associated with the use of septic tanks by home owners and the use of land fills by municipalities. He points out that septic tanks cannot be effectively used in impervious clays and overly permeable soils. Furthermore, septic tanks should be used only on large lots and this may pose problems in the future. Landfills provide a good means of garbage disposal provided they are located where the geological conditions are right. Regional plans are needed to identify lands suitable for landfill use.

In addition to the pollution of the land by solid wastes, more and more acres of land each year are despoiled through strip and contour mining activities. Prior to 1965, 7000 square miles were covered with unsightly piles of wastes. Lindsey points out that now more than four-fifths of the ores and solid fuels mined in the United States come from surface mines. More stringent reclamation standards must be applied to restore mined lands to minimal standards and to prevent much of the pollution caused by acid and sediment as a result of these mining activities.

Lindsey also considered the problems and dangers associated with the proposed Alaskan pipe lines.

At the end of Chapter 10, Lindsey discusses the land ethic. Aldo Leopold in 1949 called for the acceptance of ethical obligations to land and all of its living things. This means that economic considerations should be tempered by those of ecology. The social conscience must be extended from people to land.

If modern man could develop a general consciousness of his obligations to land and all its living things, many of our present problems could be solved more easily. Our present attitudes toward land as personal property will be the most difficult to change. One has to recognize that air and water are ambient and that the idea of personal ownership of air and water is difficult to defend. On the other hand we have long associated with the ownership of land the right to do with it what one chooses. This attitude has changed some in recent years because in many places land is zoned for certain uses. This relinquishment of total rights over land will have to be extended in the future if we are to solve the problems caused by pollution and population growth.

Problems of the Cities

As pointed out by Paul Sanford Salter in Chapter 11, the destruction and deterioration of our environment is concentrated in our urban regions. It is estimated that the population of the metropolitan centers in the United States will grow from 113 million in 1960 to 178 million by 1985. This means that 71 percent of all Americans will be living in metropolitan regions in 1985 as compared to 63 percent in 1960. The greatest problems of air, water, and land pollution are found in these regions.

After discussing the early studies in human ecology and the concepts of urban growth, Salter turns to the present problems of cities and to the things that must be done to correct these problems. A major problem of urban growth is the spreading of cities, the so-called urban sprawl, with its unplanned use of land.

This has resulted in the growing together of cities—the formation of megalopolises—such as is now found on the Atlantic coast from Boston to Washington. The early cities were separated and compact. The slow methods of transportation made this a necessity. A number of changes in transportation —railroads with their spur industrial sites, rapid transit facilities, and finally the tremendous use of automobiles—are primarily responsible for the changes in patterns of urban growth in recent years. Now many of the city workers live far from the central city. Shopping plazas and industrial parks located away from the central city are common. The urban fringes are occupied by unplanned suburbs and many industries, some of which may be environmentally harmful. The central city has deteriorated and the slum regions have spread there. A most unfortunate feature of this trend is that the people who are left in these regions of the central city are unable to keep up the area.

All cities are plagued with pollution problems—air, garbage, trash, sewage disposal, industrial waste, and so on. Salter refers to the "Indian Equivalent" concept of W.H. Davis which says that the damage a single American contributes is greater than the environmental deterioration caused by 25 people in India. There is now evidence that the air pollution of some cities is causing climatic changes in adjacent areas. Noise pollution is of primary concern in urban areas. Salter discusses the effects of noise on humans, indicates ways to reduce noise, and points out that we have just started to formulate laws that will make noise controls possible.

Salter points out that the great congestion in our cities is causing density effects on man. Among these are higher incidences of emotional and mental problems, higher incidences of lung cancer and many other diseases, more drug addiction, and more violence and vandalism.

The solution of the problems of the cities involves many changes, and Salter points out some of the things that must be done. Many new, well-planned cities must be built. A small start has been made in this direction. However, the old cities will have to be changed and used too. Residential areas must be separated from industrial areas, many small parks must be created, each urban center should have some uniqueness, housing in the central regions must be improved, educational opportunities must be improved, the racial issues must be solved—that is, conditions should be made such that people will want to return to the cities.

Much has happened in recent years that has caused many people to demand that the problems of the cities be resolved, but it is still not clear that the old basic attitudes of man about his environment—that nature was made for man to dominate and subdue—have changed enough to make this possible.

One of the biggest problems in making our cities more livable for man is the cost. Salter points out that we have been relying on property taxes to pay all of the urban bills, but that we can no longer do this. In addition to the property tax we should introduce a system of urban pricing. This would include in part tolls or permits for city streets during rush hours, tolls on expressways during rush hours, higher taxes on undeveloped urban land, higher taxes on land developers who create urban sprawl by skipping over more expensive land to develop cheaper outlying districts, and higher charges on such things as municipal golf courses and city operated marinas. In other words, it is time that we started making individuals who use certain facilities more or who profit greatly from present practices pay a good part of the load instead of charging it all to the general public.

International Aspects

The problems of the environment are worldwide and the International Biological Program is quite significant. It represents the first worldwide effort of biologists to pool their research in order to seek answers to the environmental questions of great importance to man's continued survival on this planet. In Chapter 12 W. Frank Blair outlines this program and points out some of its accomplishments.

The program was first conceived in 1961 and became a reality in 1964. Fifty-eight countries, including the United States, are participating in this massive, cooperative research effort. The goals of the program involve a worldwide study of (1) organic production of the land, fresh waters, and seas, so that adequate estimates may be made of the potential yield of new, as well as existing, natural resources, and (2) human adaptability to changing conditions. A secretariat is maintained in London for the program.

Blair points out that there was some concern on the part of United States representatives about the preoccupation with problems of increasing productivity rather than the overall problem of population numbers and the ability of the earth's resources to support a freely growing population.

He describes the organization of the IBP, how it is coordinated and the details of some of the projects. The most significant accomplishment of the IBP seems certain to be its fostering of multidisciplinary, integrated research efforts to solve complex environmental problems. Involved in these efforts are many breeds of scientists, ranging from ecologists, systematists, pedologists, and hydrologists to specialists in remote sensing, data processing, systems analysis, and ecological modeling. The purpose is to develop an understanding of major ecosystems such that it will be possible to predict the effects of specific human manipulation and alteration of these systems. Such information will provide much-needed guidelines for future environmental management.

In assessing the accomplishments to date of the IBP, Blair calls attention to these things: (1) It has provided a prototype for the kind of multidisciplinary studies that are essential in providing baselines for environmental management. (2) It has provided a focus on environmental problems at a time when policy makers all over the world have become more concerned about environmental pollution and means of restoring and maintaining environmental quality. (3) The program of IBP provides a logical precursor for the National Institute of Ecology that has been proposed by the Ecological Society of America. (4) The IBP has spawned a major international effort to monitor the quality of the world's environments. The "Man and the Biosphere" program of the United Nations has been copied almost completely from the IBP program. "The Scientific Committee on Problems of the Environment" (SCOPE) of the International Council of Scientific Unions has objectives very similar to those of IBP. It should be mentioned also that the United Nations planned a Conference on the Human Environment which was held in Stockholm in the summer of 1972.

Government and Academia

In Chapter 13 Edward J. Kormondy asks what government and academia can do to provide solutions for our environmental problems. He points out that this question must be viewed against the background attitude of so many men—that more of everything is the desired way of life.

The ultimate responsibility for protecting our environment rests jointly with the Legislative, Executive, and Judicial branches of our government. In Congress, varying aspects of environmental management are in the jurisdictions of 14 principal committees. This results in many conflicting jurisdictions. For instance, in the Senate water pollution may be considered by Interior and Insular Affairs, or by Public Works, or by Science and Astronautics. Kormondy points out that in the 89th Congress significant environmental measures were introduced in 8 of the 16 standing committees of the Senate and in 11 of the 21 standing committees of the House.

Kormondy indicates that he has been impressed by the sincere commitment with which several of the key congressional leaders face their responsibilities on environmental problems. Congress has not chosen to reorganize its committees with reference to the complex environmental problems but many of its members are becoming informed on the basic underlying principles of man-environment relationships. Members of Congress are now making great use of the Environmental Policy Division of the Legislative Reference Service and of the Environmental Clearing House.

The National Environmental Policy Act of 1969, which was signed by President Nixon on January 1, 1970, states the policy of the United States with respect to environmental quality. One of the requirements of the Act is that every recommendation or report on proposals for legislation significantly affecting the quality of the human environment must be accompanied by a detailed statement on the environmental impact of the proposed action. The Act also established in the White House the Council on Environmental Quality whose responsibility it is to transmit to the Congress an annual report on status and trends in environmental affairs and to suggest remedies for deficiencies. The reports for 1970, 1971, and

1972 are now available from the U.S. Government Printing Office in Washington, D.C. for the interested reader.[2]

If the handling of environmental problems by committees in Congress seems overlapping and cumbersome, the situation in the Executive branch has been much worse. Some 100 separate agencies, boards, committees, and commissions have been involved. These groups have been located in several different departments and bureaus. Under President Nixon this bureaucracy was reshuffled in the summer of 1970 and two new super-agencies were created: the Environmental Protection Administration (EPA) and the National Oceanographic and Atmospheric Administration (NOAA). EPA has responsibility for all programs of air and water pollution and solid wastes. NOAA is charged with exploring and predicting what is happening to our oceans and climate, and with taking appropriate remedial steps. There are strong hopes that these two agencies will be able to bring about many of the needed changes in environmental management and control.

After pointing out that universities in the past have considered that their two major roles are that of transmitting knowledge from the past and that of investigating what is in the present, Kormondy says that new roles must be assumed—the universities must be more future-oriented and more concerned with the transmission of value judgments. In his opinion the university has largely failed to instill those time-honored values that constitute the warp and woof of American society because of compartmentalization of knowledge and specialization of disciplines. There is need of conveying an holistic view of man. He points out in this connection that a number of universities have recently recognized the need for a shift in their functions.

The university must develop new programs for generating information about environmental change—information that is broad in scope, not merely about forests and fields, but about cities and major aquatic systems. Because the environment is so complex, individuals trained in many disciplines will be required for generating this kind of information. Kormondy says that the popular solution to this is to have an interdisciplinary study but that he prefers to prepare specialists who are generalists in outlook and who are able to work effectively with other individuals on a particular problem.

The university has always considered the dissemination of information as one

[2]The reports on Environmental Quality make interesting and informative reading. While the EPA has made quite a bit of progress in the past two years, it has been hampered by the failure of Congress to enact proper legislation. In three messages in 1970, 1971, and 1972, the President made 31 different environmental legislative proposals to Congress. As of October 1972, only six of these have been enacted and only one—the Clean Air Act Amendments of 1970—was a major proposal. Bills now pending before Congress deal with water quality, pesticides, noise, ocean dumping, toxic substances, land use, strip and underground mining, power plant siting, predator control, endangered species, sediment control, land and water conservation, environmental financing, marine pollution, and oil pollution.

of its responsibilities. This has been done, for the most part, through the medium of the learned and professional journals. Now the university community must assume a greater overall responsibility for public education. The public must be made aware of the relationships of environmental quality to human welfare, and this calls for environmental education from the lower grades through college.

He calls for changes in existing educational programs at all levels which will provide for an examination of man's ecology in its total perspective. What we need is not only an applied environmental science in all its dimensions, but also the totality of man in all his dimensions: his economics, his sociology, his psychology, his aesthetics, his theology—as well as his science, basic and applied.

The universities are responding to the situation. In 1958 only 20 universities had multidisciplinary programs dealing with the environment. By 1969, 500 schools were planning them and 121 viable programs were identified. Kormondy points out, however, that only a few of these programs involved the social sciences and virtually none, the humanities.

Only by providing such programs can we have the manpower needed for the proper control and management of the environment and an understanding citizenry necessary for the support of such control and management.

A Humanist Approach to the Problems

In Chapter 14 Leo Marx discusses American institutions and the ecological ideal. He points out that classic American writers such as Cooper, Emerson, Thoreau, Melville, Whitman, and Twain began to measure the quality of American life against something like an ecological ideal.

Until recently, in this country, the idea of the ecological ideal has been fostered primarily by a few voluntary organizations loosely known as the "conservation movement." According to Marx, one unfortunate result of this has been to identify the protection of the environment with special interests of private citizens—the rich man eager to protect the sanctity of his private retreat or the fellow with enough time and money to enjoy hunting, fishing, skiing, or mountain climbing. Only recently has environmental conservation become pertinent to the poor, the non-whites, and the residents of cities. Past governmental programs with reference to national parks, the Forest Service, and soil conservation have not caught the attention of the average citizen. Most Americans have not listened to the scientific writings on this subject by such men as Marsh and Osborne during the past century.

But by the end of the sixties things had changed. Environmental devastation had become so great that people from many sectors became concerned and the mass media began to spread the alarm. Scientists like Commoner and Ehrlich began talking about the possible doom facing our civilization. Marx asks the question, "What is one to believe?" and points out that much of the organized scientific community has not been of help. In answering the question, why do many scientists go about business as usual, he says that perhaps they feel that the campaign to restore the environment is already on—that the mass media and the President have stirred the nation to action. But, he points out, public awareness does not solve a problem, as has been illustrated in the problems of the Asian war and the plight of the blacks. It will take more than just awareness to solve the environmental problems, and scientists should know this.

Marx recalls Veblen's expectation that the scientific and technological professions would help to control the ravening economic appetites whetted by America's natural abundance. But until now, he says, this leadership has not appeared. He refers to the writing of Mailer and Gregg, who have compared the uncontrolled growth of American society to the uncontrolled spread of cancer cells in a living organism.

Marx states that there have been criticisms of this expansionary tendency by many and that there have been many benefits to society. However, the point he is interested in making is that if there is a single native institution which has consistently criticized American life from a vantage like that of ecology, it is the institution of letters.

Much of the imaginative literature in America has been involved with ecological problems. The pastoral impulse is the urge to retreat from society's increasing power and complexity in the direction of nature. Men like Cooper, Emerson, Thoreau, and Frost have taken us back to nature. Marx says that there is more than an undesirable element of escapism in this. The movement to nature calls into question a society dominated by a mechanistic system of values, keyed to perfecting the routine means of existence, yet oblivious to its meaning or purpose. The pastoral interlude must be taken, not as representing a program to be copied, but as a symbolic action which embodies alternative values, attitudes, modes of thought and feeling to those which characterize the dynamic, expansionary life style of modern America. The chief goal is not to increase the collective power, but an ideal more like the pursuit of happiness. In economic terms, the distinction is between the desire to maximize profits and the concept of material sufficiency. From a psychological standpoint, the pastoral retreat affirmed the possibility of maintaining man's mental equilibrium by renewed emphasis on his inner needs.

What Marx does here is point out the striking convergence of the views of the pastoral writers and those of ecologists. Our literature contains a deep intuition of the gathering environmental crisis and its causes.

Marx concludes his chapter with a number of significant statements. It would be folly, he says, to discount the urgency of the situation. We cannot rely on technology alone because the essential problems are political in nature. The political campaign to save our environment must take into account the deep-seated institutional origins of our distress. The environmentalists should join forces, when common aims can be found, with other groups concerned to effect basic institutional changes. They will find it necessary to confront many of the institutions which have prevented us from ending the Asian war earlier or giving justice to black Americans. It is essential that many more scientists and technologists become involved because only they can tell the general public which claims about environmental degradation should be taken seriously. As long as most scientists go about their business as usual and as long as they seem unperturbed by the urgent appeals of their colleagues, it is likely that most laymen, including our political leaders, will remain skeptical. Marx suggests that the American Association for the Advancement of Science should form a panel of highly qualified men, representing as many scientific disciplines as possible, to serve as a national screening committee for ecological information. The committee should assume the responsibility for locating and defining crucial problems and making public recommendations whenever feasible. In replying to those who may say that the AAAS should not be drawn into politics, Marx says that American scientists and technologists have already been involved in politics. On this issue there was convergence of opinion between Marx and Kormondy. Kormondy expressed the same opinion.

A Few Last Words

The problems of population and environmental pollution are complex and very difficult to solve. In many of the chapters reference has been made to some of the things that are being done. While we have a long way to go in effecting satisfactory solutions, it seems in order to mention a few other things that are happening—things that indicate the picture is not one of complete gloom.

On the population side there is evidence that the concept of "zero population" is gaining ground. In the past year there has been a great increase in the number of voluntary sterilizations, both male and female. Many states have revised their abortion laws and many more abortions have resulted.[3] In August of 1971 former Senator Joseph D. Tydings and Dr. Milton Eisenhower of Johns Hopkins University launched the Coalition for a National Population Policy which will work toward a zero population growth goal. In commenting on this Dr. Eisenhower said that the coalition will work to inform the public that "a rapidly increasing population is guaranteeing human misery and degradation, inadequate education, poor health facilities, political instability and a lack of human progress."

In discussing the government and pollution, Kormondy points out that in spite of our many agencies for dealing with pollution, we had none to prevent the situation that has developed in Lake Erie. Now the United States and Canada

[3]It should be noted that the Supreme Court has now ruled that state laws that prohibit a pregnant woman from deciding for herself to have an abortion during the first three months of pregnancy are unconstitutional.

have agreed on a joint program to eliminate water pollution in the Great Lakes by 1975—the most far-reaching environmental agreement ever signed by two nations. The second report of the Council on Environmental Quality, although indicating again that we have a tremendous job to do, did point out a few bright spots. Although air pollution has generally increased since 1969, emissions from automobiles apparently have reached the peak level and may now be on the decline as older cars are replaced by newer ones equipped with antipollution devices. Monitoring of the environment has been sharply stepped up. Now there are 10,000 water-quality stations for constantly checking the fresh waters of the nation. Although solid wastes are still an increasing problem, more and more industries are recycling and reusing their waste products. (It is proper to mention, also, that an examination of the reports to stockholders from many of the corporations of the country indicates that many of them are now quite aware of what they have been doing to the environment and are spending large sums of money to bring pollution under control.) Although the inner cities present the worst concentration of nearly every kind of pollution, there are now signs of local political action by inner-city residents themselves. In several large cities environmental groups have been formed to help battle the worst problems. The second Council report estimated that it will take about 105 billion dollars on the part of industry and government to bring pollution under control. President Nixon, in commenting on the report, said: "We must recognize that the goal of a cleaner environment will not be achieved by rhetoric or moral dictation alone. It will not be cheap or easy and the cost will have to be borne by each citizen, consumer, and taxpayer."

References

Environmental Quality. First Annual Report of the Council on Environmental Quality, 1970, U.S. Government Printing Office, Washington, D.C. 20402 ($1.75).

Environmental Quality. Second Annual Report of the Council on Environmental Quality, 1971, U.S. Government Printing Office, Washington, D.C. 20402 ($2.00).

Environmental Quality. Third Annual Report of the Council on Environmental Quality, 1972, U.S. Government Printing Office, Washington, D.C. 20402 ($2.00).

Index